Universitex

Springer
Berlin
Heidelberg
New York
Hong Kong
London
Milan
Paris
Tokyo

Fernando Q. Gouvêa

p-adic Numbers

An Introduction

Second Edition 1997
With 15 Figures

Springer

Fernando Q. Gouvêa
Colby College
Department of Mathematics
Waterville, ME 04901
USA
e-mail: fqgouvea@colby.edu
URL: http://www.colby.edu/personal/f/fqgouvea/

Cataloging-in-Publication Data applied for

Die Deutsche Bibliothek - CIP-Einheitsaufnahme
Gouvêa, Fernando Q.:
P-adic numbers : an introduction / Fernando Gouvêa. - 2. ed. - Berlin
; Heidelberg ; New York ; Barcelona ; Budapest ; Hong Kong ;
London ; Milan ; Paris ; Santa Clara ; Singapore ; Tokyo : Springer,
1997
   (Universitext)
   ISBN 3-540-62911-4

Corrected 3rd printing 2003

ISBN 3-540-62911-4 Springer-Verlag Berlin Heidelberg New York

---

Mathematics Subject Classification (2000): 11-01, 11S-XX

Reprint from English language edition:
p-adic Numbers: An Introduction Second Edition
by Fernando Q. Gouvêa
Copyright © Springer Berlin Heidelberg 1993, 1997
Springer Berlin Heidelberg is a part of Springer Science+Business Media
All Rights Reserved

This reprint has been authorized by Springer Science & Business Media for distribution in
China Mainland only and not for export therefrom.

Contents

Introduction

In the course of their undergraduate careers, most mathematics majors see little beyond "standard mathematics:" basic real and complex analysis, abstract algebra, some differential geometry, etc. There are few adventures in other territories, and few opportunities to visit some of the more exotic corners of mathematics. The goal of this book is to offer such an opportunity, by way of a visit to the p-adic universe. Such a visit offers a glimpse of a part of mathematics which is both important and fun, and which also is something of a meeting point between algebra and analysis.

Over the last century, p-adic numbers and p-adic analysis have come to play a central role in modern number theory. This importance comes from the fact that they afford a natural and powerful language for talking about congruences between integers, and allow the use of methods borrowed from calculus and analysis for studying such problems. More recently, p-adic numbers have shown up in other areas of mathematics, and even in physics.

For all their importance, p-adic numbers are not an extremely difficult concept; in fact, they are quite accessible to an undergraduate audience. The goal of these notes is to realize this possibility, taking its readers for a short promenade along the p-adic path. Our aim is sightseeing, rather than a scientific expedition, so we will not worry too much if we fail to emphasize a subtle point here and there, nor if our theorems are less general than they could be, nor, in fact, if we do not learn all there is to know. Rather, our goal is to introduce the reader to the rather strange world of the p-adic numbers, and to begin to make it feel familiar. What we will cover will not be sufficient for those students which will need to use p-adic numbers as a research tool. For them, a lot more reading will be necessary (and in an appendix we discuss some of the texts that are available for further reading). Instead, we try to touch a lot of bases, and set the stage for further study.

There are many ways to begin our task. Of the available options, I chose to go with the theory of absolute values on fields, and to view the p-adic numbers as directly analogous to the real numbers. In this approach, the main ingredient is a change of attitude about absolute values. It starts with the observation that from an *algebraic* point of view there is no reason to view the usual absolute value on the field $\mathbb{Q}$ of rational numbers as a given. Rather, any function satisfying the same basic properties should be just as good. If we start with the usual absolute value and look for a completion[1] of $\mathbb{Q}$ as a metric space, we get the real numbers; starting with a different absolute value, we get something else. What that something else is, and why it is interesting, is the subject of this book.

Besides its importance, the study of p-adic numbers is attractive because it blends together so many parts of mathematics. While it is certainly a part of number theory, its language is often the language of analysis, and its theorems are often analogous to, but slightly different from, those found in calculus textbooks. Both the analogy and the differences are fascinating, so that at times one gets the feeling that things are slightly out of whack, and p-adic analysis seems like classical analysis in a distorting mirror. I have tried to include many examples of this sort of thing, and I hope they are convincing.

I have done much less to convince the reader that p-adic numbers are actually useful. For the most part, I have limited myself to stating that certain things are true or that certain methods are fruitful. In every case, developing the details of the application would make this book much harder than it is supposed to be. Once again, a lot can be learned from other texts, and the student who wants to know should go to the references mentioned in the text and in the appendix.

One final pointer to Appendix B is in order. While they are working through the book, some readers may enjoy being able to work with a computer software package that can handle p-adic numbers. There aren't many of these, but a note about those that do exist has been added to the discussion of other sources of information.

Some business: the pre-requisites for reading this book are a basic knowledge of algebra and number theory, and a few courses in calculus or analysis. To be a bit more precise, the reader should be familiar with the language of congruences, with the basic theory of fields and rings, and with basic concepts about point-set topology, continuity, and infinite series. I have tried to provide as many definitions (and also informal descriptions) as I could, consistent with the requirement that the result not be too ungainly. I hope that this approach may be useful both to refresh students' memories of other subjects and to display the unity and interconnectedness of mathematics in a dramatic way.

The use of the topics mentioned above as pre-requisites is not uniform. Most students will know enough to read the first few chapters without needing to run back to their textbooks from other courses. The analysis requirements become more serious beginning in Chapter 4, and the algebraic requirements come in more strongly in Chapter 5. Even so, the whole book remains[2] well within the reach of undergraduate mathematics majors.

---

[1]If you're wondering what a "completion" is, the definition will be met later, in full gory detail. Don't worry about it yet.

[2]I hope!

There are many kinds of books about mathematics, from encyclopedic treatises to brief surveys, from dry as dust to boringly chatty. This book is closer to being a survey than to being encyclopedic, and is intended to be easy to read, but not as bed-time reading: the reader is expected to do some work. (Maybe even a lot of work.) To this end, I have included a great many problems throughout. The problems are meant to be solved, or at least attempted, at about the time when they are met in the text.[3] Most of them offer an opportunity to work with concepts that have just been introduced, and it is the author's fond hope that such problems will help create familiarity with the material. The majority of these problems ask the reader to work out the details of arguments which have been only sketched in the text, or to supply the proofs for statements given in the text (for the most part, this is only done when the proof is straightforward, and even then hints are often given). Other problems stretch out to mention matters not touched upon in the text, to indicate to the reader that there are many themes we have not had time to discuss. Finally, many are intended to prepare the reader for the discussion to follow. Such problems will often become trivial in the light of what comes later (they may be special cases or simple corollaries of theorems we will prove); leaving them for later will only render them boring.

Besides offering practice and a chance of active interaction with the material, the many problems are intended to stimulate the reader to read in a certain way. In many mathematics textbooks, one finds proofs that are "left to the reader" or dismissed as "clear" and throwaway lines mentioning interesting sidelines to the material being discussed. The experienced mathematical reader knows that these are signals to dig out pencil and paper and verify what has been said. In this book, I have tried to make sure that most such signals are followed by explicit problems. My hope is that this will help my less experienced readers gain experience of how to interact with mathematical texts.

A note to the specialists: this book is intended as a pedagogical tool. It is *not* intended as a replacement for the standard references (it is much too sketchy for that), nor as a model of elegant or detailed treatment of this (or any other) subject. Rather, I have tried to make it fun to work with, demanding, and ample. I have often spent time discussing interesting mathematics (the point-set topology, for example, or the various definitions of the field norm) just because it was interesting. I welcome any comments, and ask students in particular to tell me their reactions.

Every writer creates in his or her mind an imaginary audience for his or her text. In the case of this book, what I imagined was an upper-level undergraduate course for mathematics majors. It would include honest-to-goodness

---

[3]I realize this is very different from what most of my readers are used to. Try to think of this text as a workbook rather than a textbook: each thing should be done in turn, before going on to the next step.

undergraduates and not only graduate-level students who just happen not to have finished their undergraduate degrees yet. (In other words, this is not only for hot-shots, though hot-shots should be welcome too.) The course would very likely use an approach where students are asked to read the text, attempt the problems, and discuss the results with each other and with their instructor. The many problems asking the reader to "make a conjecture," or to attempt something ("Can you...") presuppose such a situation. For emergencies, I have provided some hints and comments on the problems. These are usually not complete solutions, but rather attempts to jump-start the solution process; they should be used only after some meditation on the problem, or they may spoil the fun.

This book grew from a set of notes for a mini-course given at the "17° Colóquio Brasileiro de Matemática," the 1989 edition of the bi-annual congress of Brazilian mathematicians. It has since been used in a course (much like the one described above) at Colby College. I would like to thank the organizers of the "Colóquio" for their invitation, and also to thank the students who sat through preliminary versions of this material for their interest and for their patience with its shortcomings. Many shortcomings will undoubtedly remain, and I would like to hear about them (who knows, there may even be a second edition someday). Please drop me a note if you have any comments.

During the final stages of the writing of this book, the author's research was partially supported by NSF grant number DMS–9203469. The writing was done in three phases, at the Universidade de São Paulo, at Queen's University at Kingston, Ontario, and at Colby College. I would like to thank NSF and all three universities for their support; Colby College, where most of the work was done, and whose computer equipment is responsible for the physical existence of this book, deserves special thanks for providing pleasant and fruitful working conditions.

This book was typeset in LaTeX using several different kinds of computers and a large number of standard macro packages. It depends, thus, on the work of many people who have given of their talents to the community of TeX users. I thank you all.

Finally, I would also like to thank César Polcino, of the Universidade de São Paulo, who first put a book on p-adic numbers in my hands, and Noriko Yui, of Queen's University, who insisted that I develop the original notes into this book; the project would not have been undertaken without them.

Note on the second edition: I am grateful to Springer-Verlag for giving me the opportunity to revise the book for this new printing. The largest changes happened in chapter four. I'd like to thank the various people who made comments and suggestions, including Silvio Levy, Alain Robert, and especially Keith Conrad.

Note on the second printing of the second edition: The need for a new printing has given me the opportunity to correct several typos, update references, and make a few small changes in the text.

Note on the third printing of the second edition: The main change for this new printing was to correct the numbering of the solutions to the problems, which was incorrect in the previous printing. I apologize to those who were inconvenienced by that mistake. Other than that, I have only made a few minor changes.

I'd like to thank the many people who found typos, made suggestions and comments, and generally gave me useful feedback. You are all encouraged to keep at it!

τι ποιεῖτε, πάντα εἰς δόξαν θεοῦ ποιεῖτε

1 Apéritif

The idea of considering new ways to measure the "distance" between two rational numbers, and then of considering the corresponding completions, did not arise merely from some desire to generalize, but rather from several concrete situations involving problems from algebra and number theory. The new "metrics" on $\mathbb{Q}$ will be each connected to a certain prime, and they will "codify" a great deal of arithmetic information related to that prime. The goal of this first chapter is to offer an *informal* introduction to these ideas. Thus, we proceed without worrying too much about mathematical rigor[1] or precision, but rather emphasizing the ideas that are behind what we are trying to accomplish. Then, in the next chapter, we will begin to develop the theory in a more formal way.

1.1 Hensel's Analogy

The p-adic numbers were first introduced by the German mathematician K. Hensel (though they are foreshadowed in the work of his predecessor E. Kummer). It seems that Hensel's main motivation was the analogy between the ring of integers $\mathbb{Z}$, together with its field of fractions $\mathbb{Q}$, and the ring $\mathbb{C}[X]$ of polynomials with complex coefficients, together with its field of fractions $\mathbb{C}(X)$. To be specific, an element of $f(X) \in \mathbb{C}(X)$ is a "rational function," i.e., a quotient of two polynomials:

$$f(X) = \frac{P(X)}{Q(X)},$$

with $P(X), Q(X) \in \mathbb{C}[X]$, $Q(X) \neq 0$; similarly, any rational number $x \in \mathbb{Q}$ is a quotient of two integers:

$$x = \frac{a}{b},$$

with $a, b \in \mathbb{Z}$, $b \neq 0$. Furthermore, the properties of the two rings are quite similar: both $\mathbb{Z}$ and $\mathbb{C}[X]$ are rings where there is *unique factorization*: any integer can be expressed uniquely as ± 1 times a product of primes, and any polynomial can be expressed uniquely as

$$P(X) = a(X - \alpha_1)(X - \alpha_2)\ldots(X - \alpha_n),$$

---

[1] which always runs the risk of becoming mathematical *rigor mortis*...

where a and $\alpha_1, \alpha_2, \ldots \alpha_n$ are complex numbers. This gives us the main point of the analogy Hensel explored: *The primes $p \in \mathbb{Z}$ are analogous to the linear polynomials $X - \alpha \in \mathbb{C}[X]$*

So far, we have nothing that is really notable, but Hensel noticed that the analogy, as it stands, goes a little deeper. Suppose we are given a polynomial $P(X)$ and a particular $\alpha \in \mathbb{C}$. Then it is possible (for example, using a Taylor expansion) to write the polynomial in the form

$$P(X) = a_0 + a_1(X - \alpha) + a_2(X - \alpha)^2 + \cdots + a_n(X - \alpha)^n$$
$$= \sum_{i=0}^{n} a_i(X - \alpha)^i$$

with $a_i \in \mathbb{C}$. Now this also works naturally for integers (let's stick to positive integers for now): given a positive integer m and a prime[2] p, we can write it "in base p," that is, in the form

$$m = a_0 + a_1 p + a_2 p^2 + \cdots + a_n p^n = \sum_{i=0}^{n} a_i p^i$$

with $a_i \in \mathbb{Z}$ and $0 \leq a_i \leq p - 1$.

The reason such expansions are interesting is that they give "local" information: the expansion in powers of $(X - \alpha)$ will show, for example, if $P(X)$ vanishes at α, and to what order. Similarly, the expansion "in base p" will show if m is divisible by p, and to what order. For example, expanding 72 in base 3 gives

$$72 = 0 + 0 \times 3 + 2 \times 3^2 + 2 \times 3^3,$$

which shows at once that 72 is divisible by 3^2.

Now, for polynomials and their quotients, one can in fact push this much further. Taking $f(X) \in \mathbb{C}(X)$ and $\alpha \in \mathbb{C}$, there is always an expansion

$$f(X) = \frac{P(X)}{Q(X)} = a_{n_0}(X - \alpha)^{n_0} + a_{n_0+1}(X - \alpha)^{n_0+1} + \ldots$$
$$= \sum_{i \geq n_0} a_i(X - \alpha)^i$$

This is just the Laurent expansion from complex analysis, but in our case it can be very easily obtained by simply doing long division with the expansions of $P(X)$ and of $Q(X)$. Notice that it is a much more complicated object than the preceding expansion:

- We can have $n_0 < 0$, that is, the expansion can begin with a negative exponent; this would signal that α is a root of $Q(X)$ and not of $P(X)$ (more precisely, that its multiplicity as a root of $Q(X)$ is bigger). In the language of analysis, we would say that $f(X)$ has a *pole* at α, of order $-n_0$.

---

[2]Remember that in the analogy choosing $(X - \alpha)$ corresponds to choosing a prime.

- The expansion will usually not be finite. In fact, it will only be finite if when we write $f(X) = P(X)/Q(X)$ in lowest terms then $Q(X)$ happens to be a power of $(X - \alpha)$ (can you prove that?). In other words, this is usually an infinite series, and it can be shown that the series for $f(\lambda)$ will converge whenever λ is close enough (but not equal to) α. However, since we want to focus on the algebraic structure here, we will treat the series as a *formal* object: it is just there, and we do not care about convergence.

Here's an example. Take the rational function

$$f(X) = \frac{X}{X - 1},$$

and let's look at the expansions for different α. (This is a calculus exercise.) If $\alpha = 0$, we get

$$\frac{X}{X - 1} = -X - X^2 - X^3 - X^4 - \ldots$$

which shows that $f(0) = 0$ with multiplicity one. For $\alpha = 1$, we get

$$\frac{X}{X - 1} = \frac{1 + X - 1}{X - 1} = (X - 1)^{-1} + 1$$

which highlights the pole of order one at $\alpha = 1$ (and also gives an example of an expansion that is finite). Finally, if we take, say, $\alpha = 2$, where there is neither pole nor zero, we get

$$\frac{X}{X - 1} = 2 - (X - 2) + (X - 2)^2 - (X - 2)^3 + \ldots$$

Problem 1 Refresh your calculus skills by checking these three equalities. Can you find the region of convergence? (Hint: all you need to remember is the geometric series.)

Problem 2 Suppose $f(X) = P(X)/Q(X)$ is in lowest terms, so that $P(X)$ and $Q(X)$ have no common zeros. Show that the expansion of $f(X)$ in powers of $(X - \alpha)$ is finite if and only if $Q(X) = (X - \alpha)^m$ for some $m \geq 0$.

The punchline is that any rational function can be expanded into a series of this kind in terms of each of the "primes" $(X-\alpha)$. (The quotes aren't really necessary, since the ideals generated by the elements of the form $(X - \alpha)$ are exactly the prime ideals of the ring $\mathbb{C}[X]$, so that $(X - \alpha)$ is a rightful bearer of the title of "prime." But all that comes later.) On the other hand, not all such series come from rational functions. In fact, we have already met examples in our calculus courses: the series for $\sin(X)$, say, or the series for e^X, which cannot be expansions of any rational function (calculus exercise: why not?).

Now, from an algebraic point of view, here's how to read the situation. We have two fields: the field $\mathbb{C}(X)$ of all rational functions, and another field which consists of all Laurent series in $(X - \alpha)$. (The next exercise asks you to check that it is indeed a field.) Let's denote the second by $\mathbb{C}((X - \alpha))$. Then the function

$$f(X) \mapsto \text{expansion around } (X - \alpha)$$

defines an *inclusion* of fields

$$\mathbb{C}(X) \hookrightarrow \mathbb{C}((X - \alpha)).$$

There are, of course, infinitely many of these (one for each α), and each one contains "local" information about how rational functions behave near α.

Problem 3 Let $\mathbb{C}((X - \alpha))$ be the set of all finite-tailed Laurent series (with complex coefficients) in $(X - \alpha)$

$$f(X) = \sum_{i \geq n_0} a_i (X - \alpha)^i.$$

Define the sum and product of two elements of $\mathbb{C}((X - \alpha))$ in the "obvious" way, and show that the resulting object is a field. Show that one may in fact take the coefficients to be in any field, with the same result.

Hensel's idea was to extend the analogy between $\mathbb{Z}$ and $\mathbb{C}[X]$ to include the construction of such expansions. Recall that the analogue of choosing α is choosing a prime number p. As we have already seen, we already know the expansion for a positive integer m: it is just the "base p" representation of m:

$$m = a_0 + a_1 p + a_2 p^2 + \cdots + a_n p^n,$$

with $a_i \in \mathbb{Z}$, $0 \leq a_i \leq p - 1$. As in the case of polynomials, this is a finite expression[3].

To pass from positive integers to positive rationals, we simply do exactly as in the other case, that is, we expand both numerator and denominator in powers of p, and then *divide formally*. The only thing one has to be careful with is that one may have to "carry." The sum of two of our a_i, for example, may be larger than p, and one has to do the obvious thing. It's probably easier to go straight to an example.

Let's take $p = 3$, and consider the rational number $24/17$. Then we have

$$a = 24 = 0 + 2 \times 3 + 2 \times 3^2 = 2p + 2p^2$$

and

$$b = 17 = 2 + 2 \times 3 + 1 \times 3^2 = 2 + 2p + p^2.$$

---

[3]The condition $0 \leq a_i \leq p - 1$ may seem to break the analogy with the complex case. But not so! The point is that the quotient of $\mathbb{C}[X]$ by the ideal generated by $(X - \alpha)$ is isomorphic to $\mathbb{C}$, and the constants in $\mathbb{C}[X]$ give a "canonical" choice of coset representatives. Similarly, the numbers between 0 and $p - 1$ are a choice of coset representatives for the quotient of $\mathbb{Z}$ by the ideal generated by p.

(Though of course $p = 3$, it's probably less confusing to write p, because one is less tempted to "add it all up." The point is to operate *formally* with our expansions.)

Then the expansion of $a/b = 24/17$ is

$$\frac{a}{b} = \frac{24}{17} = \frac{2p + 2p^2}{2 + 2p + p^2} = p + p^3 + 2p^5 + p^7 + p^8 + 2p^9 + \dots$$

To check that this is correct, all we need to do is to multiply it by (the expansion of) 17, remembering that $p = 3$:

$$(2 + 2p + p^2)(p + p^3 + 2p^5 + p^7 + p^8 + 2p^9 + \dots) =$$

$$= 2p + 2p^2 + \underbrace{p^3 + 2p^3}_{} + 2p^4 + p^5 + 4p^5 + 4p^6 +$$
$$+ 2p^7 + 2p^7 + 2p^8 + 2p^8 + p^9 + 2p^9 + 4p^9 \dots$$

since $p = 3$, we get $p^3 + 2p^3 = 3p^3 = p^4$, so

$$= 2p + 2p^2 + \underbrace{p^4 + 2p^4}_{} + p^5 + 4p^5 + 4p^6 + 2p^7 + 2p^7 +$$
$$+ 2p^8 + 2p^8 + p^9 + 2p^9 + 4p^9 + \dots$$

$$= 2p + 2p^2 + \underbrace{p^5 + p^5 + 4p^5}_{} + 4p^6 + 2p^7 + 2p^7$$
$$+ 2p^8 + 2p^8 + p^9 + 2p^9 + 4p^9 + \dots$$

$$= 2p + 2p^2 + \underbrace{2p^6 + 4p^6}_{} + 2p^7 + 2p^7 + 2p^8 + p^9 + 2p^8 + 2p^9 + 4p^9 + \dots$$

$$= 2p + 2p^2 + \underbrace{2p^7 + 2p^7 + 2p^7}_{} + 2p^8 + 2p^8 + p^9 + 2p^9 + 4p^9 + \dots$$

$$= 2p + 2p^2 + \underbrace{2p^8 + 2p^8 + 2p^8}_{} + p^9 + 2p^9 + 4p^9 + \dots$$

$$= \dots$$

$$= 2p + 2p^2$$

so that the higher powers of p disappear "to the right," leaving us with $2p + 2p^2 = 24$! (The reader will probably feel something has been shoved under the rug, and in fact there is something to prove here. But the point is that, at least formally, it works.)

Provided that one can treat the whole process formally, it is easy to check that this always works, and that the resulting series reflects the properties of the rational number $x = a/b$ as regards the prime number p (we will get into the habit of saying "locally at p" or even "near p," to emphasize the analogy). Thus, for each prime p, we can write any (positive, for now) rational number a/b in the form

$$x = \frac{a}{b} = \sum_{n \geq n_0} a_n p^n,$$

and, for example, we have $n_0 \geq 0$ if and only if $p \nmid b$, and $n_0 > 0$ if and only if $p \nmid b$ and $p \mid a$ (assuming a/b is in lowest terms). In fact, the number n_0 (which is something like the order of a zero or pole) reflects the "multiplicity" of p in a/b; it is characterized by the equation

$$x = p^{n_0} \frac{a_1}{b_1} \qquad \text{with} \quad p \nmid a_1 b_1.$$

It remains to see how to get the negative rational numbers, but since our power series in p can clearly (see Problem 5) be multiplied, it is enough to get an expansion for -1. Keeping in mind that we are working formally, and with a little imagination, that is not too hard to do. We find, for any p, that

$$-1 = (p-1) + (p-1)p + (p-1)p^2 + (p-1)p^3 + \cdots,$$

since, if we add 1, we get

$$\underbrace{1 + (p-1)} + (p-1)p + (p-1)p^2 + (p-1)p^3 + \cdots =$$

$$= \underbrace{p + (p-1)p} + (p-1)p^2 + (p-1)p^3 + \cdots$$

$$= \underbrace{p^2 + (p-1)p^2} + (p-1)p^3 + \cdots$$

$$= \cdots$$

$$= 0.$$

The conclusion is that, at least in a formal sense, every rational number x can be written as a "finite-tailed Laurent series in powers of p"

$$x = a_{n_0} p^{n_0} + a_{n_0+1} p^{n_0+1} + \cdots$$

("finite-tailed" refers, of course, to the fact that the expansion is finite *to the left*; it is usually infinite to the right). We will call this the *p-adic expansion* of x; remember that if x is a positive integer, it is just its expansion "in base p."

It is not too hard to show that the set of *all* finite-tailed Laurent series in powers of p (i.e., of all p-adic expansions) is a field (Problem 5 again), just as $\mathbb{C}((X - \alpha))$ is a field. We will denote this field by $\mathbb{Q}_p$, and call it the *field of p-adic numbers*. As before, we can describe much of what we have done by saying that the function

$$x \mapsto p\text{-adic expansion of } x$$

gives an inclusion of fields

$$\mathbb{Q} \hookrightarrow \mathbb{Q}_p.$$

(We have not yet shown that $\mathbb{Q}_p$ is strictly bigger than $\mathbb{Q}$; the next section will show that this is true.)

The definition of a p-adic number as a formal object (a finite-tailed Laurent expansion in powers of p) is of course rather unsatisfactory according to the tastes of today. We will remedy this in Chapter 3, where we will show how to construct the field $\mathbb{Q}_p$ as an analogue of the field of real numbers. For now, note only that whatever the "real" definition is, it must allow our series to converge, so that powers p^n must get *smaller* as n grows. This is pretty strange, so let's give ourselves time to get used to the idea. The problems in this section are intended to help the reader feel a little more comfortable with p-adic expansions.

Problem 4 Consider a p-adic number

$$x = a_0 + a_1 p + a_2 p^2 + a_3 p^3 + \cdots .$$

What is $-x$? (This means: what is its p-adic expansion?)

Problem 5 Show that $\mathbb{Q}_p$ is indeed a field. (You will have to begin by making explicit what the operations are, and this is a bit tricky because of "carrying." For example, the coefficient of a given power of p in the sum of two expansions depends on the coefficients of *all* the lower powers in the summands; however, this is still a finite rule.) Then show that the map $\mathbb{Q} \longrightarrow \mathbb{Q}_p$ given by sending each rational number to its expansion is a homomorphism.

Problem 6 By analogy with the real numbers, it's natural to guess that every rational number will have a periodic (or eventually periodic) p-adic expansion, and that conversely any such expansion represents a rational number. Show that this is in fact correct. (Just follow the proof for real numbers.)

Problem 7 When one deals with real numbers, one uses the notation $3.14159\ldots$ to refer to the infinite series

$$3 + \frac{1}{10} + \frac{4}{10^2} + \frac{1}{10^3} + \frac{5}{10^4} + \frac{9}{10^5} + \cdots$$

Devise a similar notation for p-adic numbers, and explain how to sum and multiply numbers expressed in your notation. Using your notation, re-do some of the examples we gave above.

Problem 8 (Some Abstract Algebra required!) Another point at which our analogy seems to break down is the fact that rational functions $f(X) \in \mathbb{C}(X)$ are really *functions*: one can really compute their value at a complex number α. This problem explains a highfalutin' way of interpreting rational numbers as functions too.

   i) First of all, show that we can identify the set of complex numbers α with the set of maximal ideals in $\mathbb{C}[X]$ via the correspondence $\alpha \leftrightarrow (X - \alpha)$.

  ii) Fix a complex number α. Show that the map $f \mapsto f(\alpha)$ defines a homomorphism from the ring $\mathbb{C}[X]$ to $\mathbb{C}$, whose kernel is exactly the ideal $(X - \alpha)$.

 iii) Now let $f(X)$ be a rational function. Show that the map $f \mapsto f(\alpha)$ still makes sense provided the denominator of f is not divisible by $X - \alpha$. If the denominator is divisible by $(X - \alpha)^n$ but not by $(X - \alpha)^{n+1}$, explain why this means that f has a pole of order n at α.

  iv) Now take $x = a/b \in \mathbb{Q}$, and choose a prime $p \in \mathbb{Z}$. If p does not divide b, define the *value of x at p* to be $a/b \pmod{p}$, which means $ab' \pmod{p}$, where b' is an integer satisfying $bb' \equiv 1 \pmod{p}$. We think of this value as an element of $\mathbb{F}_p$, the field with p elements. If p does divide b, we say that x has a pole at p. Explain how to define the order of the pole. This interprets the elements of $\mathbb{Q}$ as a sort of "function" on the primes $p \in \mathbb{Z}$. It is a bit weird, because this "function" doesn't have a "range:" the value at each p belongs to a different field $\mathbb{F}_p$.

   v) Discuss whether this way of thinking of rational numbers as functions is reasonable. Does it make the analogy any tighter?

1.2   Solving Congruences Modulo p^n

The "p-adic numbers" we have just constructed are closely related to the problem of solving congruences modulo powers of p. We will look at some examples of this.

Let's start with the easiest possible case, an equation which has solutions in $\mathbb{Q}$, such as

$$X^2 = 25.$$

We want to consider it modulo p^n for every n, i.e., to solve the congruences

$$X^2 \equiv 25 \pmod{p^n}.$$

Now, of course, our equation has solutions already in the integers: $X = \pm 5$. This automatically gives a solution of the congruence for every n; just put $X \equiv \pm 5 \pmod{p^n}$ for every n.

Problem 9 Check that these are the only solutions, up to congruence, of $X^2 \equiv 25 \pmod{p^n}$, at least when $p \neq 2, 5$. What happens in these special cases?

Let's try to understand these solutions a little better from the p-adic point of view. To make our life easier, we take $p = 3$ once again. We begin by rewriting our solutions using residue class representatives between 0 and $3^n - 1$ for the solutions modulo 3^n. The first solution, $X = 5$, gives:

$$X \equiv 2 \pmod 3$$
$$X \equiv 5 = 2 + 3 \pmod 9$$
$$X \equiv 5 = 2 + 3 \pmod{27}$$
$$\text{etc.}$$

which never changes any more, and therefore just gives the 3-adic expansion of this solution:

$$X = 5 = 2 + 1 \times 3.$$

For $X = -5$, the results are a little more interesting; the representatives are

$$X \equiv -5 \equiv 1 \pmod 3$$
$$X \equiv -5 \equiv 4 = 1 + 3 \pmod 9$$
$$X \equiv -5 \equiv 22 = 1 + 3 + 2 \times 9 \pmod{27}$$
$$X \equiv -5 \equiv 76 = 1 + 3 + 2 \times 9 + 2 \times 27 \pmod{81}$$
$$\text{etc.}$$

Again, continuing this gives the 3-adic expansion of the solution, which is a bit more interesting because it is infinite:

$$X = -5 = 1 + 1 \times 3 + 2 \times 3^2 + 2 \times 3^3 + 2 \times 3^4 + \cdots$$

(Check this against your answer in Problem 4.)

Notice that the two systems of solutions are "coherent," in the sense that when we look at, say, $X = 76$ (which is the solution modulo 3^4) and reduce it modulo 3^3, we get $X = 22$ (which is the solution modulo 3^3). Let's give this a formal definition:

Definition 1.2.1 *Let p be a prime. We say a sequence of integers α_n such that $0 \le \alpha_n \le p^n - 1$ is* coherent *if, for every $n \ge 1$, we have*

$$\alpha_{n+1} \equiv \alpha_n \pmod{p^n}.$$

If we need to emphasize the choice of prime p, we will say the sequence is p-adically coherent.

We can picture our two coherent sequences of solutions as branches in a tree (see figure 1.1). Of course this is all rather painfully obvious in the case we are considering, since the sequences of solutions are coherent simply because they "are" solutions in $\mathbb{Z}$ (76 is congruent to 22 just because both are congruent to -5). The only real bit of information we have obtained is the connection between expressing the roots as a coherent sequence and obtaining their p-adic expansions.

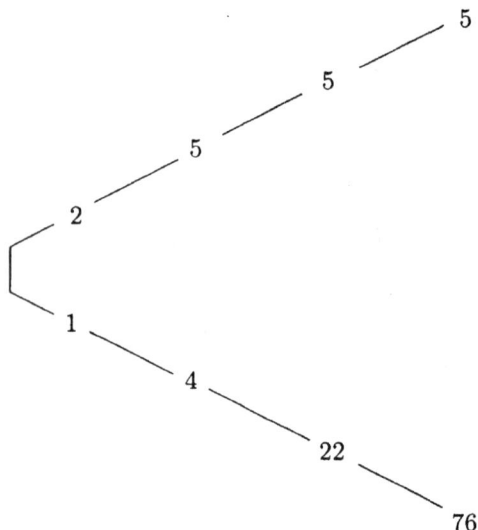

Figure 1.1: Solutions of $X^2 \equiv 25 \pmod{3^n}$

Problem 10 Before we go on to something more interesting, do a couple of similar examples on your own, say with $X^2 = 49$ and $p = 5$, and $X^3 = 27$ and $p = 2$.

Problem 11 Things already get slightly more interesting if we take $p = 2$ and the equation $X^2 = 81$. In this case, the "tree" of solutions modulo 2^n is much more complex: there are two infinite branches that correspond to the solutions $X = \pm 9$, but there are also lots of finite branches (solutions modulo 2^n that do not "lift" to solutions modulo 2^{n+1}). We will later consider what is special about this situation.

Things become much more interesting if we follow the same process with an equation that does *not* have rational roots. For example, take the system of congruences

$$X^2 \equiv 2 \pmod{7^n}, \qquad n = 1, 2, 3, \ldots$$

For $n = 1$, the solutions are $X \equiv 3 \pmod 7$ and $X \equiv 4 \equiv -3 \pmod 7$. To find the solutions for $n = 2$, note that their reductions modulo 7 must be solutions for $n = 1$. Hence we set $X = 3 + 7k$ or $X = 4 + 7k$ and solve for k:

$$(3 + 7k)^2 \equiv 2 \pmod{49}$$
$$9 + 42k \equiv 2 \pmod{49}$$

(notice that the term involving $(7k)^2$ is congruent to zero)

$$7 + 42k \equiv 0 \pmod{49}$$
$$1 + 6k \equiv 0 \pmod 7$$
$$k \equiv 1 \pmod 7$$

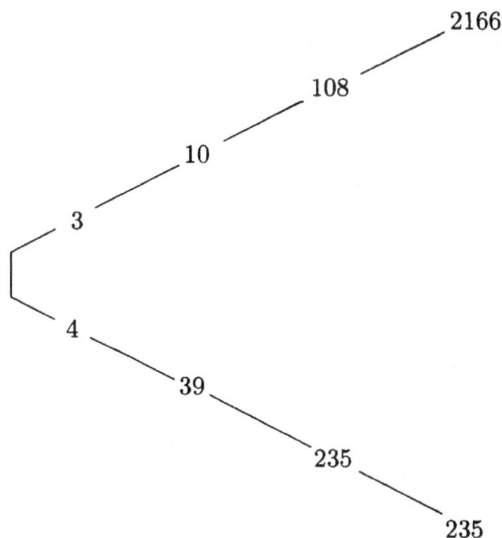

Figure 1.2: Solutions of $X^2 = 2 \pmod{7^n}$

which, since $X = 3 + 7k$, gives the solution $X \equiv 10 \pmod{49}$. Using $X = 4 + 7k$ gives the other solution $X \equiv 39 \equiv -10 \pmod{49}$.

Problem 12 Prove that for each n there can be at most two solutions. (All you need is $p \neq 2$.)

Problem 13 Show that the process above can be continued indefinitely, that is, that given a solution α_n of the congruence $X^2 \equiv 2 \pmod{7^n}$, there always exists a unique solution α_{n+1} of $x^2 \equiv 2 \pmod{7^{n+1}}$ satisfying $\alpha_{n+1} \equiv \alpha_n \pmod{7^n}$. Find a few more terms in each of the sequences of solutions above.

Again, the solutions can be represented as branches in a tree (see figure 1.2). This time, however, we can't predict *a priori* what the numbers that appear will be; instead, all we can do is convince ourselves that the process will continue as long as we want it to. The fact that one can continue finding roots indefinitely shows that there are two coherent sequences of solutions:

$$x_1 = (3, 10, 108, 2166, \ldots)$$

and

$$x_2 = (4, 39, 235, 235 \ldots) = (-3, -10, -108, -2166 \ldots) = -x_1.$$

Just as before, we can expand each number in each sequence 7-adically. The fact that the sequence is coherent means that the expansion of each root is the truncation of the expansion of the following root, so that, for example,

$$
\begin{aligned}
3 &= 3 \\
10 &= 3 + 1 \times 7 \\
108 &= 3 + 1 \times 7 + 2 \times 49
\end{aligned}
$$

This gives us two 7-adic numbers:

$$x_1 = 3 + 1 \times 7 + 2 \times 49 + 6 \times 343 + \cdots$$

and

$$x_2 = 4 + 5 \times 7 + 4 \times 49 + 0 \times 343 + \cdots = -x_1.$$

It probably bears repeating: we are not claiming that we can predict the pattern here. All we know is that we can *continue* the pattern for as long as necessary, if we have enough time and patience. It's just like finding the decimal expansion of the square root of two: we can get as close as we like, and we can *prove* that, though we can't predict what the expansion will actually be like.

In any case, we do get two 7-adic numbers, and they are indeed roots of the equation $X^2 = 2$ in $\mathbb{Q}_7$, in the usual sense:

Problem 14 Show that the 7-adic number x_1 obtained as above satisfies $x_1^2 = 2$ in $\mathbb{Q}_7$. Conclude that the field $\mathbb{Q}_7$ is strictly bigger than $\mathbb{Q}$.

The tie between solving sequences of congruences modulo higher and higher powers of p and solving the corresponding equation in $\mathbb{Q}_p$ is quite close, as the problems below try to emphasize. It is also one of the more important reasons for using p-adic methods in number theory.

Problem 15 Check that $X^2 = 2$ has no solutions in the field $\mathbb{Q}_5$. (Begin by expressing the putative solution as a 5-adic expansion. Show that it must be of the form $a_0 + a_1 5 + a_2 5^2 + \ldots$, and conclude that a_0 must satisfy a congruence modulo 5. Finally, check that the congruence you obtained has no solutions modulo 5.) Notice that this shows (in a very roundabout way) that 2 has no square root in $\mathbb{Q}$, since any square root in $\mathbb{Q}$ would be a square root in any of the $\mathbb{Q}_p$ (remember that there is an inclusion $\mathbb{Q} \hookrightarrow \mathbb{Q}_p$), hence in particular in $\mathbb{Q}_5$.

Problem 16 Check that $X^2 + 1 = 0$ has a solution in $\mathbb{Q}_5$, but not in $\mathbb{Q}_7$.

Problem 17 Show that a p-adic number

$$x = a_0 + a_1 p + a_2 p^2 + a_3 p^3 + \cdots$$

is a solution in $\mathbb{Q}_p$ of an equation $X^2 = m$ if and only if the sequence

$$(a_0, a_0 + a_1 p, a_0 + a_1 p + a_2 p^2, \ldots)$$

is a coherent sequence of solutions of the congruences $X^2 \equiv m \pmod{p^n}$. (Hint: compute x^2 up to a certain power of p, and compare it with m to read off a congruence modulo that power of p.)

We have already mentioned that there is some analogy between p-adic numbers and real numbers. The next problem gives an example of this. Over $\mathbb{R}$, there is a simple condition that determines whether the equation $X^2 = m$ has a solution (just check the sign of m). In $\mathbb{Q}_p$, the condition is also simple:

Problem 18 Let m be any integer, and suppose that the congruence $X^2 \equiv m$ (mod p) has a solution; show that if $p \neq 2$ and $p \nmid m$ it is always possible to "extend" this solution to a full coherent sequence of solutions of $X^2 \equiv m \pmod{p^n}$. Use this to find a necessary and sufficient condition for the equation $X^2 = m$ to have a root in $\mathbb{Q}_p$ for $p \neq 2$. What is special about $p = 2$?

Problem 19 Show that for every p, the inclusion $\mathbb{Q} \hookrightarrow \mathbb{Q}_p$ is strict, that is, some p-adic numbers are not (expansions of) rational numbers. (Hint: find an equation that has a root in $\mathbb{Q}_p$ but not in $\mathbb{Q}$; the root will be a p-adic number that is not rational. The basic work has all been done; just be careful with $p = 2$.)

Problem 20 In the same spirit as the previous problem, show that $\mathbb{Q}_p$ is never algebraically closed; more precisely, for each p one can find an algebraic equation with rational coefficients that has no roots in $\mathbb{Q}_p$.

1.3   Other Examples

Working with p-adic numbers is useful in all sorts of contexts. We round off this chapter by giving two rather whimsical examples.

Consider the equation $X = 1 + 3X$. This is of course easy to solve, but let's try something strange and look at it as a fixed-point problem, i.e, as the problem of finding a solution for $f(x) = x$ for some function $f(x)$. Such problems are often solved by iteration, plugging in an arbitrary initial value, then computing $f(x)$ over and over in the hope that we will get closer and closer to a fixed point. To try that in our case, we take $x_0 = 1$ and iterate, so that $x_{n+1} = 1 + 3x_n$. Here's what we get:

$$
\begin{aligned}
x_0 &= 1 \\
x_1 &= 1 + 3x_0 = 1 + 3 \\
x_2 &= 1 + 3x_1 = 1 + 3 + 3^2 \\
&\quad \cdots \\
x_n &= 1 + 3 + 3^2 + \cdots + 3^n.
\end{aligned}
$$

In $\mathbb{R}$, this is a divergent sequence, and we were all taught in calculus classes never to have any dealings with them. On the other hand, it is the sequence of partial sums of a geometric series, and we all know that

$$
1 + a + a^2 + a^3 + \cdots = \frac{1}{1-a}
$$

(Well, we know it for $|a| < 1$, but what the heck...) Plugging in blindly gives $x = 1/(1 - 3) = -1/2$, which is (surprise!) the correct answer.

This dubious playing around with divergent sequences is clearly illegal in calculus class, but it works. Here's one way to understand why. While the sequence is certainly divergent in $\mathbb{R}$, there is nothing to keep us from looking at the sequence in $\mathbb{Q}_3$ (the elements in the sequence are in $\mathbb{Q}$, which is contained in both $\mathbb{R}$ and $\mathbb{Q}_3$). Now, in $\mathbb{Q}_3$, the sequence is obviously convergent, to the 3-adic number

$$1 + 3 + 3^2 + \cdots + 3^n + \cdots.$$

One then easily checks (by the same argument used over $\mathbb{R}$!) that this is equal to $-1/2$.

Of course it is silly to solve a linear equation in such a roundabout way, but the remarkable fact here is that an argument that was either dubious or outright illegal at first sight turns out to work perfectly well in the p-adic context. The series we used is divergent only if we insist of thinking of it as a series of real numbers. Once we put it in the "right" context, it becomes quite nice. In fact, we will see in the next chapter that there is an absolute value in $\mathbb{Q}_3$, and that with respect to the notion of size determined by that absolute value our series is convergent.

The point, then, is that introducing the p-adic fields broadens our world in such a way as to allow arguments that were previously impossible. This toy example points the way to many analogous situations where considering the p-adic numbers simplifies matters tremendously.

Problem 21 Show that, for any prime p, the formula

$$1 + p + p^2 + p^3 + \cdots = \frac{1}{1-p}$$

is true in $\mathbb{Q}_p$.

The next example is perhaps even more interesting. It shows that sometimes introducing p-adic ideas allows a more conceptual proof of a fact that seems obscure (and hard to prove) otherwise. This example is a bit more advanced, and we will take for granted things that we will prove only later, but the reader should be able to follow it. We will work with $p = 2$, that is, in the field $\mathbb{Q}_2$ of 2-adic numbers.

Consider the usual MacLaurin series for the logarithm of $1 + X$:

$$\log(1 + X) = X - \frac{X^2}{2} + \frac{X^3}{3} - \frac{X^4}{4} + \cdots.$$

Since powers of 2 are "small" in $\mathbb{Q}_2$, it turns out that we can plug in $X = -2$ to compute the logarithm of -1:

$$\log(-1) = \log(1 - 2) = -\left(2 + \frac{2^2}{2} + \frac{2^3}{3} + \frac{2^4}{4} + \cdots\right)$$

(This is of course wildly divergent in $\mathbb{R}$, but it turns out to be convergent in $\mathbb{Q}_2$; this is not completely obvious because of the denominators, but it does work—see ahead.) Now, if the series converges, it must converge to zero, by the usual properties of the logarithm:

$$2 \log(-1) = \log(1) = 0.$$

This means that the partial sums

$$2 + \frac{2^2}{2} + \frac{2^3}{3} + \frac{2^4}{4} + \cdots + \frac{2^n}{n}$$

must get closer and closer to zero as n grows. Remember that what this means is that the terms in the 2-adic expansion "disappear to the right," that is, that the partial sums, written in base 2, begin with longer and longer stretches of zeros. Here's the upshot:

Fact 1.3.1 *For each integer $M > 0$ there exists an n such that the partial sum*

$$2 + \frac{2^2}{2} + \frac{2^3}{3} + \frac{2^4}{4} + \cdots + \frac{2^n}{n}$$

is divisible by 2^M.

Problem 22 Can you give a direct proof of that fact?

What this example points out is that using p-adic methods, and in particular the methods of the calculus in the p-adic context, we can often prove facts about divisibility by powers of p which are otherwise quite hard to understand. The proofs are often, as in this case, "cleaner" than any direct proof would be, and therefore easier to understand. We will look at many more examples of this before we are done.

2   Foundations

The goal of this chapter is to begin to lay a solid foundation for the theory we described informally in Chapter One. The main idea will be to introduce a different absolute value function on the field of rational numbers. This will give us a different way to measure distances, hence a different calculus. Once we have that, we will use it (in Chapter 3) to construct the p-adic numbers.

To get the p-adic numbers, we need to start with the field $\mathbb{Q}$ of rational numbers. However, rather than deal exclusively with $\mathbb{Q}$, we will devote this chapter to studying absolute values on fields in general. Of course, the main example we will have in mind will be $\mathbb{Q}$, but the general theory is easy enough that it would be a waste to specialize to rational numbers too soon. (Later, when the generality would cost us some effort, we will speedily go back to the special case of the rationals.)

So, for this chapter, $\Bbbk$ will be an arbitrary field, and we will be interested in constructing an abstract theory of absolute values on $\Bbbk$. We will do this by starting from the basic properties of the absolute values we already know and love, and then looking for other functions with similar properties.

One thing to notice from the start is that we will want to think of our new absolute values as giving alternative ways to measure the "size" of things. This can feel rather strange at first, so it's wise to keep many concrete examples in mind as we go.

2.1   Absolute Values on a Field

Let $\Bbbk$ be a field and let $\mathbb{R}_+ = \{x \in \mathbb{R} : x \geq 0\}$ be the set of all non-negative real numbers. We begin by defining an absolute value on $\Bbbk$ and exploring the possibilities implicit in the definition. The definition just tries to capture what seem to be the most important properties of the everyday absolute value.

Definition 2.1.1 *An* absolute value *on* $\Bbbk$ *is a function*

$$| \ | : \Bbbk \longrightarrow \mathbb{R}_+$$

that satisfies the following conditions:

   i) $|x| = 0$ *if and only if* $x = 0$

   ii) $|xy| = |x|\,|y|$ *for all* $x, y \in \Bbbk$

iii) $|x + y| \leq |x| + |y|$ *for all* $x, y \in \mathbb{k}$.

We will say an absolute value on $\mathbb{k}$ is non-archimedean *if it satisfies the additional condition:*

iv) $|x + y| \leq \max\{|x|, |y|\}$ *for all* $x, y \in \mathbb{k}$;

otherwise, we will say that the absolute value is archimedean.

Note that condition *(iv)* implies condition *(iii)*, since $\max\{|x|, |y|\}$ is certainly smaller than the sum $|x| + |y|$. We will later discuss in more detail why non-archimedean absolute values are important, and where their name comes from; for now, let's just mention that they are quite common.

EXAMPLES:
1. The most obvious example, of course, is our model: take $\mathbb{k} = \mathbb{Q}$, and take the usual absolute value $|\ |$ defined by

$$|x| = \begin{cases} x & \text{if } x \geq 0 \\ -x & \text{if } x < 0 \end{cases}$$

A more sophisticated way of describing this absolute value is to say that it is actually the absolute value on the field $\mathbb{R}$ of real numbers, applied to $\mathbb{Q}$ via the inclusion $\mathbb{Q} \hookrightarrow \mathbb{R}$. It is easy to see that this absolute value is *archimedean*. (Take $x = y = 1$ to see that condition *(iv)* does not hold.) For reasons that we will discuss later, this absolute value is usually called the *infinite absolute value* on $\mathbb{Q}$, or the *absolute value at infinity*, and is written as $|\ |_\infty$.
2. The most boring example is the one we get by setting $|x| = 1$ if $x \neq 0$ and $|0| = 0$. This works for any field $\mathbb{k}$, and defines a non-archimedean absolute value. It is known, for obvious reasons, as the *trivial absolute value*. It will often have to be excluded in the theorems to follow.

There are many simple properties that one can deduce quickly from the conditions above. We will try to develop them systematically in the next section. For now, let's try to be as concrete as we can. First of all, it's worth pointing out that for the special case of *finite* fields, the whole theory is trivial:

Problem 23 Let k be a finite field. Show that the only absolute value on k is the trivial absolute value.

We now go on to introduce the example that we will focus on for most of this book. Take $\mathbb{k} = \mathbb{Q}$, and choose any prime $p \in \mathbb{Z}$. Any integer $n \in \mathbb{Z}$ can be written as $n = p^v n'$, with $p \nmid n'$, and this representation is unique. Since v is determined by p and n, it make sense to define a function v_p by setting $v_p(n) = v$, so that $v_p(n)$ is just the multiplicity of p as a divisor of n. Formally:

Definition 2.1.2 *Fix a prime number $p \in \mathbb{Z}$. The p-adic valuation on $\mathbb{Z}$ is the function*

$$v_p : \mathbb{Z} - \{0\} \longrightarrow \mathbb{R}$$

defined as follows: for each integer $n \in \mathbb{Z}$, $n \neq 0$, let $v_p(n)$ be the unique positive integer satisfying

$$n = p^{v_p(n)}\, n' \qquad \text{with} \quad p \nmid n'.$$

We extend v_p to the field of rational numbers as follows: if $x = a/b \in \mathbb{Q}^{\times}$, then

$$v_p(x) = v_p(a) - v_p(b).$$

It is often convenient to set $v_p(0) = +\infty$, with the usual conventions on how to handle this symbol. The reasoning here is that we can certainly divide 0 by p, and the answer is 0, which we can divide by p, and the answer is 0, which we can divide by p...

Problem 24 Check that for any $x \in \mathbb{Q}$, the value of $v_p(x)$ does not depend on its representation as a quotient of two integers. In other words, if $a/b = c/d$, then $v_p(a) - v_p(b) = v_p(c) - v_p(d)$.

It is in fact easy to see that the p-adic valuation of any $x \in \mathbb{Q}^{\times}$ is determined by the formula

$$x = p^{v_p(x)} \cdot \frac{a}{b} \qquad p \nmid ab.$$

Problem 25 Compute a few examples, to get a feel for the thing. For example, determine $v_5(400)$, $v_7(902)$, $v_2(621)$, $v_3(123/48)$, $v_5(180/3)$.

The basic properties of the p-adic valuation v_p are the following:

Lemma 2.1.3 *For all x and $y \in \mathbb{Q}$, we have*

i) $v_p(xy) = v_p(x) + v_p(y)$

ii) $v_p(x + y) \geq \min\{v_p(x), v_p(y)\}$,

with the obvious conventions with respect to $v_p(0) = +\infty$.

Problem 26 Prove Lemma 2.1.3. (Hint: the first property is easy to see by writing out factorizations of x and y; the second comes from the fact that common powers of p can be factored out from a sum.)

Now here comes the really tricky thing: if we compare the two properties in this lemma with conditions (ii) and (iv) in the definition of absolute values, we see that they are very similar, except that the product in the first has been turned into a sum (as when taking a logarithm) and that the inequality in the second has been reversed. We can "unreverse" the inequality by changing the sign, and then turn the sum into a product by putting it into an exponent. This suggests the following, which is the crucial definition:

Definition 2.1.4 *For any $x \in \mathbb{Q}$, we define the p-adic absolute value of x by*

$$|x|_p = p^{-v_p(x)}$$

if $x \neq 0$, and we set $|0|_p = 0$.

Notice that the definition of $|0|_p$ matches our convention that $v_p(0) = +\infty$ if we interpret $p^{-\infty}$ in the only reasonable way. To see that our definition really does give an absolute value, we need to check that our requirements have been satisfied.

Proposition 2.1.5 *The function $|\;|_p$ is a non-archimedean absolute value on $\mathbb{Q}$.*

PROOF: Everything follows at once from Lemma 2.1.3.   □

To get a general impression about what the p-adic absolute value is doing, notice that when a number n is very divisible by our prime p the valuation $v_p(n)$ will be large, and then the absolute value $|n|_p$ will be small. (Look at that minus sign in the exponent!) So the p-adic absolute value gives, in a strange sort of way, a measure of how divisible by p a number is.

Problem 27 More practice: take $\mathbf{k} = \mathbb{Q}$, $p = 7$, and let $|\;| = |\;|_7$ be the 7-adic absolute value. Compute $|35|$, $|56/12|$, $|177553|$, $|3/686|$.

The connection between a non-archimedean absolute value and a function such as in Lemma 2.1.3 (called a *valuation*, or sometimes an *additive valuation*) is quite general. In fact, one can develop the theory taking either object (valuation or absolute value) as the primitive one. In this book, we will stick to absolute values, because they are closer to our intuition, but it is often convenient to go the other way.

Problem 28 (Some abstract algebra required) Let A be an integral domain, and let K be its field of fractions. Let $v : A - \{0\} \longrightarrow \mathbb{R}$ be a function satisfying the conditions of Lemma 2.1.3, i.e., a valuation on A. Extend v to K by setting $v(a/b) = v(a) - v(b)$. Show that the function $|\;|_v : K \longrightarrow \mathbb{R}_+$ defined by

$$|x|_v = e^{-v(x)} \qquad \text{for } x \neq 0$$

and $|0| = 0$ is a non-archimedean absolute value on K. Conversely, show that if $|\;|$ is a non-archimedean absolute value, then $-\log|\;|$ is a valuation.

Problem 29 Let $v : \mathbf{k}^\times \longrightarrow \mathbb{R}$ be a valuation. Show that the image of v is an additive subgroup of $\mathbb{R}$. This is sometimes called the *value group* of the valuation v. What is the value group of the p-adic valuation?

Though the p-adic absolute value is certainly the most interesting one from the point of view of this book, it's worth pointing out that there are other interesting absolute values on other fields. Before we go on to look at them, however, here are two problems to force another look at the p-adic absolute value.

Problem 30 Show that $|p^n|_p \to 0$ when $n \to \infty$, so that high powers of p are *small* with respect to the p-adic absolute value.

Problem 31 Show that for any $c \in \mathbb{R}$, $c > 1$, the equation $|x| = c^{-v_p(x)}$ defines a non-archimedean absolute value on $\mathbb{Q}$. Make a conjecture about the relation between this absolute value and the p-adic absolute value $|\ |_p$. Make a conjecture about why we chose $c = p$ for the p-adic absolute value.

Our final example is intended to show that the theory we are developing is indeed quite general, and in fact can be applied, almost without change, in all sorts of contexts. The example we want to consider also serves to confirm Hensel's intuition on the similarity between $\mathbb{Q}$ and fields of rational functions. So let F be any field (for example, a finite field, or $\mathbb{C}$), let F[t] be the ring of polynomials with coefficients in F, and let F(t) be the field of rational functions over F, which is the field of fractions of the form $f(t)/g(t)$ where $f(t)$ and $g(t)$ belong to F[t] (and $g(t) \neq 0$, of course). We will define several valuations (and therefore several absolute values) on F(t). The first is very specific to this situation, since it depends on the notion of degree of a polynomial; by contrast, the others are closely analogous to the p-adic absolute value.

First, for any polynomial $f(t) \in$ F[t], we set $v_\infty(f) = -\deg(f(t))$, and extend this to rational functions as before, by setting $v_\infty(0) = +\infty$ and

$$v_\infty\left(\frac{f(t)}{g(t)}\right) = v_\infty(f(t)) - v_\infty(g(t)) = \deg(g(t)) - \deg(f(t)).$$

It is easy to check that this is a valuation:

Problem 32 Check that for any $f(t)$, $g(t) \in$ F(t) we have $v_\infty(f(t)g(t)) = v_\infty(f(t)) + v_\infty(g(t))$ and also $v_\infty(f(t) + g(t)) \geq \min\{v_\infty(f(t)), v_\infty(g(t))\}$. (Is it enough to check for polynomials? Why?)

This gives us a non-archimedean absolute value just as before:

$$|f(t)|_\infty = e^{-v_\infty(f)}$$

for any $f(t) \in$ F(t). (As we hinted in Problem 31, any real number greater than one will do for the basis of the exponential; choosing e just fixes one; if F is a finite field, a nicer choice might be the number of elements in F.)

Problem 33 When is a rational function "small" with respect to $|\ |_\infty$? Is a polynomial ever small?

One can get other valuations on $F(t)$ by imitating the definition of the p-adic valuation, since $F[t]$ is a unique factorization domain. Just choose an irreducible polynomial $p = p(t)$ and proceed as before: define a valuation by counting the multiplicity of $p(t)$ as a factor.

Problem 34 Do it! For an irreducible polynomial $p(t) \in F[t]$, define the $p(t)$-adic valuation and absolute value on $F(t)$.

Problem 35 Since F is a subfield of $F(t)$, any absolute value on $F(t)$ also gives an absolute value on F. For the examples we have just constructed, describe the absolute value on F obtained in this way.

Problem 36 Suppose $F = \mathbb{C}$. What are the irreducible polynomials in this case? Are we getting close to realizing Hensel's analogy?

Problem 37 All of the absolute values we have constructed on $F(t)$ are non-archimedean. Try to construct an archimedean absolute value on some $F(t)$. (First of all, this may or may not be possible, depending on F. If you're very sneaky, it can be done for $F = \mathbb{Q}$. Can it be done in such a way that the induced absolute value on F is the trivial one?)

Problem 38 The field $F(t)$ contains the subring of polynomials $F[t]$, but it also contains the subring $F[1/t]$ of "polynomials in $1/t$." In fact, every element of $F(t)$ can be written as a quotient of elements in $F[1/t]$, so this subring serves just as well as $F[t]$ as a starting point. Very well, in $F[1/t]$ the "polynomial" $1/t$ is clearly irreducible, so we can construct, as in Problem 34, a $1/t$-adic valuation v_1. Check that v_1 is the same as the v_∞ constructed above. This means that all of the valuations we have constructed on $F(t)$ are of the "$p(t)$-adic" type.

Problem 39 Let $\mathbf{k} = \mathbb{Q}(i)$ be the field obtained by adjoining $i = \sqrt{-1}$ to the rational numbers, so that any element of $\mathbf{k}$ can be written as $a + bi$ with $a, b \in \mathbb{Q}$. The "integers" in $\mathbf{k}$ are the elements of $\mathbb{Z}[i] = \{a + bi : a, b \in \mathbb{Z}\}$. It is not too hard to check that this is a unique factorization domain, so that its properties are much like those of the usual integers[1]. The primes of $\mathbb{Z}[i]$ are of three kinds:

i) $1 + i$ is prime,

ii) if $p \in \mathbb{Z}$ is a prime number and $p \equiv 3 \pmod 4$, then p is a prime in $\mathbb{Z}[i]$,

iii) for each prime $p \in \mathbb{Z}$ which is congruent to 1 modulo 4, there are two primes $x + yi$ and $x - yi$ in $\mathbb{Z}[i]$ satisfying $(x + yi)(x - yi) = x^2 + y^2 = p$.

In each case, we can use the prime $\pi \in \mathbb{Z}[i]$ to construct a π-adic valuation v_π and (from it) π-adic absolute value $|\ |_\pi$ on $\mathbf{k}$ as before:

$$|\alpha|_\pi = c^{-v_\pi(\alpha)}$$

(can you come up with a "good" choice for the constant c?). Check that this works, and explore the resulting situation. For example, since $\mathbb{Q}$ is contained in $\mathbf{k}$, this induces an absolute value on $\mathbb{Q}$; describe the induced absolute value. In particular, for a fixed π, can you compute $v_\pi(p)$ as p ranges through the primes in $\mathbb{Z}$?

---

[1]See most introductory texts in algebra or number theory, or just take it for granted.

There is an extensive theory of how valuations extend (or not) from subfields to larger fields, and this theory turns out to be closely connected to algebraic number theory. (In fact, many texts on algebraic numbers develop the theory in terms of valuations and absolute values rather than in terms of ring theory; the best example is probably [Has80].) Some aspects of this subject are discussed in Chapter 5.

Keeping in mind this set of examples, and of course especially the p-adic absolute value, let's go on to look at absolute values in general in a more careful way.

2.2   Basic Properties

In this section, $\Bbbk$ will be an arbitrary field, and $|\ |$ will be a *non-trivial* absolute value on $\Bbbk$, which may or not be archimedean. The first few things to prove are some "obvious" facts, which we had better make sure work in a general setting.

Lemma 2.2.1 *For any absolute value* $|\ |$ *on any field* $\Bbbk$, *we have:*

  i) $|1| = 1$

  ii) If $x \in \Bbbk$ *and* $|x^n| = 1$, *then* $|x| = 1$.

  iii) $|-1| = 1$

  iv) For any $x \in \Bbbk$, $|-x| = |x|$.

  v) If $\Bbbk$ *is a finite field, then* $|\ |$ *is trivial.*

PROOF: The crucial fact is to remember that $|x|$ is a positive real number. Then, to prove the first statement, all one needs to note is that

$$|1| = |1^2| = |1|^2,$$

since the only non-zero positive real number α for which $\alpha^2 = \alpha$ is $\alpha = 1$. The remaining statements follow in a similar fashion.   □

Problem 40 Prove the remaining statements in the Lemma.

Our first serious theorem is a necessary and sufficient condition for an absolute value to be non-archimedean. We begin by noticing (or remembering) that for any field $\Bbbk$ we have a map $\mathbb{Z} \longrightarrow \Bbbk$ defined by

$$n \mapsto \begin{cases} \underbrace{1 + 1 + \cdots + 1}_{n} & \text{if } n > 0 \\[2mm] 0 & \text{if } n = 0 \\[2mm] -\underbrace{(1 + 1 + \cdots + 1)}_{-n} & \text{if } n < 0 \end{cases}$$

For example, if $k \supset \mathbb{Q}$, this is just the usual inclusion of $\mathbb{Z}$ into $\mathbb{Q}$; if k is a finite field, the image is a subfield of k, which will have a prime number of elements.

Theorem 2.2.2 *Let $A \subset k$ be the image of $\mathbb{Z}$ in k. An absolute value $|\ |$ on k is non-archimedean if and only if $|a| \leq 1$ for all $a \in A$. In particular, an absolute value on $\mathbb{Q}$ is non-archimedean if and only if $|n| \leq 1$ for every $n \in \mathbb{Z}$.*

PROOF: One part is easy: we have $|\pm 1| = 1$ always; hence, if $|\ |$ is non-archimedean, we get that

$$|a \pm 1| \leq \max\{|a|, 1\}.$$

By induction, if follows that $|a| \leq 1$ for every $a \in A$.

The converse requires some hocus-pocus: suppose that $|a| \leq 1$ for all $a \in A$. We want to prove that for any two elements $x, y \in k$, we have $|x + y| \leq \max\{|x|, |y|\}$. If $y = 0$, this is obvious. If not, we can divide through by $|y|$, and we see that this is equivalent to the inequality

$$\left| \frac{x}{y} + 1 \right| \leq \max \left\{ \left| \frac{x}{y} \right|, 1 \right\}.$$

This means that we need only prove the inequality for the case when the second summand is 1, and the general fact will follow. In other words, we want to prove that for any $x \in k$ we have

$$|x + 1| \leq \max\{|x|, 1\}.$$

Now let m be any positive integer. Then we have

$$|x + 1|^m = \left| \sum_{k=0}^{m} \binom{m}{k} x^k \right|$$

$$\leq \sum_{k=0}^{m} \left| \binom{m}{k} \right| |x^k|$$

Now, since $\binom{m}{k}$ is an integer, we have $\left| \binom{m}{k} \right| \leq 1$, so we can continue with

$$|x + 1|^m \leq \sum_{k=0}^{m} \left| \binom{m}{k} \right| |x^k|$$

$$\leq \sum_{k=0}^{m} |x^k| = \sum_{k=0}^{m} |x|^k$$

$$\leq (m + 1) \max\{1, |x|^m\}.$$

(For the last step, notice that the largest value of $|x|^k$ for $k = 0, 1, 2, \ldots m$ is equal to $|x|^m$ if $|x| > 1$ and is equal to 1 otherwise.) Taking the m-th root on both sides gives

$$|x + 1| \leq \sqrt[m]{m + 1} \; \max\{1, |x|\}.$$

Now this strange inequality holds for *every* positive integer m, no matter how large, and we know (from calculus) that

$$\lim_{m \to \infty} \sqrt[m]{m + 1} = 1.$$

Therefore, if we let $m \to \infty$ we get

$$|x + 1| \leq \max\{|x|, 1\},$$

which is what we wanted to prove.   □

This helps explain the difference between archimedean and non-archimedean absolute values. It allows us to restate things in the following way. An absolute value is *archimedean* if it has the following property:

Archimedean Property: *Given* $x, y \in \Bbbk$, $x \neq 0$, *there exists a positive integer* n *such that* $|nx| > |y|$.

This property holds for the "usual" absolute value on $\mathbb{Q}$ and in the the real numbers, and (in a slightly different form) this observation does go back to Archimedes.

It is easy to see that the Archimedean Property is equivalent to the assertion that there are arbitrarily "big" integers (translation: that there are integers whose absolute values are arbitrarily big). In other words, the archimedean property is equivalent to the assertion that

$$\sup\{|n| : n \in \mathbb{Z}\} = +\infty.$$

Put into this context, what our theorem says is that:

Corollary 2.2.3 *An absolute value* $| \; |$ *is non-archimedean if and only if* $\sup\{|n| : n \in \mathbb{Z}\} = 1$.

One can complete the circle by showing that these are the only two possibilities.

Problem 41 Show that if $\sup\{|n| : n \in \mathbb{Z}\} = C < +\infty$, then $| \; |$ is non-archimedean, and $C = 1$.

2.3 Topology

The whole point of an absolute value is that it provides us with a notion of "size." In other words, once we have an absolute value, we can use it to measure distances between numbers, that is, to put a *metric* on our field. Having the metric, we can define open and closed sets, and in general investigate what is called the *topology* of our field.[2]

The first step is measuring distances, in the obvious way:

Definition 2.3.1 *Let* $\Bbbk$ *be a field and* $|\ |$ *an absolute value on* $\Bbbk$. *We define the distance* $d(x,y)$ *between two elements* x, $y \in \Bbbk$ *by*

$$d(x,y) = |x - y|.$$

The function $d(x,y)$ is called the *metric* induced by the absolute value.

The definition of $d(x,y)$ parallels, of course, the usual way we define the distance between two real numbers. The first point we need to make is that a great many of the notions that we can define using the usual distance on $\mathbb{R}$ work just as well for any old distance.

Problem 42 Show that $d(x,y)$ has the following properties:

i) for any x, $y \in \Bbbk$, $d(x,y) \geq 0$, and $d(x,y) = 0$ if and only if $x = y$

ii) for any x, $y \in \Bbbk$, $d(x,y) = d(y,x)$

iii) for any x, y, $z \in \Bbbk$, $d(x,z) \leq d(x,y) + d(y,z)$

These are the general defining properties for a metric; the last inequality is called the *triangle inequality*, since it expresses the usual fact that the sum of the lengths of two legs of a triangle is bigger than the length of the other side. ("A line is the shortest path between two points.") A set on which a metric is defined is called a *metric space*, so we can read the statement of this last problem as saying that any field with an absolute value can be made into a metric space by defining $d(x,y) = |x - y|$. For more on metric spaces in general, check practically any book on real analysis (for example, [Rud76]) or an introductory text on general topology.

Problem 43 The point of this problem is to check that the metric $d(x,y)$ (or, equivalently, the absolute value it is derived from) relates well to the operations in the field $\Bbbk$:

i) Fix x_0, $y_0 \in \Bbbk$. Show that for any $\varepsilon > 0$ there exists a $\delta > 0$ such that, whenever $d(x,x_0) < \delta$ and $d(y,y_0) < \delta$, we have $d(x + y, x_0 + y_0) < \varepsilon$. In other words, addition is a continuous function.

---

[2]The reader who has never met topology or metric spaces before should not feel spooked; all we are doing is repeating the usual constructions of the calculus, but using our unusual absolute values.

ii) Fix x_0, $y_0 \in \mathbb{k}$. Show that for any $\varepsilon > 0$ there exists a $\delta > 0$ such that, whenever $d(x, x_0) < \delta$ and $d(y, y_0) < \delta$, we have $d(xy, x_0 y_0) < \varepsilon$. In other words, multiplication is a continuous function.

iii) Fix $x_0 \in \mathbb{k}$, $x_0 \neq 0$. Show that for any $\varepsilon > 0$ there exists a $\delta > 0$ such that, whenever $d(x, x_0) < \delta$, we have $x \neq 0$ and $d(1/x, 1/x_0) < \varepsilon$. In other words, taking inverses is a continuous function.

This shows that the metric $d(x, y)$ makes $\mathbb{k}$ a *topological field*.

The fact that the absolute value is non-archimedean can also be expressed in terms of the metric:

Lemma 2.3.2 *Let $|\ |$ be an absolute value on a field $\mathbb{k}$, and define a metric by $d(x, y) = |x - y|$. Then $|\ |$ is non-archimedean if and only if for any x, y, $z \in \mathbb{k}$, we have*

$$d(x, y) \leq \max\{d(x, z), d(z, y)\}.$$

PROOF: To go one way, apply the non-archimedean property to the equation

$$(x - y) = (x - z) + (z - y).$$

For the converse, take $y = -y_1$ and $z = 0$ in the inequality satisfied by $d(\cdot, \cdot)$. □

Problem 44 Give the details of the proof of the lemma. Prove also that the inequality in the Lemma implies the triangle inequality from Problem 42.

This inequality is known as the "ultrametric inequality," and a metric for which it is true is sometimes called an "ultrametric." A space with an ultrametric is called an "ultrametric space." Such spaces have rather curious properties, and we will spend the rest of this section exploring them[3]. The main point in what follows is that, once we have a way to measure distances, we can do geometry. Since our way to measure distances is rather strange, the geometry is also rather strange.

Proposition 2.3.3 *Let $\mathbb{k}$ be a field and let $|\ |$ be a non-archimedean absolute value on $\mathbb{k}$. If x, $y \in \mathbb{k}$ and $|x| \neq |y|$, then*

$$|x + y| = \max\{|x|, |y|\}.$$

---

[3]Ultrametric spaces sound like the sort of thing only a mathematician would dream up. Surprisingly, they have recently turned up in physics (in the theory of "spin glasses"). This may be one more example of the "unreasonable effectiveness of mathematics in the physical sciences"—see [Wig64].

PROOF: Exchanging x and y if necessary, we may suppose that $|x| > |y|$. Then we know that

$$|x + y| \le |x| = \max\{|x|, |y|\}.$$

On the other hand, $x = (x + y) - y$, so that

$$|x| \le \max\{|x + y|, |y|\}.$$

Since we know that $|x| > |y|$, this inequality can hold only if

$$\max\{|x + y|, |y|\} = |x + y|.$$

This gives the reverse inequality $|x| \le |x + y|$, and from it (using our first inequality) we can conclude that $|x| = |x + y|$. $\square$

This has an interesting corollary that captures in a memorable statement a property that ends up having a big role later on:

Corollary 2.3.4 *In an ultrametric space, all "triangles" are isosceles.*

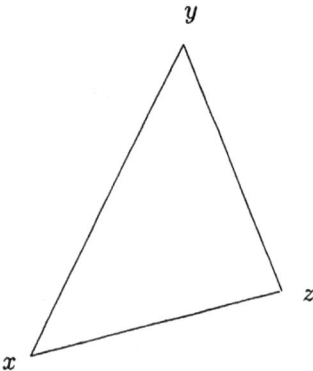

Figure 2.1: All isosceles!

PROOF: Let x, y and z be three elements of our space (the vertices of our "triangle"). The lengths of the sides of the "triangle" are the three distances

$$d(x, y) = |x - y|$$
$$d(y, z) = |y - z|$$
$$d(x, z) = |x - z|$$

Now, of course,

$$(x - y) + (y - z) = (x - z),$$

so that we can invoke the proposition to show that if $|x - y| \ne |y - z|$, then $|x - z|$ is equal to the bigger of the two. In any case, two of the "sides" are equal. $\square$

This is a rather unintuitive result (and it will have an enormous impact on the topology on our field). Thus, rather than simply barging on, it may be worth a brief look at the case of the p-adic absolute value to try to understand what is behind the truth of the proposition. As before, we put $|x| = p^{-v_p(x)}$. Since we're looking for insight, not for proof, we will only look at the case where $x, y \in \mathbb{Z}$. Say that $v_p(x) = n$ and $v_p(y) = m$, so that

$$x = p^n x' \qquad y = p^m y' \qquad p \nmid x'y'.$$

Translating into the absolute values, we get

$$|x| = p^{-n} \quad \text{and} \quad |y| = p^{-m}.$$

We will have $|x| > |y|$ when $n < m$; say $m = n + \varepsilon$, with $\varepsilon > 0$. Then

$$x + y = p^n x' + p^{n+\varepsilon} y' = p^n (x' + p^\varepsilon y').$$

Now, since $p \nmid x'$, we have $p \nmid (x' + p^\varepsilon y')$, and therefore $v_p(x + y) = n$, which means $|x + y| = p^{-n} = |x|$, as the proposition states.

On the other hand, suppose that $|x| = |y|$, that is, $n = m$. Then we get

$$x + y = p^n (x' + y')$$

with $p \nmid x'$ and $p \nmid y'$, and it is perfectly possible that $p | (x' + y')$. If so, the most we can say is that $v_p(x + y) \geq n = \min\{v_p(x), v_p(y)\}$, which translates to

$$|x + y| \leq \max\{|x|, |y|\} = |x| = |y|.$$

Notice that in either case two of the three absolute values $|x|$, $|y|$ and $|x + y|$ are equal.

Problem 45 Give $\mathbb{Q}$ the 5-adic topology, and consider the triangle whose vertices are $x = 2/15$, $y = 1/5$, $z = 7/15$; what are the lengths of the three sides?

In metric spaces, more important than triangles are the "balls" or "disks." These also turn out to be pretty strange in the case of an ultrametric.

Definition 2.3.5 *Let* $\Bbbk$ *be a field with an absolute value* $| \ |$. *Let* $a \in \Bbbk$ *be an element and* $r \in \mathbb{R}_+$ *be a real number. The* open ball *of radius* r *and center* a *is the set*

$$B(a, r) = \{x \in \Bbbk : d(x, a) < r\} = \{x \in \Bbbk : |x - a| < r\}.$$

The closed ball *of radius* r *and center* a *is the set*

$$\overline{B}(a, r) = \{x \in \Bbbk : d(x, a) \leq r\} = \{x \in \Bbbk : |x - a| \leq r\}.$$

These are standard definitions in any metric space. The open balls are the prototypes of the open sets, and the closed balls of the closed sets[4].

Problem 46 Show that open balls are always open sets, and that closed balls are always closed sets. (This is true for any absolute value, archimedean or not.)

---

[4]Here are the definitions: a set U is open if any element in U belongs to a (usually small) an open ball that is contained in U; a set is closed if its complement is an open set. A point x is a boundary point of a set S if any open ball with center x contains points that are in S and points that are not in S. S is closed exactly when it contains all of its boundary points.

For non-archimedean absolute values, we get some surprising properties:

Proposition 2.3.6 *Let* $\Bbbk$ *be a field with a non-archimedean absolute value.*

i) *If* $b \in B(a, r)$, *then* $B(a, r) = B(b, r)$; *in other words, every point that is contained in an open ball is a center of that ball.*

ii) *If* $b \in \overline{B}(a, r)$, *then* $\overline{B}(a, r) = \overline{B}(b, r)$; *in other words, every point that is contained in a closed ball is a center of that ball.*

iii) *The set* $B(a, r)$ *is both open and closed.*

iv) *If* $r \neq 0$, *the set* $\overline{B}(a, r)$ *is both open and closed.*

v) *If* $a, b \in \Bbbk$ *and* $r, s \in \mathbb{R}_+^\times$, *we have* $B(a, r) \cap B(b, s) \neq \emptyset$ *if and only if* $B(a, r) \subset B(b, s)$ *or* $B(a, r) \supset B(b, s)$; *in other words, any two open balls are either disjoint or contained in one another.*

vi) *If* $a, b \in \Bbbk$ *and* $r, s \in \mathbb{R}_+^\times$, *we have* $\overline{B}(a, r) \cap \overline{B}(b, s) \neq \emptyset$ *if and only if* $\overline{B}(a, r) \subset \overline{B}(b, s)$ *or* $\overline{B}(a, r) \supset \overline{B}(b, s)$; *in other words, any two closed balls are either disjoint or contained in one another.*

PROOF: Most of this is easy. The weird parts all depend on the fact that "all triangles are isosceles;" drawing pictures may help understand what is going on.

i) By the definition, $b \in B(a, r)$ if and only if $|b - a| < r$. Now, taking any x for which $|x - a| < r$, the non-archimedean property tells us that

$$|x - b| \leq \max\{|x - a|, |b - a|\} < r,$$

so that $x \in B(b, r)$; this shows that $B(a, r) \subset B(b, r)$. Switching a and b, we get the opposite inclusion, so that the two balls are equal.

ii) Replace $<$ with $\leq$ in the proof of *(i)*.

iii) The open ball $B(a, r)$ is always an open set in any metric space. (Here's a one-line[5] proof: any x in $B(a, r)$ is in $B(a, r)$ which is contained in $B(a, r)$!) What we need to show is that in our non-archimedean case, it is also closed. So take an x in the boundary of $B(a, r)$; this means that any open ball centered in x must contain points that are in $B(a, r)$. Choose a number $s \leq r$, and look at the open ball $B(x, s)$ with center x and radius s. Now, since x is a boundary point, $B(a, r) \cap B(x, s) \neq \emptyset$, so that there exists an element

$$y \in B(a, r) \cap B(x, s).$$

This means that $|y - a| < r$ and $|y - x| < s \leq r$. Applying the non-archimedean inequality, we get

$$|x - a| \leq \max\{|x - y|, |y - a|\} < \max\{s, r\} \leq r,$$

---

[5]This is a kind of logical skullduggery which delights mathematicians. Look closely at the definition, and you'll see it's correct...

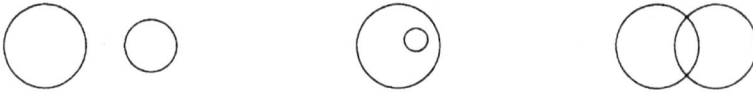

These are allowed...   but not this!

Figure 2.2: Balls for non-archimedean absolute values

so that $x \in B(a, r)$. This shows that any boundary point of $B(a, r)$ belongs to $B(a, r)$, which means that $B(a, r)$ is a closed set.

iv) This is a lot like *(iii)*.

v) We can assume that $r \leq s$ (otherwise switch them around). If the intersection is not empty, there exists a $c \in B(a, r) \cap B(b, s)$. Then we know, from (i), that $B(a, r) = B(c, r)$ and $B(b, s) = B(c, s)$. Hence

$$B(a, r) = B(c, r) \subset B(c, s) = B(b, s),$$

as claimed.

vi) Identical to the preceding, using *(ii)*.   □

Problem 47 Supply the missing portions of the proof (parts *(iv)* and *(vi)*). Why is the condition $r \neq 0$ necessary for closed, but not for open, balls?

The geometry of the balls in a non-archimedean space seems very strange at first sight; getting a good feeling for it may be the most important initial step toward understanding the p-adic absolute value. The next problems are intended to help with that goal.

Problem 48 Describe the closed ball of radius 1 around the point $x = 0$ in $\mathbb{Q}$ with respect to the p-adic absolute value. Describe the open ball of radius 1 around $x = 3$; which integers belong to this ball?

Problem 49 Let $\mathbb{k} = \mathbb{Q}$ and $|\ | = |\ |_p$. Show that the closed ball $\overline{B}(0, 1)$ can be written as a disjoint union of open balls, as follows:

$$\overline{B}(0, 1) = B(0, 1) \cup B(1, 1) \cup B(2, 1) \cup \cdots \cup B(p - 1, 1)$$

(both the equality and the disjointness need to be checked). This gives another proof that the closed unit ball is open, since unions of open sets are always open.

Problem 50 Take the 5-adic absolute value on $\mathbb{Q}$. Show that $B(1, 1) = B(1, 1/2) = \overline{B}(1, 1/5)$. What is going on here?

Problem 51 Under the hypotheses of the proposition, show that for $a \in \mathbb{k}$ and $r \in \mathbb{R}_+$, $r \neq 0$, the "sphere" $\{x \in \mathbb{k} : |x - a| = r\}$ is both an open and a closed set. (Notice that the "sphere" is *not* the boundary of the open ball. In fact, show that the boundary of the open ball is empty.)

Sets that are both open and closed are rather rare in the usual calculus, but are very common when we are dealing with non-archimedean absolute values. (As we've just seen!) So we give them a name.

Definition 2.3.7 *Let* $\Bbbk$ *be a field with an absolute value* $|\ |$ *(or, more generally, any metric space). We say a set* $S \subset \Bbbk$ *is* clopen *if it is both an open and a closed set.*

The fact that there are so many clopen sets around makes the topology of fields with non-archimedean valuations rather strange. For example, recall that a set S is called *disconnected* if one can find two open sets U_1 and U_2 such that

- $U_1 \cap U_2 = \emptyset$

- $S = (S \cap U_1) \cup (S \cap U_2)$

- neither $S \cap U_1$ nor $S \cap U_2$ is empty.

The idea, of course, is that such an S is made up of two "pieces" (namely, the intersections with each of the open sets). Sets which *cannot* be divided in this way are called *connected*.

Problem 52 Show that a set S is disconnected if and only if we can write it as a union $S = A \cup B$ of two sets satisfying the condition

$$\overline{A} \cap B = A \cap \overline{B} = \emptyset,$$

where, for a set X, $\overline{X}$ means the *closure* of X, that is, the union of X and all of its boundary points.

Problem 53 What are the connected sets in $\mathbb{R}$? (Hint: they appear all the time in elementary calculus.)

Problem 54 Show that in a field with a (non-trivial) non-archimedean valuation every closed ball with radius $r > 0$ is disconnected. Is the same true for open balls?

If we take a point $x \in \Bbbk$, we define the *connected component of* x to be the union of all the connected sets that contain x. Since the union of two non-disjoint connected sets is connected, this is a connected set, so we can describe it as the largest connected set containing x. For example, if $\Bbbk = \mathbb{R}$ is the real numbers, then the connected component of any point $x \in \mathbb{R}$ is all of $\mathbb{R}$ (simply because $\mathbb{R}$ is connected). Things are quite different in the non-archimedean case:

Proposition 2.3.8 *In a field* $\Bbbk$ *with a non-archimedean absolute value, the connected component of any point* $x \in \Bbbk$ *is the set* $\{x\}$ *consisting of only that point.*

Problem 55 Prove the Proposition. In the language of general topology, this says that **k** is a *totally disconnected* topological space.

What this says is that there are really no interesting connected sets in **k**: only the sets with only one element are connected. On the other hand, provided the absolute value on **k** is non-trivial, the set $\{x\}$ is *not* open (if every set $\{x\}$ were open, the topology on **k** would be *discrete*, i.e., every set would be open, which only happens with the trivial absolute value).

Problem 56 Take the *usual* absolute value on $\mathbb{Q}$, which of course is archimedean. Are there any clopen sets in $\mathbb{Q}$ with respect to this absolute value? Is $\mathbb{Q}$ totally disconnected with respect to this absolute value?

The same questions make sense in the real numbers, of course. Are there any clopen sets in $\mathbb{R}$?

Problem 57 Take the p-adic absolute value on $\mathbb{Q}$. Show that with respect to this absolute value every open ball is the disjoint union of open balls. (So that open balls are disconnected in a rather dramatic way.) Do you think this is true for any field with a non-archimedean absolute value? If not, can you come up with a counter-example?

2.4 Algebra

So far, we have mostly concentrated on the geometry we obtain from an absolute value on a field **k**. In this section, we take a more algebraic point-of-view, and look for connections between (non-archimedean) absolute values and the algebraic structure[6] of the underlying field. These connections turn out to be quite serious. In fact, they point to a tight connection between geometric and algebraic properties of such fields. (This section necessarily requires a little more background in abstract algebra than the preceding ones, but shouldn't be very hard to manage.)

To begin with, every non-archimedean absolute value is attached to a subring of the field **k**, and this subring has some rather nice properties:

Proposition 2.4.1 *Let* **k** *be a field, and let* $|\ |$ *be a non-archimedean valuation on* **k**. *The set*

$$\mathcal{O} = \overline{B}(0,1) = \{x \in \mathbf{k} : |x| \leq 1\}$$

is a subring of **k**. *Its subset*

$$\mathfrak{P} = B(0,1) = \{x \in \mathbf{k} : |x| < 1\}$$

is an ideal of $\mathcal{O}$. *Furthermore,* $\mathfrak{P}$ *is a maximal ideal in* $\mathcal{O}$, *and every element of the complement* $\mathcal{O} - \mathfrak{P}$ *is invertible in* $\mathcal{O}$.

---

[6]in the sense of abstract algebra

Problem 58 Prove the Proposition. It is all a matter of using the definitions directly, and remembering that the absolute value is non-archimedean. Notice that the statement about the complement of $\mathfrak{P}$ implies at once that $\mathfrak{P}$ is a maximal ideal.

Rings that contain a unique maximal ideal whose complement consists of invertible elements are called *local rings*. The Proposition, then, shows us how to attach to any non-archimedean absolute value on $\mathbb{k}$ a subring of $\mathbb{k}$ which is a local ring. Let's give it a name:

Definition 2.4.2 *Let* $\mathbb{k}$ *be a field and* $|\ |$ *be a non-archimedean absolute value on* $\mathbb{k}$. *The subring*

$$\mathcal{O} = \overline{B}(0,1) = \{x \in \mathbb{k} : |x| \leq 1\} \subset \mathbb{k}$$

is called the valuation ring *of* $|\ |$. *The ideal*

$$\mathfrak{P} = B(0,1) = \{x \in \mathbb{k} : |x| < 1\} \subset \mathcal{O}$$

is called the valuation ideal *of* $|\ |$. *The quotient*

$$\kappa = \mathcal{O}/\mathfrak{P}$$

is called the residue field *of* $|\ |$.

(For the residue field, remember that the quotient of a ring by a maximal ideal is always a field.)

It is natural to expect that many of the properties of the absolute value are connected to algebraic properties of its associated valuation ring. In fact, one can develop the theory by concentrating on this side of things (so that finding an absolute value on a field gets translated into finding a subring with certain properties). Exactly what properties characterize the rings that arise in this way is a question that will be touched upon in one of the problems for this section.

Since we'll be mostly interested in the p-adic absolute values, let's record what we get in that case:

Proposition 2.4.3 *Let* $\mathbb{k} = \mathbb{Q}$ *and let* $|\ | = |\ |_p$ *be the p-adic absolute value. Then:*

   i) the associated valuation ring is $\mathcal{O} = \mathbb{Z}_{(p)} = \{a/b \in \mathbb{Q} : p \nmid b\}$;

   ii) the valuation ideal is $\mathfrak{P} = p\mathbb{Z}_{(p)} = \{a/b \in \mathbb{Q} : p \nmid b \text{ and } p \mid a\}$;

   iii) the residue field is $\kappa = \mathbb{F}_p$ *(the field with p elements).*

PROOF: All we need to do is remember the definitions. We have

$$\left|\frac{a}{b}\right| = p^{-v} \qquad \text{when} \qquad \frac{a}{b} = p^v \frac{a_1}{b_1} \quad \text{with } p \nmid a_1 b_1.$$

So we get that $a/b \in \mathcal{O}$ if and only if $v \geq 0$. If a/b is in lowest terms, this just means $p \nmid b$, as claimed. Similarly, $a/b \in \mathfrak{P}$ happens when $v > 0$, hence when $p \nmid b$ and $p|a$. The last statement is an easy exercise in quotient rings.$\square$

Problem 59 Prove the last statement in the Proposition. (Hint: the jazzy proof begins with the inclusion $\mathbb{Z} \hookrightarrow \mathbb{Z}_{(p)}$, and checks that it induces a map on quotient rings.)

Problem 60 Compute the valuation ring, valuation ideal, and the residue field for the non-archimedean valuations on $F(t)$ introduced above.

One could go further in exploring these connections between absolute values and algebraic structure, but we will stop here, at least for now. As we go along, we will develop a clearer feeling for how the connection works by finding out more and more about the specific case of the p-adic absolute value. The following problems use a little more background from abstract algebra.

Problem 61 Consider $\mathbb{Q}$ with a p-adic absolute value, and let $a \in \mathbb{Z}$. Describe the open ball $B(a, 1)$ with center a and radius 1 in terms of the algebraic structure. Use your description to interpret algebraically the fact (Problem 49) that the closed ball $\overline{B}(0, 1)$ is the disjoint union of open balls of radius 1.

Problem 62 In the case of the p-adic absolute value, the valuation ideal is a *principal* ideal, that is, it is the set of multiples of an element of $\mathcal{O}$ (to wit, the element p). Is this always the case for the examples we have considered? Make a conjecture as to whether it will always be the case for any non-archimedean absolute value. (Hint: if so, it shouldn't be too hard to prove in general. . .)

Problem 63 Let k be a field, and let $|\ |$ be an absolute value on k. Define a valuation v on k by

$$v(x) = -\log|x|$$

for $x \neq 0$ and $v(0) = +\infty$. Check that if $|\ |$ is non-archimedean then this is indeed a valuation (i.e., it has the properties listed in Lemma 2.1.3).

  i) If $|\ |$ is the p-adic absolute value, how does v relate to the p-adic valuation v_p? What is the image of v in this case?

  ii) Show that the valuation ideal of $|\ |$ is a principal ideal if and only if the image of v is a *discrete* additive subgroup of $\mathbb{R}$. (We showed above that the image is a subgroup; the point here is the discreteness, which means that each element of the subgroup is contained in an open interval that does not contain any other elements of the subgroup.)

  iii) Show that if the image of v is a discrete subgroup of $\mathbb{R}$ then the valuation ring $\mathcal{O}$ is a principal ideal domain whose only prime ideals are 0 and $\mathfrak{P}$. (For example, check that this happens for the p-adic absolute values.)

3 p-adic Numbers

Having built our foundation, we can now apply the general theory to the specific case of the field $\mathbb{Q}$ of rational numbers. Extending our scope to include all fields of algebraic numbers (i.e., finite extensions of $\mathbb{Q}$), or even to include what the experts call "global fields" in general, would not be very hard. Nevertheless, we have preferred to stick, at first, to the most concrete example available. In a later chapter, we will consider some aspects of the problem of extending valuations from $\mathbb{Q}$ to larger fields. More details about the theory of valuations on global fields can be found in several of the references.

3.1 Absolute Values on $\mathbb{Q}$

We have already found a few examples of absolute values on the field $\mathbb{Q}$ of rational numbers. The next step will be to show that these are essentially all the possible absolute values; for that we will need to introduce a refined notion of what it means for two absolute values to be "the same." Up to that notion of equivalence, we will be able to show that the absolute values we have are the complete list of possible absolute values on $\mathbb{Q}$. Finally, we will prove the *product formula* as an initial example of how all the absolute values work together in the arithmetic of $\mathbb{Q}$.

We begin by recording what has been achieved so far, namely that we have constructed the following absolute values on the field $\mathbb{Q}$:

- the trivial absolute value;

- the "usual" absolute value $|\ |_\infty$, which we have called the "absolute value at infinity," and which is associated to the real numbers;

- for each prime p, the p-adic absolute value $|\ |_p$.

Notice that, except for the trivial absolute value (which we will tend to ignore), we have written all of these in the form $|\ |_p$, where p is either a prime or ∞. It turns out to be convenient to think of the symbol ∞ as some sort of prime number in $\mathbb{Z}$, and refer to it as "the infinite prime," and to the corresponding absolute value as the "∞-adic" absolute value. This will allow us to say things like "$|\ |_p$ for all primes $p \leq \infty$." Though there are some reasons[1] for doing this, at this point we will use it only as a notational convenience.

To be able to state our main theorem in this section, we must first make a good definition of when two absolute values are "the same." The main idea here is that we use absolute values on a field k to introduce a topology (open and closed sets, connectedness, etc.) on k. So it is reasonable to define:

Definition 3.1.1 *Two absolute values* $|\ |_1$ *and* $|\ |_2$ *on a field* k *are called equivalent if they define the same topology on* k, *that is, if every set that is open with respect to one is also open with respect to the other.*

This is easier to say than to check, so we had better find a more accessible criterion:

Lemma 3.1.2 *Let* $|\ |_1$ *and* $|\ |_2$ *be absolute values on a field* k. *The following statements are equivalent:*

i) $|\ |_1$ *and* $|\ |_2$ *are equivalent absolute values;*

ii) for any $x \in$ k *we have* $|x|_1 < 1$ *if and only if* $|x|_2 < 1$;

iii) there exists a positive real number α *such that for every* $x \in$ k *we have*

$$|x|_1 = |x|_2^\alpha.$$

PROOF: We follow the usual method of proving a circle of implications

$$(i) \Longrightarrow (ii) \Longrightarrow (iii) \Longrightarrow (i)$$

(1) First, suppose (i), i.e., that $|\ |_1$ and $|\ |_2$ are equivalent. Then any sequence that converges with respect to one absolute value must also converge in the other (because the topologies are the same). But, given any $x \in$ k, it is easy to see that

$$\lim_{n \to \infty} x^n = 0$$

with respect to the topology induced by an absolute value $|\ |$ if and only if $|x| < 1$. This gives (ii).

(2) We leave it to the reader to prove that (ii) implies (iii), not because it is easy, but because it is the hardest part of the theorem, and the convoluted

---

[1]The reasons hinge on the close connection between primes and absolute values that we are about to establish. If all the other absolute values correspond to primes, then so should the usual absolute value. As to why it should be called the *infinite* prime, that is far less clear. In fact, John H. Conway has been heard to argue quite vigorously that the "usual" absolute value should be attached to the "prime" -1, and this does seem to make more sense. (Think of the ± 1 that appears in prime factorizations.) Unfortunately, number theorists are too used to talking of "primes at infinity" for this to change easily, and we have preferred to go along with convention.

argument that one ends up resorting to can only be appreciated after one has become convinced that easier methods don't work. The next problem includes some hints.

(3) If we assume (iii), we get that

$$|x - a|_1 < r \iff |x - a|_2^\alpha < r \iff |x - a|_2 < r^{1/\alpha},$$

so that any open ball with respect to $|\ |_1$ is also an open ball (albeit of different radius) with respect to $|\ |_2$. This is enough to show that the topologies defined by the two absolute values are identical.   □

Problem 64 Prove step (2) above. The first hurdle is finding the number α. For that, just choose any appropriate x_0 and choose α to be the unique real number that will make $|x_0|_1 = |x_0|_2^\alpha$. The proof will be done if you can show that the same equation will hold for every $x \in k$; it is here that you have to find a way to use condition (ii). (This is quite hard, but worth a try. The argument suggested in Appendix A is quite sophisticated, and it will be hard to understand why it is needed unless some effort has been expended to do it in an easier way.)

Problem 65 Let $|\ |_1$ and $|\ |_2$ be two absolute values on a field k. If every open ball with respect to one of these is also an open ball with respect to the other, show that the induced topologies are identical, i.e., that every set that is open with respect to the one is open with respect to the other. (Hint: this only requires a straightforward reading of the definition.)

Problem 66 Show that we can add the following condition to the list in the proposition:

iv) for any $x \in \mathbf{k}$, we have $|x|_1 \leq 1$ if and only if $|x|_2 \leq 1$.

Problem 67 Suppose that $|\ |$ is an absolute value that is equivalent to the trivial absolute value. Must it *be* the trivial absolute value? Do we need to change the definition of "nontrivial"?

Problem 68 Show that if p and q are two different primes, the p-adic and the q-adic absolute values are not equivalent. Do the same when p is a prime and $q = \infty$.

Problem 69 Show that in general a non-archimedean absolute value cannot be equivalent to an archimedean absolute value.

As an example, recall that we considered, in Problem 31, an absolute value defined by

$$|x| = c^{-v_p(x)},$$

where $c > 1$ was a real number. Now we can check that this is equivalent to the p-adic absolute value—just choose α so that $c^\alpha = p$. We will see later that the choice $c = p$ is dictated by "global" considerations (namely, the product formula).

Now we come to the main theorem in this section. It says that we have already found all the absolute values on $\mathbb{Q}$.

Theorem 3.1.3 (Ostrowski) *Every non-trivial absolute value on $\mathbb{Q}$ is equivalent to one of the absolute values $|\ |_p$, where either p is a prime number or $p = \infty$.*

PROOF: Let $|\ |$ be a non-trivial absolute value on $\mathbb{Q}$. We will consider the possible cases.

a) Suppose, first, that $|\ |$ is *archimedean*. We want to show, in this case, that it is equivalent to the "usual" (∞-adic) absolute value. Let n_0 be the least positive integer for which $|n_0| > 1$ (there has to be one, because otherwise $|\ |$ would be non-archimedean). Now of course we can find a positive real number α so that

$$|n_0| = n_0^{\alpha}.$$

(Finding a formula for α is an easy exercise on logarithms.) We claim that this α will do, that is, that it will realize the equivalence between $|\ |$ and $|\ |_{\infty}$. That means that we want to prove that for every $x \in \mathbb{Q}$ we have $|x| = |x|_{\infty}^{\alpha}$. Given the known properties of absolute values, this will follow if we know it for positive integers, that is, if we show that $|n| = n^{\alpha}$ for any positive integer n. (Check this!)

We know that the equality holds for $n = n_0$. To prove it in general, we use a little trick. Take an arbitrary integer n, and write "in base n_0," i.e., in the form

$$n = a_0 + a_1 n_0 + a_2 n_0^2 + \cdots + a_k n_0^k,$$

with $0 \le a_i \le n_0 - 1$ and $a_k \ne 0$. Notice that k is determined by the inequality $n_0^k \le n < n_0^{k+1}$, which says that

$$k = \left\lfloor \frac{\log n}{\log n_0} \right\rfloor,$$

where $\lfloor x \rfloor$ denotes the "floor" of x, that is, the largest integer that is less than or equal to x. Now take absolute values. We get

$$|n| = |a_0 + a_1 n_0 + a_2 n_0^2 + \cdots + a_k n_0^k|$$
$$\le |a_0| + |a_1| n_0^{\alpha} + |a_2| n_0^{2\alpha} + \cdots + |a_k| n_0^{k\alpha}$$

Since we chose n_0 to be the *smallest* integer whose absolute value was greater than 1, we know that $|a_i| \le 1$, so that we get

$$|n| \le 1 + n_0^{\alpha} + n_0^{2\alpha} + \cdots + n_0^{k\alpha}$$

$$= n_0^{k\alpha}\left(1 + n_0^{-\alpha} + n_0^{-2\alpha} + \cdots + n_0^{-k\alpha}\right)$$

$$\le n_0^{k\alpha} \sum_{i=0}^{\infty} n_0^{-i\alpha}$$

$$= n_0^{k\alpha} \frac{n_0^{\alpha}}{n_0^{\alpha} - 1}.$$

If we set $C = n_0^\alpha/(n_0^\alpha - 1)$ (which is, the reader will note, a positive number), we can read this as saying that

$$|n| \leq Cn_0^{k\alpha} \leq Cn^\alpha.$$

Now we use a dirty trick. This formula applies for every n (since the one we chose was arbitrary); applying it to an integer of the form n^N we get

$$|n^N| \leq Cn^{N\alpha}$$

(the crucial point is that the number C does not depend on n—check its definition above!). Taking N-th roots, we get

$$|n| \leq \sqrt[N]{C}\, n^\alpha.$$

Since any N will do, we can let $N \to \infty$, which makes $\sqrt[N]{C} \to 1$, and so gives an inequality: $|n| \leq n^\alpha$. This is half of what we want.

Now we need to show the inequality in the opposite direction. For that, we go back to the expression in base n_0

$$n = a_0 + a_1 n_0 + a_2 n_0^2 + \cdots + a_k n_0^k.$$

Since $n_0^{k+1} > n \geq n_0^k$, we get

$$n_0^{(k+1)\alpha} = |n_0^{k+1}| = |n + n_0^{k+1} - n| \leq |n| + |n_0^{k+1} - n|,$$

so that

$$|n| \geq n_0^{(k+1)\alpha} - |n_0^{k+1} - n| \geq n_0^{(k+1)\alpha} - (n_0^{k+1} - n)^\alpha,$$

where we have made use of the inequality proved in the previous paragraph. Now since $n \geq n_0^k$, it follows that

$$|n| \geq n_0^{(k+1)\alpha} - (n_0^{k+1} - n_0^k)^\alpha$$

$$= n_0^{(k+1)\alpha}\left(1 - \left(1 - \frac{1}{n_0}\right)^\alpha\right)$$

$$= C' n_0^{(k+1)\alpha}$$

$$> C' n^\alpha,$$

and once again $C' = 1 - (1 - 1/n_0)^\alpha$ does not depend on n and is positive. Using precisely the same trick as before, we get the reverse inequality $|n| \geq n^\alpha$, and hence $|n| = n^\alpha$. This proves that $|\ |$ is equivalent to the "usual" absolute value $|\ |_\infty$, as claimed.

b) Now suppose $|\ |$ is *non-archimedean*. Then, as we have shown, we have $|n| \leq 1$ for every integer n. Since $|\ |$ is non-trivial, there must exist a smallest integer n_0 such that $|n_0| < 1$.

The first thing to see is that n_0 must be a prime number. To see that, suppose that $n_0 = a \cdot b$ with a and b both smaller than n_0. Then, by our choice for n_0, we would have $|a| = |b| = 1$ and $|n_0| < 1$, which cannot be. Thus, n_0 is prime, so let's call it by a prime-like name; set $p = n_0$. Now, of course, we want to show that $|\ |$ is equivalent to the p-adic absolute value, where p is this particular prime.

The next step is to show that if $n \in \mathbb{Z}$ is not divisible by p, then $|n| = 1$. This is not too hard. If we divide n by p we will have a remainder, so that we can write

$$n = rp + s$$

with $0 < s < p$. By the minimality of p (see the preceding paragraph), we have $|s| = 1$. We also have $|rp| < 1$, because $|r| \leq 1$ (because $|\ |$ is non-archimedean) and $|p| < 1$ (by construction). Since $|\ |$ is non-archimedean (and therefore "all triangles are isosceles"), it follows that $|n| = 1$.

Finally, given any $n \in \mathbb{Z}$, write it as $n = p^v n'$ with $p \nmid n'$. Then

$$|n| = |p|^v |n'| = |p|^v = c^{-v},$$

where $c = |p|^{-1} > 1$, so that $|\ |$ is equivalent to the p-adic absolute value, as claimed. $\square$

Problem 70 There's one fishy thing about the first part of the proof: once we have the conclusion we know $n_0 = 2$, but while we're proving we have to consider the possibility that n_0 is large. So we might have $n < n_0$, which would make the k in the expansion in base n_0 equal to zero. In other words, if $n < n_0$ its expansion in base n_0 is just n. Do we need to modify the proof to account for this case?

This theorem is the main reason for thinking of the "usual" absolute value $|\ |_\infty$ (or of the inclusion $\mathbb{Q} \hookrightarrow \mathbb{R}$ from which it comes) as some sort of "prime" of $\mathbb{Q}$. The point is that then it is true that every absolute value of $\mathbb{Q}$ "comes from" a (finite or infinite) prime.

There are lots of contexts in arithmetic where it is useful to work with "all of the primes," that is, to use information obtained from all of the absolute values of $\mathbb{Q}$. In terms of general "feeling," the real absolute value records information related to *sign*, while the other absolute values record information related to the various primes. Here is the most fundamental example of this:

Proposition 3.1.4 (Product Formula) *For any* $x \in \mathbb{Q}^\times$, *we have*

$$\prod_{p \leq \infty} |x|_p = 1,$$

where $p \leq \infty$ *means that we take the product over all of the primes of* $\mathbb{Q}$, *including the "prime at infinity."*

PROOF: It is easy to see that we only need to prove the formula when x is a positive integer, and that the general case will then follow. So let x be a positive integer, which we can factor as $x = p_1^{a_1} \cdot p_2^{a_2} \cdots p_k^{a_k}$. Then we have

$$\begin{cases} |x|_q = 1 & \text{if } q \neq p_i \\ |x|_{p_i} = p_i^{-a_i} & \text{for } i = 1, 2, \ldots, k \\ |x|_\infty = p_1^{a_1} \cdot p_2^{a_2} \cdots p_k^{a_k} \end{cases}$$

The result then follows. □

This formula establishes a close relation between the absolute values of $\mathbb{Q}$; for example, it says that if we know all but one of the absolute values of a number $x \in \mathbb{Q}$, then we can determine the missing one. This turns out to be surprisingly important in many applications (for example, the theory of heights on algebraic varieties).

A similar result is true for finite extensions of $\mathbb{Q}$, except that in that case we must use *several* "infinite primes" (one for each different inclusion into $\mathbb{R}$ or $\mathbb{C}$). Of course, we also need an extension of Ostrowski's theorem for this to make sense, and a correct notion of a "prime" in such a field. It is because of these technicalities that we have chosen to deal only with the theory over $\mathbb{Q}$. See the references for the general case.

3.2 Completions

We are now ready to construct the p-adic fields $\mathbb{Q}_p$. The main point will be to pursue the idea that all of the absolute values on $\mathbb{Q}$ are "equally important," and hence should be treated equally. We first need to recall three important concepts from basic topology (we only state them in the context of fields with absolute values, but they are really general concepts for metric spaces).

Definition 3.2.1 *Let* $\Bbbk$ *be a field and let* $|\ |$ *be an absolute value on* $\Bbbk$.

i) *A sequence of elements* $x_n \in \Bbbk$ *is called a* Cauchy *sequence if for every* $\varepsilon > 0$ *one can find a bound* M *such that we have* $|x_n - x_m| < \varepsilon$ *whenever* $m, n \geq M$.

ii) *The field* $\Bbbk$ *is called* complete *with respect to* $|\ |$ *if every Cauchy sequence of elements of* $\Bbbk$ *has a limit.*

iii) *A subset* $S \subset \Bbbk$ *is called* dense *in* $\Bbbk$ *if every open ball around every element of* $\Bbbk$ *contains an element of* S; *in symbols, if for every* $x \in \Bbbk$ *and every* $\varepsilon > 0$ *we have*

$$B(x, \varepsilon) \cap S \neq \emptyset.$$

The reader has probably met these concepts in a course on real analysis, since one of the big things about the field $\mathbb{R}$ of real numbers is that it is a *complete* field, i.e., that every Cauchy sequence converges. In intuitive terms, a Cauchy sequence is a sequence that "ought to" have a limit, because its terms get crowded into smaller and smaller balls (think of choosing a sequence of smaller and smaller values for ε). In other words, a field is complete if sequences that ought to converge do converge.

Problem 71 Show that $\mathbb{Q}$ is not complete with respect to the usual absolute value $|\ |_\infty$. (This was done in real analysis, too; one way is to construct a Cauchy sequence whose limit, if it existed, would have to be the square root of 2. Since 2 has no square root in $\mathbb{Q}$, there can be no limit.)

The following problem is intended to deal with a very common misunderstanding (which the reader also probably met in her course on real analysis). It is especially important to get this straight now, because things will get confusing for non-archimedean absolute values.

Problem 72 Show that the condition

$$\lim_{n \to \infty} |x_{n+1} - x_n| = 0$$

is *not the same* as the Cauchy condition, by showing that there exists a sequence of real numbers that satisfies this condition but is not a Cauchy sequence. In informal terms, the Cauchy condition is *stronger* than the assertion that successive terms of the sequence get closer and closer together. (Hint: one example of such a sequence was met in Calculus, in the portion on series...)

Our reason for recalling these notions is that, as our theory now stands, the archimedean absolute value $|\ |_\infty$ is different from all the rest, because there exists an inclusion $\mathbb{Q} \hookrightarrow \mathbb{R}$ of $\mathbb{Q}$ into a field $\mathbb{R}$ (yes, we do mean the real numbers) which satisfies the following conditions:

- the absolute value $|\ |_\infty$ extends to $\mathbb{R}$,

- $\mathbb{R}$ is *complete* with respect to the metric given by this absolute value, and

- $\mathbb{Q}$ is *dense* in $\mathbb{R}$ (with respect to the metric given by $|\ |_\infty$).

This is all probably well-known to the reader (see the standard references for proofs). We summarize that list of properties by saying that $\mathbb{R}$ is the *completion* of $\mathbb{Q}$ with respect to the absolute value $|\ |_\infty$. The point is that $\mathbb{R}$ is the smallest field containing $\mathbb{Q}$ which is complete with respect to this absolute value. We can see this because any such field would have to include the limit of any Cauchy sequence of elements of $\mathbb{Q}$, and, since $\mathbb{Q}$ is dense in $\mathbb{R}$, any element of $\mathbb{R}$ is a limit of such a sequence.

Problem 73 Can you prove the assertions of the preceding paragraph?

Our main goal in this section is to restore the parity between the absolute values on $\mathbb{Q}$, by constructing, for each of the other absolute values, a completion analogous to $\mathbb{R}$. That is, we want to show that for each prime p there exists some field to which we can extend the p-adic absolute value, which is then complete with respect to the extended absolute value, and in which $\mathbb{Q}$ is dense. The existence of such a field is a general theorem about metric spaces, which the reader may have met in another context; if so[2], she may prefer to skip directly to the end of this section. In any case, this section is for those who wish to see the full construction[3] of such a completion.

Problem 74 Should we bother trying to construct a completion of $\mathbb{Q}$ with respect to the trivial absolute value?

For the rest of this section, we let $|\ | = |\ |_p$ be the p-adic absolute value on $\mathbb{Q}$, for some prime p. The first useful thing to note is that Cauchy sequences can be characterized much more simply when the absolute value is non-archimedean.

Lemma 3.2.2 *A sequence (x_n) of rational numbers is a Cauchy sequence with respect to a non-archimedean absolute value $|\ |$ if and only if we have*

$$\lim_{n \to \infty} |x_{n+1} - x_n| = 0.$$

PROOF: If $m = n + r > n$, we get

$$|x_m - x_n| = |x_{n+r} - x_{n+r-1} + x_{n+r-1} - x_{n+r-2} + \cdots + x_{n+1} - x_n|$$

$$\leq \max\{|x_{n+r} - x_{n+r-1}|, |x_{n+r-1} - x_{n+r-2}|, \ldots, |x_{n+1} - x_n|\},$$

because the absolute value is non-archimedean. The result then follows at once. $\qquad\qquad\qquad\square$

This makes analysis much simpler when the field is non-archimedean, as we will see later. We should insist, once again, that this Lemma is false for archimedean absolute values, as Problem 72 shows.

The next step is to show that $\mathbb{Q}$ is not complete with respect to the p-adic absolute values, so that the completion process is really going to accomplish something.

Lemma 3.2.3 *The field $\mathbb{Q}$ of rational numbers is not complete with respect to any of its nontrivial absolute values.*

---

[2]or if she is willing to grant the existence of a completion

[3]One remark is important: as in the case of the construction of the real numbers, the method of constructing our completion is less important than the properties of the resulting field. In other words, the construction itself is important *only* because it establishes the existence of a completion. It will not be of any further use after that, so that skipping this section is a real possibility.

PROOF: Given Ostrowski's Theorem (3.1.3), we need to check this for $|\ |_p$ for $p \leq \infty$. That $\mathbb{Q}$ is not complete for $|\ |_\infty$ is well known (and left as a Problem above), so we look at the p-adic absolute values.

If we take $|\ | = |\ |_p$ for some prime p, we need to construct a Cauchy sequence in $\mathbb{Q}$ which does not have a limit in $\mathbb{Q}$. This was essentially the content of a problem from Chapter 1. To construct the necessary Cauchy sequence, we need only find a coherent sequence of solutions modulo p^n of an equation that has no solution in $\mathbb{Q}$. We work this out in the case $p \neq 2$, and leave the case $p = 2$ to the reader.

Thus, suppose $p \neq 2$ is a prime. Choose an integer $a \in \mathbb{Z}$ such that

- a is not a square in $\mathbb{Q}$;

- p does not divide a;

- a is a quadratic residue modulo p, i.e., the congruence $X^2 \equiv a \pmod{p}$ has a solution.

For example, we might take any square in $\mathbb{Z}$ and add a multiple of p to get a suitable a. Now we can construct a Cauchy sequence (with respect to $|\ |_p$) in the following way:

- choose x_0 to be any solution of $x_0^2 \equiv a \pmod{p}$;

- choose x_1 so that $x_1 \equiv x_0 \pmod{p}$ and $x_1^2 \equiv a \pmod{p^2}$ (the existence of x_1 was proved in one of the problems in Chapter 1, and is easy to see in any case);

- in general, choose x_n so that

$$x_n \equiv x_{n-1} \pmod{p^n} \quad \text{and} \quad x_n^2 \equiv a \pmod{p^{n+1}}$$

(the same remark applies as to existence).

It was in fact checked in Problem 18 that such sequences do exist whenever the initial element x_0 exists (it is here that we need to know that $p \neq 2$).

The next step is to check that we really have a Cauchy sequence. It is clear from the construction that we have

$$|x_{n+1} - x_n| = |\lambda p^{n+1}| \leq p^{-(n+1)} \to 0,$$

which shows, together with Lemma 3.2.2, that the sequence of the x_n is indeed a Cauchy sequence. On the other hand, we also know that

$$|x_n^2 - a| = |\mu p^{n+1}| \leq p^{-(n+1)} \to 0,$$

so that the limit, if it existed, would have to be a square root of a. Since a is not a square, there can be no limit, which shows $\mathbb{Q}$ is not complete with respect to $|\ |_p$.   □

Problem 75 Finish the proof, by showing that $\mathbb{Q}$ is also not complete with respect to the 2-adic absolute value. (Hint: the easiest way is probably to use cube roots instead of square roots...)

Since $\mathbb{Q}$ is not complete, we need to construct a completion. There are several ways to do so. We will follow the path of least resistance. What we want to do is to "add to $\mathbb{Q}$ the limits of all the Cauchy sequences." Since at first no such limits exist, one cannot literally do that. What we do instead is to use a standard mathematician's ruse, replacing the limit we do not have with the sequence we do have (so that in the end the sequence will be sort of like a limit of itself!). To do that, we begin with the set of all Cauchy sequences as the basic object, then use the algebraic operations on $\mathbb{Q}$ to handle the resulting object. (The construction uses some notions from abstract algebra; these can be avoided, but doing so would make our life much harder.)

Definition 3.2.4 *Let* $|\ | = |\ |_p$ *be a non-archimedean absolute value on* $\mathbb{Q}$. *We denote by* $\mathcal{C}$, *or* $\mathcal{C}_p(\mathbb{Q})$ *if we want to emphasize* p *and* $\mathbb{Q}$, *the set of all Cauchy sequences of elements of* $\mathbb{Q}$:

$$\mathcal{C} = \mathcal{C}_p(\mathbb{Q}) = \{(x_n) : (x_n) \text{ is a Cauchy sequence with respect to } |\ |_p\}.$$

The first thing to check is that $\mathcal{C}$ has a natural ring structure, using the "obvious" definitions for the sum and product of two sequences.

Proposition 3.2.5 *Defining*

$$(x_n) + (y_n) = (x_n + y_n)$$

$$(x_n) \cdot (y_n) = (x_n y_n)$$

makes $\mathcal{C}$ *a commutative ring with unity.*

PROOF: Easy; the only thing that really needs checking is that the sequences on the right hand side are Cauchy. $\qquad\qquad\square$

Problem 76 Check that the sum and product of two Cauchy sequences, as defined above, are also Cauchy sequences.

Problem 77 What is the zero element of the ring $\mathcal{C}$? What is the unit element? Can you decide which elements are invertible?

The ring $\mathcal{C}$ is not a field (as is clear from the previous exercise, since not all non-zero elements are invertible). In fact, it contains "zero divisors," i.e., non-zero elements whose product is zero.

Problem 78 Find two non-zero Cauchy sequences (say, with respect to the p-adic absolute value, but it doesn't really matter) whose product is the zero sequence.

We should check at once that this huge ring does contain the field of rational numbers, since, after all, the point of the whole exercise is to construct something which extends $\mathbb{Q}$. In fact, all we need to do is notice that if $x \in \mathbb{Q}$ is any number, the sequence

$$x,\ x,\ x,\ x,\ \ldots$$

is certainly Cauchy; we will call it the *constant sequence associated to* x and denote it by (x). Then we have

Lemma 3.2.6 *The map $x \mapsto (x)$ is an inclusion of $\mathbb{Q}$ into $\mathcal{C}$.*

PROOF: This is clear from the definitions.   □

The main problem with $\mathcal{C}$ is that it does not yet capture the idea of "adding all limits of all Cauchy sequences," because different Cauchy sequences whose terms get close to each other "ought" to have the same limit, but they are different objects in $\mathcal{C}$. This sort of situation calls for identifying two sequences which "ought" to have the same limit, which means we must pass to a quotient[4] of $\mathcal{C}$.

It is here that the algebraic structure helps us, because it makes it easy to describe when it is that two sequences "ought" to have the same limit: this should happen when their terms get close to each other, i.e., when the difference of the sequences tends to zero. So we begin by looking at the set of sequences that tend to zero.

Definition 3.2.7 *We define $\mathcal{N} \subset \mathcal{C}$ to be the ideal*

$$\mathcal{N} = \{(x_n) : x_n \to 0\} = \{(x_n) : \lim_{n \to \infty} |x_n|_p = 0\}$$

of sequences that tend to zero with respect to the absolute value $|\ |_p$.

Problem 79 Check that $\mathcal{N}$ is in fact an ideal of $\mathcal{C}$. (This is really already known from way back when in Calculus class.)

Lemma 3.2.8 $\mathcal{N}$ *is a maximal ideal of $\mathcal{C}$.*

PROOF: Let $(x_n) \in \mathcal{C}$ be a Cauchy sequence that does not tend to zero (i.e., does not belong to $\mathcal{N}$), and let I be the ideal generated by (x_n) and $\mathcal{N}$. What we want to show is that I must be all of $\mathcal{C}$. We will do that by showing that the unit element (1) (i.e., the constant sequence corresponding to 1) is in I.

---

[4]The operation of passing to a quotient to identify objects is one of those absolutely basic ideas that one meets over and over in mathematics. In every case, one has to introduce an equivalence relation of some sort, then identify equivalent elements. In our situation, we will take advantage of the machinery of abstract algebra to do this, since $\mathcal{C}$ is a commutative ring.

This is enough, because any ideal that contains the unit element must be the whole ring.

Now, since (x_n) does not tend to zero and is a Cauchy sequence, it must "eventually" be away from zero, that is, there must exist a number $c > 0$ and an integer N such that $|x_n| \geq c > 0$ whenever $n \geq$ N. (If this is not clear, the reader should find a proof!) Now in particular this means that $x_n \neq 0$ for $n \geq$ N, so that we may define a new sequence (y_n) by setting $y_n = 0$ if $n <$ N and $y_n = 1/x_n$ if $n \geq$ N.

The first thing to check is that (y_n) is a Cauchy sequence. But that is clear because if $n \geq$ N we have

$$|y_{n+1} - y_n| = \left| \frac{1}{x_{n+1}} - \frac{1}{x_n} \right| = \frac{|x_{n+1} - x_n|}{|x_n x_{n+1}|} \leq \frac{|x_{n+1} - x_n|}{c^2} \longrightarrow 0,$$

which shows $(y_n) \in \mathcal{C}$ because $|\ |$ is non-archimedean. (One can modify the argument slightly so that it works also if $|\ |$ is archimedean, but this is easier.)

Now notice that,

$$x_n y_n = \begin{cases} 0 & \text{if } n < \text{N} \\ 1 & \text{if } n \geq \text{N} \end{cases}$$

This means that the product sequence $(x_n)(y_n)$ consists of a finite number of 0's followed by an infinite string of 1's. In particular, if we subtract it from the constant sequence (1), we get a sequence that tends to zero (in fact, which goes to zero and then stays there). In other words

$$(1) - (x_n)(y_n) \in \mathcal{N}.$$

But this says that (1) can be written as a multiple of (x_n) plus an element of $\mathcal{N}$, and hence belongs to I, as we had claimed.   □

Problem 80 To make sure that you understand the proof, check that it works just as well for any field **k** with an absolute value $|\ |$. (The only catch is to supply a version of the check that the "almost inverse" sequence is Cauchy that does not depend on $|\ |$ being non-archimedean. But this is easy: the use of Lemma 3.2.2 is really a red herring.)

We want to identify sequences that differ by elements of $\mathcal{N}$, on the grounds that they ought to have the same limit. This is done in the standard way, by taking the quotient of the ring $\mathcal{C}$ by the ideal $\mathcal{N}$. To make things even nicer, taking a quotient of a ring by a maximal ideal gives a field.

Definition 3.2.9 *We define the field of p-adic numbers to be the quotient of the ring $\mathcal{C}$ by its maximal ideal $\mathcal{N}$:*

$$\mathbb{Q}_p = \mathcal{C}/\mathcal{N}.$$

Notice that two different constant sequences never differ by an element of $\mathcal{N}$ (their difference is just another constant sequence...). Hence, we still have an inclusion

$$\mathbb{Q} \hookrightarrow \mathbb{Q}_p$$

by sending $x \in \mathbb{Q}$ to the equivalence class of the constant sequence (x).

Very well: we now have a field, and an inclusion of $\mathbb{Q}$ into the field. It remains to check that it has the stated properties of the completion. The first is that the absolute value $|\ |_p$ extends to $\mathbb{Q}_p$. This follows easily from the following lemma.

Lemma 3.2.10 *Let* $(x_n) \in \mathcal{C}$, $(x_n) \notin \mathcal{N}$. *The sequence of real numbers* $|x_n|_p$ *is eventually stationary, that is, there exits an integer* N *such that* $|x_n|_p = |x_m|_p$ *whenever* $m, n \geq N$.

PROOF: Since (x_n) is a Cauchy sequence which does not tend to zero, we can (as in the previous lemma) find c and N_1 such that

$$n \geq N_1 \implies |x_n| \geq c > 0.$$

On the other hand, there also exists an integer N_2 for which

$$n, m \geq N_2 \implies |x_n - x_m| < c.$$

We want both conditions to be true at once, so set $N = \max\{N_1, N_2\}$. Then we have

$$n, m \geq N \implies |x_n - x_m| < \max\{|x_n|, |x_m|\},$$

which gives $|x_n| = |x_m|$ by the non-archimedean property ("all triangles are isosceles"). □

This means that the following definition makes sense:

Definition 3.2.11 *If* $\lambda \in \mathbb{Q}_p$ *is an element of* $\mathbb{Q}_p$, *and* (x_n) *is any Cauchy sequence representing* λ, *we define*

$$|\lambda|_p = \lim_{n \to \infty} |x_n|_p.$$

(Recall that we have defined $\mathbb{Q}_p$ as a quotient, so that elements of $\mathbb{Q}_p$ are equivalence classes of Cauchy sequences.) There are several things to check here, but they are all quite easy to verify, so we leave them to the reader.

Problem 81 Let $\lambda \in \mathbb{Q}_p$. Explain why the limit defining $|\ |_p$ exists.

Problem 82 Let $\lambda \in \mathbb{Q}_p$. Show that $|\lambda|_p$, as defined above, does not depend on the choice of the sequence (x_n) representing λ. In other words, show that if we replace (x_n) by an equivalent sequence $(\tilde{x}_n)$ (which means, recall, that the difference $(x_n - \tilde{x}_n)$ is a sequence that tends to zero), then

$$\lim_{n \to \infty} |x_n|_p = \lim_{n \to \infty} |\tilde{x}_n|_p.$$

(One can either do this directly, or note that the definition of $| \ |_p$ defines a function on $\mathcal{C}$ which maps $\mathcal{N}$ to zero, and hence descends to the quotient.)

Problem 83 Let $\lambda \in \mathbb{Q}_p$. Show that $|\lambda|_p = 0$ if and only if $\lambda = 0$. (You will need to remember what it means for an element to equal zero in the quotient.)

Problem 84 Show that the function $| \ |_p : \mathbb{Q}_p \longrightarrow \mathbb{R}_+$ is a non-archimedean absolute value.

Problem 85 Let $x \in \mathbb{Q}$, and let (x) be the constant sequence which is the image of x in $\mathbb{Q}_p$. Check that the two definitions of $| \ |_p$ are consistent, that is, that $|(x)|_p = |x|_p$. (Yes, this is essentially obvious.)

These problems, taken together, show that we have indeed defined an absolute value on $\mathbb{Q}_p$ which extends the p-adic absolute value on $\mathbb{Q}$. There is one more important fact which should be recorded, which is that the set of values is the same for both fields.

Problem 86 Show that the image of $\mathbb{Q}$ under $| \ |_p$ is equal to the image of $\mathbb{Q}_p$ under $| \ |_p$. In other words, for any $\lambda \in \mathbb{Q}_p$ which is different from zero, there exists $n \in \mathbb{Z}$ such that $|\lambda|_p = p^{-n}$.

To check that we have indeed obtained the completion, we must now check the remaining two requirements: that $\mathbb{Q}$ is dense in $\mathbb{Q}_p$, and that $\mathbb{Q}_p$ is complete. The first is easy:

Proposition 3.2.12 *The image of $\mathbb{Q}$ under the inclusion $\mathbb{Q} \hookrightarrow \mathbb{Q}_p$ is a dense subset of $\mathbb{Q}_p$.*

PROOF: We need to show that any open ball around an element $\lambda \in \mathbb{Q}_p$ contains an element of (the image of) $\mathbb{Q}$, i.e., a constant sequence. So fix a radius ε. We will show that there is a constant sequence belonging to the open ball $B(\lambda, \varepsilon)$.

First of all, let (x_n) be a Cauchy sequence representing λ, and let ε' be a number slightly smaller than ε. By the Cauchy property, there exists a number N such that $|x_n - x_m| < \varepsilon'$ whenever $n, m \geq N$. Let $y = x_N$ and consider the constant sequence (y). We claim that

$$(y) \in B(\lambda, \varepsilon),$$

i.e., that $|\lambda - (y)| < \epsilon$. To see this, recall that $\lambda - (y)$ is represented by the sequence $(x_n - y)$, and that we have defined

$$|(x_n - y)| = \lim_{n \to \infty} |x_n - y|.$$

But for any $n \geq N$ we have

$$|x_n - y| = |x_n - x_N| < \varepsilon'$$

so that, in the limit, we get

$$\lim_{n\to\infty} |x_n - y| \leq \varepsilon' < \varepsilon,$$

so that (y) does indeed belong to $B(\lambda, \varepsilon)$, and we are done.   □

Problem 87 Why does $<$ become $\leq$ in the limit? Do we really need the business of decreasing ε slightly to ε'?

It remains to show that $\mathbb{Q}_p$ is complete, i.e., that every Cauchy sequence in $\mathbb{Q}_p$ converges to an element of $\mathbb{Q}_p$. This seems almost obvious, until one realizes that a Cauchy sequence of elements of $\mathbb{Q}_p$ amounts to a sequence of Cauchy sequences, which seems to make everything very confusing. In fact, it is not so hard if one keeps one's wits, but it is the sort of thing best done in the privacy of one's own home, so we leave it to the reader:

Problem 88 Show that $\mathbb{Q}_p$ is complete with respect to $|\ |_p$. To do this, follow these steps:

i) Let $\lambda_1, \lambda_2, \ldots, \lambda_n, \ldots$ be a Cauchy sequence of elements of $\mathbb{Q}_p$ (so that each λ "is" a Cauchy sequence of elements of $\mathbb{Q}$, taken up to equivalence). Use the fact that the image of $\mathbb{Q}$ is dense in $\mathbb{Q}_p$ to show that it is possible to find rational numbers $y^{(1)}, y^{(2)}, \ldots, y^{(n)}, \ldots$ such that we have

$$\lim_{n\to\infty} |\lambda_n - (y^{(n)})| = 0.$$

(Read carefully! This says that the absolute value of the difference between λ_n and the *constant sequence* $(y^{(n)})$ tends to zero. The absolute value is taken in $\mathbb{Q}_p$, not in $\mathbb{Q}$.)

ii) Show that the rational numbers $y^{(1)}, y^{(2)}, \ldots, y^{(n)}, \ldots$ themselves form a Cauchy sequence in $\mathbb{Q}$. Let λ denote the element of $\mathbb{Q}_p$ corresponding to this sequence.

iii) Show that
$$\lim_{n\to\infty} \lambda_n = \lambda.$$

iv) Conclude that $\mathbb{Q}_p$ is complete.

Putting it all together, we have proved the following theorem:

Theorem 3.2.13 *For each prime $p \in \mathbb{Z}$ there exists a field $\mathbb{Q}_p$ with a non-archimedean absolute value $|\ |_p$, such that:*

i) there exists an inclusion $\mathbb{Q} \hookrightarrow \mathbb{Q}_p$, and the absolute value induced by $|\ |_p$ on $\mathbb{Q}$ via this inclusion is the p-adic absolute value;

ii) the image of $\mathbb{Q}$ under this inclusion is dense in $\mathbb{Q}_p$ (with respect to the absolute value $|\ |_p$); and

iii) $\mathbb{Q}_p$ is complete with respect to the absolute value $|\ |_p$.

The field $\mathbb{Q}_p$ satisfying (i), (ii) and (iii) is unique up to unique isomorphism preserving the absolute values.

PROOF: We've done it all except the uniqueness statement. To get that, suppose we have another such field K. Then we can think of the inclusion $\mathbb{Q} \hookrightarrow K$ as a map defined on a dense subset of $\mathbb{Q}_p$. Since this map has to preserve the absolute values of any element of $\mathbb{Q}$, it is continuous. Now, any map defined on a dense subset which is continuous can be extended uniquely to the whole field, so that we get a map $\mathbb{Q}_p \longrightarrow K$ which is the unique continuous extension of the inclusion of $\mathbb{Q}$ in K. It is now easy to check that it is an isomorphism that preserves the absolute values, and its uniqueness is clear by construction.   □

Problem 89 Fill in the gaps in the uniqueness proof:

   i) Since the inclusion preserves the operations on $\mathbb{Q}$, and these operations are continuous, show that the extended map is a homomorphism of fields (and hence is injective).

   ii) Perform precisely the same construction in reverse to get a map in the opposite direction, and show that the resulting map is the inverse of the first. (Hint: the composition is a continuous map which restricts to the identity on $\mathbb{Q}$!)

   iii) Check that the isomorphism thus constructed preserves absolute values. (Hint: is the absolute value function itself continuous?)

   The strong uniqueness statement is important because it says we can now forget the construction of $\mathbb{Q}_p$, and work only with the properties specified in the theorem. This is precisely what we will do.

Problem 90 Why is it important that something be "unique up to unique isomorphism"? Can you give an example of some mathematical object that is unique up to isomorphism, but not up to unique isomorphism?

3.3 Exploring $\mathbb{Q}_p$

The goal of this section is to explore the field $\mathbb{Q}_p$ which we have just constructed. The basic idea for the whole section is to get away from the explicit construction we gave above. Since, as we showed, the field $\mathbb{Q}_p$ is entirely determined by its properties, we can forget the construction, and begin the exploration from the list of properties we have just obtained:

- there is an absolute value $|\ | = |\ |_p$ on $\mathbb{Q}_p$, and $\mathbb{Q}_p$ is complete with respect to this absolute value;

- there is an inclusion $\mathbb{Q} \hookrightarrow \mathbb{Q}_p$ whose image is dense in $\mathbb{Q}_p$, and the restriction of the absolute value $|\ |_p$ to (the image of) $\mathbb{Q}$ coincides with the p-adic absolute value;

- the set of values of $\mathbb{Q}$ and of $\mathbb{Q}_p$ under $|\ |_p$ is the same; specifically, the two sets

$$\{x \in \mathbb{R}_+ : x = |\lambda|_p \text{ for some } \lambda \in \mathbb{Q}\}$$

and

$$\{x \in \mathbb{R}_+ : x = |\lambda|_p \text{ for some } \lambda \in \mathbb{Q}_p\}$$

are both equal to the set $\{p^n : n \in \mathbb{Z}\} \cup \{0\}$ of powers of p, together with 0.

As the wording suggests, we will from now on *identify* $\mathbb{Q}$ with its image under the inclusion in $\mathbb{Q}_p$, that is, we will think of $\mathbb{Q}$ as a subfield of $\mathbb{Q}_p$. The last property turns out to be very useful, so we will re-state it as a lemma.

Lemma 3.3.1 *For each $x \in \mathbb{Q}_p$, $x \neq 0$, there exists an integer $n \in \mathbb{Z}$ such that $|x|_p = p^{-n}$.*

Another way of saying this is in terms of the p-adic valuation v_p. Remember that for $x \in \mathbb{Q}$ we had $|x|_p = p^{-v_p(x)}$; so what the lemma says is:

Lemma 3.3.2 *For each $x \in \mathbb{Q}_p$, $x \neq 0$, there exists an integer $v_p(x)$ such that $|x|_p = p^{-v_p(x)}$. In other words, the p-adic valuation v_p extends to $\mathbb{Q}_p$.*

As before, we extend v_p to all of $\mathbb{Q}_p$ by setting $v_p(0) = +\infty$. Later in this section (when we have a good way to describe elements of $\mathbb{Q}_p$) we will be able to describe v_p in a more precise way.

Now we begin to explore the structure of $\mathbb{Q}_p$. Since $\mathbb{Q}_p$ is a field with a non-archimedean valuation, we can consider the corresponding valuation ring, as in Chapter 2. The resulting ring has a name of its own:

Definition 3.3.3 *The* ring of p-adic integers *is the valuation ring*

$$\mathbb{Z}_p = \{x \in \mathbb{Q}_p : |x|_p \leq 1\}.$$

Of course, $\mathbb{Z}_p$ is also the closed unit ball with center 0, so we already know a few things about it. For example, it is a clopen set in $\mathbb{Q}_p$, because every ball is. Here is a much more precise description:

Proposition 3.3.4 *The ring $\mathbb{Z}_p$ of p-adic integers is a local ring whose maximal ideal is the principal ideal $p\mathbb{Z}_p = \{x \in \mathbb{Q}_p : |x|_p < 1\}$. Furthermore,*

i) $\mathbb{Q} \cap \mathbb{Z}_p = \mathbb{Z}_{(p)} = \left\{ \dfrac{a}{b} \in \mathbb{Q} : p \nmid b \right\}.$

ii) The inclusion $\mathbb{Z} \hookrightarrow \mathbb{Z}_p$ has dense image. In particular, given $x \in \mathbb{Z}_p$ and $n \geq 1$, there exists $\alpha \in \mathbb{Z}$, $0 \leq \alpha \leq p^n - 1$, such that $|x - \alpha| \leq p^{-n}$. The integer α with these properties is unique.

iii) For any $x \in \mathbb{Z}_p$, there exists a Cauchy sequence α_n converging to x, of the following type:

- *$\alpha_n \in \mathbb{Z}$ satisfies $0 \leq \alpha_n \leq p^n - 1$*
- *for every n we have $\alpha_n \equiv \alpha_{n-1} \pmod{p^{n-1}}$.*

The sequence (α_n) with these properties is unique.

PROOF: Most of this follows directly from things we have already checked. To begin with, $\mathbb{Z}_p$ is a valuation ring, hence is a local ring. To see that the valuation ideal is indeed generated by p, we use Lemma 3.3.1:

$$|x| < 1 \implies |x| \leq \frac{1}{p} \implies \left|\frac{x}{p}\right| \leq 1 \implies x \in p\mathbb{Z}_p.$$

This shows that the valuation ideal is contained in $p\mathbb{Z}_p$, but that is enough, since the valuation ideal is a *maximal* ideal, and $p\mathbb{Z}_p \neq \mathbb{Z}_p$. Now to the other statements:

(i) is clear, because we already know that $\mathbb{Z}_{(p)}$ is the valuation ring in $\mathbb{Q}$ corresponding to the p-adic valuation.

To check (ii), choose $x \in \mathbb{Z}_p$ and $n \geq 1$. Since $\mathbb{Q}$ is dense in $\mathbb{Q}_p$, one can certainly find $a/b \in \mathbb{Q}$ which is close enough to x so that

$$\left|x - \frac{a}{b}\right| \leq p^{-n} < 1.$$

The point is to show that we can in fact choose an *integer*. But notice that for a/b as above, we will have

$$\left|\frac{a}{b}\right| \leq \max\left\{|x|, \left|x - \frac{a}{b}\right|\right\} \leq 1,$$

which says that $a/b \in \mathbb{Z}_{(p)}$, that is, $p \nmid b$. Now recall that, from the elementary theory of congruences, if $p \nmid b$ there exists an integer $b' \in \mathbb{Z}$ such that $bb' \equiv 1 \pmod{p^n}$, which implies (the reader will check) that

$$\left|\frac{a}{b} - ab'\right| \leq p^{-n},$$

and of course $ab' \in \mathbb{Z}$. Finally, we need to check that we can find an integer between zero and $p^n - 1$, but this is clear from the connection between congruences modulo powers of p and the p-adic absolute value: choosing α to be the unique integer such that

$$0 \leq \alpha \leq p^n - 1 \qquad \text{and} \qquad \alpha \equiv ab' \pmod{p^n}$$

gives $|x - \alpha| \leq p^{-n}$ (check it!), which is what we want.

Finally, (iii) follows directly from (ii); just use (ii) for a sequence of integers $n = 1, 2, \ldots$. $\qquad \square$

This proposition says several important things (and implies a bunch of others—see the next few Corollaries). For example, it says that $\mathbb{Z}_p$ is the completion of $\mathbb{Z}$ with respect to the p-adic absolute value, which would be another way to begin the whole story. Notice, too, that the sequence (iii) is exactly one of our "coherent sequences" from Chapter 1, so that things are coming together rather nicely. Here are some more consequences.

Corollary 3.3.5 $\mathbb{Q}_p = \mathbb{Z}_p[1/p]$, that is, for every $x \in \mathbb{Q}_p$ there exists $n \geq 0$ such that $p^n x \in \mathbb{Z}_p$. The map $\mathbb{Q}_p \longrightarrow \mathbb{Q}_p$ given by $x \mapsto px$ is a homeomorphism. (This means that it is a continuous map with a continuous inverse, so it preserves the topology of $\mathbb{Q}_p$.) The sets $p^n \mathbb{Z}_p$, $n \in \mathbb{Z}$ form a fundamental system of neighborhoods of $0 \in \mathbb{Q}_p$ which covers all of $\mathbb{Q}_p$.

PROOF: If $x \in \mathbb{Q}_p$, we can compute its valuation $v_p(x)$. If $v_p(x) \geq 0$, then x is already an element of $\mathbb{Z}_p$. Otherwise, $v_p(x)$ is negative, and we have

$$v_p(p^{-v_p(x)}x) = -v_p(x) + v_p(x) = 0,$$

which means that $p^{-v_p(x)}x \in \mathbb{Z}_p$, as claimed. That multiplication by p is a homeomorphism is immediate from the fact that the field operations are continuous functions. The remaining statements will be checked in the next problem. □

Problem 91 Prove the corollary. Recall that a *neighborhood* of a point x is a set containing an open ball around x, and that a *fundamental system of neighborhoods* is a bunch of neighborhoods with the property that any other neighborhood contains one of them. Finally, a collection of sets *covers* a set X if the union of all the sets in the collection contains (or is) the set X.

It may be useful to remember that a map is continuous exactly when the inverse image of any open set is an open set (this is often easier to work with than the ε-δ definition).

Problem 92 (Just to keep us awake.) Describe a fundamental system of neighborhoods of 0 in $\mathbb{R}$ which also covers $\mathbb{R}$.

Recall that we pointed out that the p-adic valuation v_p can be extended to $\mathbb{Q}_p$, because for any $x \in \mathbb{Q}_p$ there exists an integer $v_p(x)$ such that $|x|_p = p^{-v_p(x)}$. The last corollary allows us to understand this a little better:

Problem 93 Show that we can give the following more natural description of $v_p(x)$: by the corollary, x belongs to some $p^n \mathbb{Z}_p$; let n_0 be the *largest* n for which this is true; then $v_p(x) = n_0$. (Be careful: n_0 may very well be negative.)

Hence, for example, $v_p(x) = 0$ if $x \in \mathbb{Z}_p$ but $x \notin p\mathbb{Z}_p$, so that $n_0 = 0$. This agrees, of course, with the original definition, since $v_p(x) = 0$ means $|x| = 1$.

One of the main points of these results is that the topology (neighborhoods, open sets,...) of $\mathbb{Q}_p$ is closely connected to its algebraic structure (multiplication by p, subrings). For example, it is very useful to burn into one's brain that for $x, y \in \mathbb{Q}_p$ we have

$$|x - y| \leq p^{-n} \qquad \text{if and only if} \qquad x - y \in p^n \mathbb{Z}_p.$$

The next few results forge ahead in this direction.

Corollary 3.3.6 *For any $n \geq 1$, the sequence*

$$0 \longrightarrow \mathbb{Z}_p \xrightarrow{p^n} \mathbb{Z}_p \longrightarrow \mathbb{Z}/p^n\mathbb{Z} \longrightarrow 0,$$

where the map $\mathbb{Z}_p \longrightarrow \mathbb{Z}_p$ is given by $x \mapsto p^n x$, is exact, and the maps are continuous (where we give $\mathbb{Z}/p^n\mathbb{Z}$ the discrete topology). In particular,

$$\mathbb{Z}_p/p^n\mathbb{Z}_p \cong \mathbb{Z}/p^n\mathbb{Z}.$$

Recall that a sequence $A \xrightarrow{f} B \xrightarrow{g} C$ is *exact* if image(f) = ker(g). A five-term sequence as above is exact when it is exact at each stage, so that the claims above are:

- the map $\mathbb{Z}_p \longrightarrow \mathbb{Z}_p$ given by multiplication by p^n is injective (its kernel is the image of zero, which is zero)

- there is a map $\mathbb{Z}_p \longrightarrow \mathbb{Z}/p^n\mathbb{Z}$ which is surjective

- the kernel of this map is precisely the image of $\mathbb{Z}_p$ under the first map, which of course is $p^n\mathbb{Z}_p$.

Recall, too, that the discrete topology is the one where *all* sets are open.

Problem 94 Check that the Corollary is true.

Problem 95 Use the Corollary (plus other facts about $\mathbb{Z}_p$) to show that if $x \in \mathbb{Z}_p$, $x \neq 0$, then $nx \neq 0$ for any $n \in \mathbb{Z}$ unless $n = 0$. In group-theoretic parlance, this says that the additive group $\mathbb{Z}_p^+$ is *torsion-free*.

The sets $a + p^n\mathbb{Z}_p$, with $a \in \mathbb{Q}$ and $n \in \mathbb{Z}$ are closed balls in $\mathbb{Q}_p$ (with center a and radius p^{-n}), hence are clopen sets. Since $\mathbb{Q}$ is dense in $\mathbb{Q}_p$, they cover all of $\mathbb{Q}_p$. As we have already shown, the space $\mathbb{Q}_p$ is totally disconnected (the connected component of any point is the set consisting of only that point). Furthermore, given any two points we can always find balls around them that do not intersect (which is a useful thing to know about a topology: points can be separated). In big words:

Corollary 3.3.7 $\mathbb{Q}_p$ *is a totally disconnected Hausdorff topological space.*

A more interesting topological property is compactness, which plays a big role in classical analysis. A subset X of a topological space is called *compact* if it has the following property:

- any collection of open sets which covers X has a *finite* subcollection which also covers X.

This is a rather unintuitive definition, but it turns out to be quite important. For example, the compact sets in $\mathbb{R}$ are precisely the closed and bounded sets, which play a big role in real analysis.

Problem 96 Read up on compactness in any introductory book on general topology. In particular, prove, or find out how to prove, the following:

   i) A closed interval in $\mathbb{R}$ is compact.

   ii) The image of a compact set by a continuous map is a compact set.

   iii) Any sequence of points contained in a compact set has a subsequence which is convergent.

   iv) In a metric space, a set X will be compact if it is complete (every Cauchy sequence in X converges to a point in X) and totally bounded (for every ε, there exists a *finite* covering of X by balls of radius ε).

Problem 97 A space is called *locally compact* when every point has a neighborhood which is a compact set. Show that $\mathbb{R}$ is locally compact. (This property is very important in classical analysis.)

Problem 98 If **k** is a field with an absolute value, show that **k** is locally compact if and only if there exists a neighborhood of zero that is compact. (Hint: if a set X is compact, the set $\{a + x : x \in X\}$ is the image of X under a continuous map, hence is also compact.)

Corollary 3.3.8 $\mathbb{Z}_p$ *is compact, and* $\mathbb{Q}_p$ *is locally compact.*

PROOF: Since $\mathbb{Z}_p$ is a neighborhood of zero, proving that it is compact is enough to prove that $\mathbb{Q}_p$ is locally compact, so that the second statement follows from the first.

To prove the first statement, remember that we already know that $\mathbb{Z}_p$ is complete (because it is a closed set in a complete field), so that (using one of the statements above), what we need to prove is that it is totally bounded, that is, that for any ε one can cover $\mathbb{Z}_p$ with finitely many balls of radius ε. It is enough to check this for every $\varepsilon = p^{-n}$, $n \geq 0$. But remember that

$$\mathbb{Z}_p / p^n \mathbb{Z}_p \cong \mathbb{Z} / p^n \mathbb{Z},$$

and that the cosets of $p^n \mathbb{Z}_p$ in $\mathbb{Z}_p$ are also balls in the *p*-adic topology. This means that as a ranges through $0, 1, \ldots, p^n - 1$ (or any other set of coset representatives), the p^n balls

$$a + p^n \mathbb{Z}_p = \{a + p^n x : x \in \mathbb{Z}_p\} = \{y \in \mathbb{Z}_p : |y - a| \leq p^{-n}\} = \overline{B}(a, p^{-n})$$

cover $\mathbb{Z}_p$, and we are done. $\qquad\square$

Problem 99 Why is it enough to check for this special family of values for ε?

One should notice that the crucial element in the compactness is the finiteness of the quotients. In fact, one can check that knowing that one quotient is finite will do the trick.

Problem 100 Let $\mathbf{k}$ be a field, $|\ |$ a non-archimedean absolute value on $\mathbf{k}$, $\mathcal{O} \subset \mathbf{k}$ the valuation ring, and $\mathfrak{P}$ the valuation ideal. Suppose that $\mathbf{k}$ is complete and that $\mathfrak{P}$ is principal. Show that $\mathbf{k}$ is locally compact if and only if the residue field $\mathcal{O}/\mathfrak{P}$ is finite. Do we really need the completeness of $\mathbf{k}$? Do we really need to know that $\mathfrak{P}$ is principal?

The elements of $\mathbb{Q}_p$ are, at this point, hard to grab hold of, because we only "know" $\mathbb{Q}_p$ via its basic properties. To counteract this a little, we will now give two different descriptions of the elements of $\mathbb{Q}_p$, both of which we have already met in Chapter 1: as "coherent sequences," and as "p-adic expansions." The description in terms of coherent sequences, which we will give first, is interesting for theoretical reasons, while the description in terms of expansions will give us the most "concrete" version of $\mathbb{Q}_p$. The first description will be stated in rather sophisticated terms, and the reader may want to skim through it rather than check all the details.

We begin from item (iii) in the last proposition: given $x \in \mathbb{Z}_p$, we can find a rather special kind of Cauchy sequence converging to x. This sequence has the property of being "coherent," which we met in Chapter 1:

- $\alpha_n \in \mathbb{Z}$, $0 \le \alpha_n \le p^n - 1$

- $\alpha_{n+1} \equiv \alpha_n \pmod{p^n}$

and in addition converges to x because $|x - \alpha_n|_p \le p^{-n}$. Finally, we checked that this sequence is unique.

On the other hand, suppose we have such a sequence (α_n). The coherence property clearly makes it a Cauchy sequence, because $|\alpha_{n+1} - \alpha_n|_p \le p^{-n}$. Hence, it must converge to some element, which will be in $\mathbb{Z}_p$ because the α's are in $\mathbb{Z}$.

Problem 101 Check that a limit of a Cauchy sequence of integers must be an element of $\mathbb{Z}_p$ (rather than merely of $\mathbb{Q}_p$).

This means that we can *identify* the elements of $\mathbb{Z}_p$ with such sequences. We will summarize this in the next proposition, but in a rather sophisticated language. To set it up, let's write φ_n for the projection on the quotient

$$\varphi_n : \mathbb{Z}_p \longrightarrow \mathbb{Z}/p^n\mathbb{Z}.$$

As an element of $\mathbb{Z}/p^n\mathbb{Z}$, we then have $\varphi_n(x) = \alpha_n \pmod{p^n}$ (just because the set of integers between 0 and $p^n - 1$ gives representatives for the cosets, and the α_n are chosen as the representatives corresponding to x). We also set

$$A_n = \mathbb{Z}/p^n\mathbb{Z}$$

and think of it as a topological ring with a discrete topology[5]. We have an obvious map $\psi_n : A_n \longrightarrow A_{n-1}$, which sends $(a \mod p^n)$ to $(a \mod p^{n-1})$. We want to consider the product of all these rings, that is, the ring of sequences (α_n) such that $\alpha_n \in A_n$. (The operations are defined in the obvious way, term by term.) There is a standard way to put a topology on this ring (it is called the *product topology*). This topology is rather tricky to describe, and we do not really need to know much about it. We just point out that the product ring will be compact with this topology.

With all that set up, we can state:

Proposition 3.3.9 *The projection maps φ_n together give an inclusion*

$$\varphi : \mathbb{Z}_p \hookrightarrow \prod_{n \geq 1} A_n$$

which identifies $\mathbb{Z}_p$, as a topological ring, with the closed subring of $\prod A_n$ consisting of the coherent sequences, i.e., those sequences (α_n) for which we have $\psi_n(\alpha_n) = \alpha_{n-1}$ for every $n > 1$.

PROOF: If all the concepts are understood, this is just a re-statement of known facts. See the next problem. □

Problem 102 Prove the proposition. Notice that we could use this to give another construction of $\mathbb{Z}_p$, with a more algebraic flavor (and a bit more subtle to handle). For example, the fact that closed subsets of a compact set are necessarily compact would provide the proof that $\mathbb{Z}_p$ is compact in this version of the theory.

It is often useful to describe how several functions are related by drawing what is called a "commutative diagram." One says a diagram of homomorphisms is *commutative* if the homomorphisms obtained by "following different routes around the diagram" always coincide. For example, the diagram

---

[5]This is mumbo-jumbo. All it means is that all the other rings will have a topology because they have absolute values. The ring A_n, on the other hand, doesn't come with such a "built-in" topology, so we just give it the simplest one of all: the one where all sets are open, which corresponds to the trivial absolute value. The point is that, this makes the important map, which is the projection from $\mathbb{Z}_p$ to A_n, be continuous.

is commutative, because reducing modulo p^n and then reducing modulo p^{n-1} is the same as reducing modulo p^{n-1} by itself, so that one can follow either path from $\mathbb{Z}$ to A_{n-1} and get the same map. Using the language of commutative diagrams, one can describe a very important property of $\mathbb{Z}_p$:

Problem 103 Show that $\mathbb{Z}_p$ has the following property, which is an instance of what are usually called *universal properties*: given any ring R plus homomorphisms $R \longrightarrow A_n$ (one for each $n \geq 1$) such that all the triangles

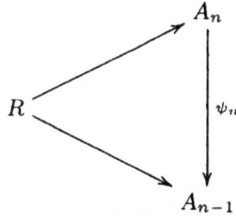

are commutative, there exists a unique homomorphism $R \longrightarrow \mathbb{Z}_p$ from which all the maps to the A_n are obtained (i.e., all the obvious triangles commute). In highfalutin' terms, this says that $\mathbb{Z}_p$ is the *inverse limit* of the A_n.

One can begin the theory from this point, and deduce all the rest from the universal property; this is the approach in [Ser73]. For ordinary mortals, however, this may all be a little too abstract, so we go on to obtain a canonical way to represent the elements of $\mathbb{Q}_p$ as "power series in p." This will finally return us to the picture we sketched in Chapter 1.

We begin with a p-adic integer $x \in \mathbb{Z}_p$. As we have just shown, there exists a coherent sequence of integers α_n converging to x such that:

- $\alpha_n \equiv x \pmod{p^n}$

- $\alpha_{n+1} \equiv \alpha_n \pmod{p^n}$

- $0 \leq \alpha_n \leq p^n - 1$.

To understand the α_n a little better, we write them in base p. The point is that for integers written in base p the process of reducing modulo p^n is very simple: just strip off all but the last n digits[6]. This means that the coherence condition

$$\alpha_{n+1} \equiv \alpha_n \pmod{p^n}$$

simply says that the last n digits of both numbers are the same. Going up the sequence, what we get is

$$
\begin{aligned}
\alpha_0 &= b_0 & 0 \leq b_0 \leq p-1 \\
\alpha_1 &= b_0 + b_1 p & 0 \leq b_1 \leq p-1 \\
\alpha_2 &= b_0 + b_1 p + b_2 p^2 & 0 \leq b_2 \leq p-1 \\
\alpha_3 &= b_0 + b_1 p + b_2 p^2 + b_3 p^3 & 0 \leq b_3 \leq p-1
\end{aligned}
$$

---

[6]Just as in base 10: to get your number modulo 10, keep the last digit only; to get it modulo 100, keep the last two, and so on.

and so on. Putting all of this together, we get an infinitely long expansion

$$x = b_0 + b_1 p + b_2 p^2 + \cdots + b_n p^n + \cdots$$

Of course, to be able to really write that equals sign with a clear conscience, we must check that the series on the right does converge to x. But that is easy:

Lemma 3.3.10 *Given any $x \in \mathbb{Z}_p$, the series*

$$b_0 + b_1 p + b_2 p^2 + \cdots + b_n p^n + \cdots$$

obtained as above converges to x.

PROOF: Remember that a series converges to x if the sequence of its partial sums converges to x. But the partial sums of our series are exactly the α_n, which we already know converge to x (we picked them that way). □

To sum up, this gives

Corollary 3.3.11 *Every $x \in \mathbb{Z}_p$ can be written in the form*

$$x = b_0 + b_1 p + b_2 p^2 + \cdots + b_n p^n + \cdots$$

with $0 \le b_i \le p - 1$, and this representation is unique.

PROOF: We have checked all but the uniqueness. To see that, notice that we already know the α_n are unique, and this implies that the b_n are too (because they are just the digits[7] in base p). □

Now, we need to get all of $\mathbb{Q}_p$. But remember that any element of $\mathbb{Q}_p$ can be written in the form y/p^m with $y \in \mathbb{Z}_p$. If we express y as a power series in p, then divide by p^m, we just get a power series in p where some of the powers may be negative. So:

Corollary 3.3.12 *Every $x \in \mathbb{Q}_p$ can be written in the form*

$$\begin{aligned}
x &= b_{-n_0} p^{-n_0} + \cdots + b_0 + b_1 p + b_2 p^2 + \cdots + b_n p^n + \cdots \\
&= \sum_{n \ge -n_0} b_n p^n
\end{aligned}$$

with $0 \le b_n \le p - 1$ and $-n_0 = v_p(x)$. This representation is unique.

PROOF: All that remains to be checked is the statement about $v_p(x)$, which is clear. □

---

[7]Should they be called pigits? Or pits?

This lands us right back in Chapter 1, and shows that one can think of an element of $\mathbb{Q}_p$, i.e., a p-adic number, as a p-adic expansion. As we noted in passing before, the coefficients b_n must be taken in a set of representatives of the classes modulo p. The numbers between 0 and $p - 1$ are only the most obvious choice for these representatives. There are situations, however, where other choices are expedient.

Problem 104 Show that the condition $0 \le b_n \le p - 1$ can be replaced by the condition $b_n \in X$, where X is any set of coset representatives for the quotient $\mathbb{Z}_p/p\mathbb{Z}_p$. (Note that the condition on X is that it be a subset of $\mathbb{Z}_p$ which gives coset representatives, so that the b_n don't even need to be integers!)

The *p-adic units* are the invertible elements of $\mathbb{Z}_p$. We will denote the set of all such elements by $\mathbb{Z}_p^\times$. Since $x \in \mathbb{Z}_p$ means $|x| \le 1$ and $x^{-1} \in \mathbb{Z}_p$ means $|x^{-1}| = |x|^{-1} \le 1$, we see that

$$\mathbb{Z}_p^\times = \{x \in \mathbb{Q}_p : |x| = 1\}.$$

It is also easy to see that

$$\mathbb{Z}_p^\times \cap \mathbb{Q} = \left\{ \frac{a}{b} \in \mathbb{Q} : p \nmid ab \right\}.$$

As in every ring, the p-adic units form a group. In our case, this group contains quite a few elements (notice that $\mathbb{Z}_p^\times \cap \mathbb{Q}$ is already quite large). We will later study its structure a little more closely.

Problem 105 Let $x \in \mathbb{Z}_p$. What condition on its p-adic expansion will guarantee that x is a p-adic unit?

Problem 106 What are the invertible elements of $\mathbb{Z}$? Of $F[t]$? Of $\mathbb{C}[[t]]$ (power series in one variable with coefficients in $\mathbb{C}$)?

Problem 107 One of the consequences of the fact that $\mathbb{Z}_p$ is compact is the fact that every infinite sequence of elements of $\mathbb{Z}_p$ has a convergent subsequence. Use the p-adic expansion to show this directly.

3.4   Hensel's Lemma

The theorem known as "Hensel's Lemma" is probably the most important algebraic property of the p-adic numbers (and of other fields like $\mathbb{Q}_p$, which are complete with respect to a non-archimedean valuation). Basically, it says that in many circumstances one can decide quite easily whether a polynomial has roots in $\mathbb{Z}_p$. The test involves finding an "approximate" root of the polynomial, and then verifying a condition on the derivative[8] of the polynomial.

---

[8]To be precise, the formal derivative, so that if $F(X) = a_0 + a_1 X + a_2 X^2 + \cdots + a_n X^n$ is the polynomial, $F'(X) = a_1 + 2a_2 X + \cdots + na_n X^{n-1}$ is its derivative. There is no limit process involved.

Theorem 3.4.1 (Hensel's Lemma) *Let* $F(X) = a_0 + a_1 X + a_2 X^2 + \cdots + a_n X^n$ *be a polynomial whose coefficients are in* $\mathbb{Z}_p$. *Suppose that there exists a p-adic integer* $\alpha_1 \in \mathbb{Z}_p$ *such that*

$$F(\alpha_1) \equiv 0 \pmod{p\mathbb{Z}_p}$$

and

$$F'(\alpha_1) \not\equiv 0 \pmod{p\mathbb{Z}_p},$$

where $F'(X)$ *is the (formal) derivative of* $F(X)$. *Then there exists a unique p-adic integer* $\alpha \in \mathbb{Z}_p$ *such that* $\alpha \equiv \alpha_1 \pmod{p\mathbb{Z}_p}$ *and* $F(\alpha) = 0$.

PROOF: We will show that the root α exists by constructing a Cauchy sequence of integers converging to it. The idea is essentially what is known as "Newton's method" in the classical theory. The attentive reader will recognize an idea that we have been using repeatedly since the first chapter.

What we will construct is a sequence of integers $\alpha_1, \alpha_2, \ldots, \alpha_n, \ldots$ such that, for all $n \geq 1$, we have

i) $F(\alpha_n) \equiv 0 \pmod{p^n}$,

ii) $\alpha_n \equiv \alpha_{n+1} \pmod{p^n}$.

It is easy to see that such a sequence will be Cauchy (in fact, it is a "coherent sequence" in our terms above), and that its limit α will satisfy $F(\alpha) = 0$ (by continuity) and $\alpha \equiv \alpha_1 \pmod{p}$ (by construction). Conversely, a root α will determine such a sequence α_n. Thus, once we have the α_n the theorem will be proved.

The main assumption in the theorem is that α_1 exists. To find α_2, we note that condition (ii) requires that

$$\alpha_2 = \alpha_1 + b_1 p$$

for some $b_1 \in \mathbb{Z}_p$. Plugging this expression into the polynomial $F(X)$ and expanding, we get

$$
\begin{aligned}
F(\alpha_2) \ &= F(\alpha_1 + b_1 p) \\
&= F(\alpha_1) + F'(\alpha_1)b_1 p + \text{terms in } p^n, \ n \geq 2 \\
&\equiv F(\alpha_1) + F'(\alpha_1)b_1 p \pmod{p^2}
\end{aligned}
$$

(This is easy to check directly, but it is probably best to think of it as a kind of formal Taylor expansion—see problem 108.) To show that one can find α_2, we have to show that one can find b_1 so that

$$F(\alpha_1) + F'(\alpha_1)b_1 p \equiv 0 \pmod{p^2}.$$

Now, we know that $F(\alpha_1) \equiv 0 \pmod{p}$, so that $F(\alpha_1) = px$ for some x. The equation then becomes

$$px + F'(\alpha_1)b_1 p \equiv 0 \pmod{p^2},$$

which gives (after we divide by p)

$$x + F'(\alpha_1)b_1 \equiv 0 \pmod{p}.$$

To solve this, notice that $F'(\alpha_1)$ is not divisible by p, and hence is *invertible* in $\mathbb{Z}_p$, so that we can (and must) take

$$b_1 \equiv -x(F'(\alpha_1))^{-1} \pmod{p}.$$

(In fact, we can choose such a b_1 in $\mathbb{Z}$, with $0 \le b_1 \le p - 1$, and then b_1 is uniquely determined.) For this choice of b_1, we set $\alpha_2 = \alpha_1 + b_1 p$, which will have the stated properties.

This shows that one can take the first step: given α_1, find α_2. But a careful inspection shows that exactly the same calculation works to get α_{n+1} from α_n. Hence, we can construct the whole sequence, and it is uniquely determined at each step. This proves the theorem.   □

Problem 108 Let $F(X)$ be a polynomial with coefficients in a field **k** of characteristic zero. Show that the Taylor formula is true for $F(X)$, i.e., that

$$F(x + h) = F(x) + F'(x)h + \frac{1}{2!}F''(x)h^2 + \frac{1}{3!}F'''(x)h^3 + \ldots$$

for any $x, h \in \mathbf{k}$.

Problem 109 Check that the calculation given in the proof does indeed work to get α_{n+1} from α_n and that α is indeed unique.

Problem 110 Why do we say that the calculation in the proof is analogous to Newton's method for finding approximate solutions to polynomial equations?

Problem 111 What happens to the calculation if we do *not* assume that we have $F'(\alpha_1) \not\equiv 0 \pmod{p}$? Can you give an example where the theorem fails because this condition does not hold? (Hint: look back at our games with the polynomial $X^2 - m$.)

One should emphasize that there are many different versions of this result, all of which tend to be referred to as "Hensel's Lemma." For example, the next problem gives a version that can be used when the hypothesis on $F'(\alpha_1)$ does not hold; later on in this section we give still another example, which will be of crucial importance later on.

Problem 112 Show that in Hensel's Lemma we can weaken the condition $F'(\alpha_1) \not\equiv 0$ by replacing it with the condition $|F(\alpha_1)| < |F'(\alpha_1)|^2$. What should replace the conclusion that $\alpha \equiv \alpha_1 \pmod{p}$? Why is this version of Hensel's Lemma more general than the first? Can you give an example where this version can be used but the original version cannot? (Hint: if you did the previous problem, you should be able to do this one too.)

A nice application of Hensel's Lemma is to determine which roots of unity can be found in $\mathbb{Q}_p$. Recall that an element ζ of a field is called an *m-th root of unity* if $\zeta^m = 1$; it is called a *primitive m-th* root of unity if in addition $\zeta^n \neq 1$ for $0 < n < m$. In $\mathbb{R}$, there are only two roots of unity, 1 and -1. On the other hand, we have already checked (in a problem long ago) that the equation $X^2 + 1 = 0$ has a root in $\mathbb{Q}_5$, and it is easy to see that its root will be a fourth root of unity. So it is interesting to try to determine which roots of unity exist.

To use Hensel's Lemma, we need a polynomial. Since we are looking for roots of unity, we will use $F(X) = X^m - 1$. Notice that $F'(X) = mX^{m-1}$, so that $F'(\lambda) = m\lambda^{m-1}$ will be congruent to zero modulo p if either p divides λ (in which case λ will not be an approximate root of $F(X)$ anyway) or p divides m. Thus, the second condition in the theorem will hold provided m is not divisible by p. For the first condition, we need to find an approximate root, and it is actually quite easy to decide when that can be done:

Problem 113 Fix a prime p and a number m not divisible by p. Show that there exists an integer α_1 such that $\alpha_1^m \equiv 1 \pmod{p}$ but $\alpha_1 \not\equiv 1 \pmod{p}$ if and only if $\gcd(m, p-1) > 1$, and that for any such α_1 the least positive integer m with this property must be a divisor of $p-1$. (Hint: $\mathbb{Z}/p\mathbb{Z}$ is a field, and the set of its invertible elements is a *cyclic* group.)

Then Hensel's Lemma yields:

Proposition 3.4.2 *For any prime p and any positive integer m not divisible by p, there exists a primitive m-th root of unity in $\mathbb{Q}_p$ if and only if m divides $p-1$.*

Problem 114 Prove the proposition. There are at least two loose ends to tie: the "only if" part, and the verification that such roots of unity must be in $\mathbb{Z}_p$, and not merely in $\mathbb{Q}_p$.

If m divides $p-1$, then any m-th root of unity is also a $(p-1)$-st root of unity, so that the upshot is that the roots of unity in $\mathbb{Q}_p$ of order prime to p are exactly the $(p-1)$-st roots. This determines all the roots of unity in $\mathbb{Q}_p$, except for the possibility that there be p^n-th roots of unity in $\mathbb{Q}_p$. These are inaccessible[9] to us by this method. It turns out, however, that they are *not*

---

[9]Can you explain why? What would using Hensel's lemma to find, say, p-th roots of unity involve?

in $\mathbb{Q}_p$ (except when $p = 2$, in which case ± 1—but no fourth roots of 1—do belong to $\mathbb{Q}_2$). Hence, we have determined *all* the roots of unity belonging to $\mathbb{Q}_p$, though we will only be able to *prove* that this is the case later on. (If you can't wait, look at page 114.)

Problem 115 Show that the set of roots of unity in $\mathbb{Q}_p$ is a subgroup of the group $\mathbb{Z}_p^\times$ of p-adic units. Show that the set of $(p-1)$-st roots of unity in $\mathbb{Q}_p$ is a cyclic group of order $(p-1)$. (The main content of the last statement is that there are $(p-1)$ p-adic roots of the polynomial $X^{p-1} - 1$. Use Hensel's Lemma.)

Another interesting application is to determine the squares in $\mathbb{Q}_p$. This is something we essentially did in Chapter 1. First we do the p-adic units:

Proposition 3.4.3 *Let $p \neq 2$ be a prime, and let $b \in \mathbb{Z}_p^\times$ be a p-adic unit. If there exists α_1 such that $\alpha_1^2 \equiv b \pmod{p\mathbb{Z}_p}$, then b is the square of an element of $\mathbb{Z}_p^\times$.*

PROOF: Apply Hensel's Lemma to $X^2 - b$, and notice that $p \neq 2$ and $b \in \mathbb{Z}_p^\times$ are enough to make sure that $2\alpha_1 \not\equiv 0 \pmod{p}$. $\square$

Then we extend to all of $\mathbb{Q}_p$, by noticing that any $x \in \mathbb{Q}_p$ can be written as $x = p^{v_p(x)}x'$ with $x' \in \mathbb{Z}_p^\times$ (in fact, that is pretty much the definition of $v_p(x)$). What the next result says is that x will be a square if $v_p(x)$ is even and x' is a square.

Corollary 3.4.4 *Let $p \neq 2$ be a prime. An element $x \in \mathbb{Q}_p$ is a square if and only if it can be written $x = p^{2n}y^2$ with $n \in \mathbb{Z}$ and $y \in \mathbb{Z}_p^\times$ a p-adic unit. The quotient group $\mathbb{Q}_p^\times/(\mathbb{Q}_p^\times)^2$ has order four. If $c \in \mathbb{Z}_p^\times$ is any element whose reduction modulo p is not a quadratic residue, then the set $\{1, p, c, cp\}$ is a complete set of coset representatives.*

PROOF: The first statement is essentially obvious (because powers of p and p-adic units "do not mix"). Applying the proposition and standard properties of quadratic residues and non-residues gives the rest. $\square$

It is interesting to compare this result to its analogue in $\mathbb{R}$, which says that a real number is a square if it is positive, and that the quotient $\mathbb{R}^\times/(\mathbb{R}^\times)^2$ is of order two, with coset representatives $\{1, -1\}$. From this point of view, the Corollary can be thought of as a p-adic version of the "rule of signs" for multiplying real numbers.

We still need to consider $p = 2$. For that, we need to use the stronger form of Hensel's Lemma given in problem 112, since $F'(\alpha_1) = 2\alpha_1$ will of course always be divisible by 2.

Problem 116 Show that if $b \in \mathbb{Z}_2$, and $b \equiv 1 \pmod{8\mathbb{Z}_2}$ (so that in particular b is a 2-adic unit), then b is a square in $\mathbb{Z}_2$. Conversely, show that any 2-adic unit which is a square is congruent to 1 modulo 8. Conclude that the group $\mathbb{Q}_2^\times/(\mathbb{Q}_2^\times)^2$ has order 8, and is generated by the classes of -1, 5, and 2, so that a complete set of coset representatives is $\{1, -1, 5, -5, 2, -2, 10, -10\}$.

To conclude this section, we prove another form of Hensel's lemma, which is more general than the first. The idea is to interpret the first form of Hensel's Lemma as saying that if a polynomial factors modulo p and one of the factors is of the form $(X - \alpha)$, so that

$$f(X) \equiv (X - \alpha)g(X) \pmod{p},$$

then there is a similar factorization in $\mathbb{Z}_p[X]$. The obvious generalization is to consider arbitrary factorizations. The condition on the derivative, above, essentially says the the root α is not a double root, that is, that the second factor $g(X)$ is not divisible by $(X - \alpha)$. For general factorizations, then, the assumption should be that the factors are relatively prime (as polynomials) modulo p. Let's make this precise:

Definition 3.4.5 *Let $g(X)$ and $h(X)$ be polynomials in $\mathbb{Z}_p[X]$. Let $\bar{g}(X)$ and $\bar{h}(X) \in \mathbb{F}_p[X]$ be the polynomials obtained by reducing the coefficients modulo p. We say $g(X)$ and $h(X)$ are* relatively prime modulo p *if $\gcd(\bar{g}, \bar{h}) = 1$ in $\mathbb{F}_p[X]$, or, equivalently, if there exist polynomials $a(X), b(X) \in \mathbb{Z}_p[X]$ such that*

$$a(X)g(X) + b(X)h(X) \equiv 1 \pmod{p},$$

where we understand congruence coefficient-by-coefficient, i.e., we say two polynomials are congruent modulo p if each coefficient of one is congruent modulo p to the corresponding coefficient of the other.

Problem 117 Is being relatively prime modulo p weaker or stronger than being relatively prime in $\mathbb{Z}_p[X]$?

The next theorem says that this idea does work.

Theorem 3.4.6 (Hensel's Lemma, Second Form) *Let $f(X) \in \mathbb{Z}_p[X]$ be a polynomial with coefficients in $\mathbb{Z}_p$, and assume that there exist polynomials $g_1(X)$ and $h_1(X)$ in $\mathbb{Z}_p[X]$ such that*

i) $g_1(X)$ is monic[10]

ii) $g_1(X)$ and $h_1(X)$ are relatively prime modulo p, and

iii) $f(X) \equiv g_1(X)h_1(X) \pmod{p}$ (understood coefficient-by-coefficient).

Then there exist polynomials $g(X), h(X) \in \mathbb{Z}_p[X]$ such that

i) $g(X)$ is monic,

ii) $g(X) \equiv g_1(X) \pmod{p}$ and $h(X) \equiv h_1(X) \pmod{p}$, and

iii) $f(X) = g(X)h(X)$.

---

[10]This means that the coefficient of the term of highest degree is one.

PROOF: This is just like the original version: we start from the "approximate" factorization, and improve the approximation more and more until, in the limit, we get a factorization over $\mathbb{Z}_p$. Notice that the conditions on $g(X)$ imply that $\deg g(X) = \deg g_1(X)$.

Let d be the degree of $f(X)$, and m be the degree of $g_1(X)$ (remember that g_1 is monic). Then we can assume that $\deg(h_1) \leq d - m$ (it could be less, because the top coefficient of f could be divisible by p). We want to construct two sequences of polynomials $g_n(X)$ and $h_n(X)$ such that

i) each g_n is monic and of degree m,

ii) $g_{n+1} \equiv g_n \pmod{p^n}$ and $h_{n+1} \equiv h_n \pmod{p^n}$

iii) $f(X) \equiv g_n(X)h_n(X) \pmod{p^n}$.

(As always, we take the congruences coefficient-by-coefficient.) If we can find such sequences, we are clearly done, since going to the limit gives the desired polynomials $g(X)$ and $h(X)$. In other words, the coefficients of, say, $g(X)$ will be the limits of the corresponding coefficients of the $g_n(X)$. (Can you see why it's important to know that the degrees of the g_n are not changing?)

We already have $g_1(X)$ and $h_1(X)$; let's describe how to get $g_2(X)$ and $h_2(X)$. Since the g's are to be congruent, we must have

$$g_2(X) = g_1(X) + p\, r_1(X)$$

for some polynomial $r_1(X) \in \mathbb{Z}_p[X]$; similarly, we must have

$$h_2(X) = h_1(X) + p\, s_1(X).$$

To show that g_2 and h_2 exist, we simply have to show that it is possible to find r_1 and s_1 such that the desired conditions are satisfied. For that, we need to solve the equation

$$f(X) \equiv g_2(X)h_2(X) \pmod{p^2},$$

which we expand to

$$f(X) \equiv (g_1(X) + p\, r_1(X))(h_1(X) + p\, s_1(X)) \pmod{p^2}.$$

Multiplying out, we get

$$f(X) \equiv g_1(X)h_1(X) + p\, r_1(X)h_1(X) + p\, s_1(X)g_1(X) + p^2\, r_1(X)s_1(X)$$

$$\equiv g_1(X)h_1(X) + p\, r_1(X)h_1(X) + p\, s_1(X)g_1(X) \pmod{p^2}$$

Now remember that $f(X) \equiv g_1(X)h_1(X) \pmod{p}$, so that we have

$$f(X) - g_1(X)h_1(X) = p\, k_1(X)$$

for some $k_1(X) \in \mathbb{Z}_p[X]$. Rearranging, we get

$$p\,k_1(X) \equiv p\,r_1(X)h_1(X) + p\,s_1(X)g_1(X) \quad (\bmod\ p^2).$$

Dividing through by p, we get

$$k_1(X) \equiv r_1(X)h_1(X) + s_1(X)g_1(X) \quad (\bmod\ p).$$

This is the equation we need to solve to determine r_1 and s_1.

The first step towards doing so is to recall that we have assumed that g_1 and h_1 are relatively prime modulo p. This means that we know that there exist $a(X)$, $b(X) \in \mathbb{Z}_p[X]$ such that $a(X)g_1(X) + b(X)h_1(X) \equiv 1 \pmod{p}$. Consider, then, the two polynomials

$$\tilde{r}_1(X) = b(X)k_1(X) \qquad \text{and} \qquad \tilde{s}_1(X) = a(X)k_1(X).$$

These will almost do the trick: they clearly will make all the congruence conditions true. The only problem is that we have no control over the degree of $\tilde{r}_1(X)$, and hence cannot guarantee that $g_1(X) + p\tilde{r}_1(X)$ is monic.

To remedy that, only a slight change is needed. We already know that

$$\tilde{r}_1(X)h_1(X) + \tilde{s}_1(X)g_1(X) \equiv k_1(X) \quad (\bmod\ p).$$

Now divide $\tilde{r}_1(X)$ by $g_1(X)$, and let $r_1(X)$ be the remainder:

$$\tilde{r}_1(X) = g_1(X)q(X) + r_1(X).$$

We know, of course, that $\deg r_1(X) < \deg g_1(X)$. But now, if we set

$$s_1(X) = \tilde{s}_1(X) + h_1(X)q(X),$$

it all works out:

$$
\begin{aligned}
r_1(X)h_1(X) + s_1(X)g_1(X) &\equiv \\
&\equiv (\tilde{r}_1(X) - g_1(X)q(X))h_1(X) + (\tilde{s}_1(X) + h_1(X)q(X))g_1(X) \\
&\equiv \tilde{r}_1(X)h_1(X) - g_1(X)h_1(X)q(X) + \tilde{s}_1(X)g_1(X) + g_1(X)h_1(X)q(X) \\
&\equiv \tilde{r}_1(X)h_1(X) + \tilde{s}_1(X)g_1(X) \\
&\equiv k_1(X) \quad (\bmod\ p),
\end{aligned}
$$

so that our congruence conditions are satisfied, and the fact that the degree of $r_1(X)$ is smaller than the degree of $g_1(X)$ is enough to guarantee that $g_1(X) + pr_1(X)$ is monic, and we are done.

This shows that g_2 and h_2 exist. Since they are congruent to g_1 and h_1 modulo p, they are also relatively prime modulo p, so that there will be no difficulty in going on to the next step.

Now we repeat the argument changing the indices and exponents to find g_3 and h_3. It is an easy exercise to show that this can always be done, and that produces the sequence whose convergence proves the theorem.   □

Problem 118 To make sure you understand that final twist in the proof, work out the details for the following example. Let $p = 2$, and consider the polynomial $f(X) = 2X^2 + X + 2$. Modulo 2, this is easy to factor: just take $g_1(X) = X$ and $h_1(X) = 1$. Follow the steps in the proof to find $g_2(X)$ and $h_2(X)$, and discuss what happens if we try to use $\tilde{r}_1$ and $\tilde{s}_1$ instead of r_1 and s_1.

Problem 119 Work out the construction of g_3 and h_3 in full detail, to convince yourself that you understand the process.

Problem 120 Fill in the last step of the proof by giving a full proof, by induction, that the g_n and h_n exist for every n.

The reader will have noticed that this argument is essentially identical to the one we gave for the first version of Hensel's Lemma. It might be interesting to check whether one can formulate a stronger version that is analogous to the one in problem 112.

3.5   Local and Global

One of the consequences of Hensel's Lemma is that, given a polynomial with integer coefficients, it is usually not too hard to decide whether it has roots in $\mathbb{Z}_p$, since it is enough to find roots modulo p. The "same" is true for $\mathbb{R}$, where we can usually decide whether there are roots by sign considerations (for example, if the polynomial has different signs at x_1 and x_2, there must be a root between these two numbers).

Suppose, however, that we want to look for roots in $\mathbb{Q}$. At least this much is easy to see: if there are roots in $\mathbb{Q}$, then there are also roots in $\mathbb{Q}_p$ for every $p \leq \infty$ (i.e., in all the $\mathbb{Q}_p$ and in $\mathbb{R}$). Hence we can certainly conclude that there are *no* rational roots if there is some $p \leq \infty$ for which there are no p-adic[11] roots.

The way to think about this situation is following Hensel's original analogy: the p-adic fields (including $\mathbb{R}$) are analogous to fields of Laurent expansions, and correspond to "local" information "near" the prime p. The fact that roots in $\mathbb{Q}$ automatically are roots in $\mathbb{Q}_p$ for every p means that a "global" root is also a "local" root at each p, i.e, "everywhere."

Much more interesting would be a converse: that "local" roots could be "patched together" to give a "global" root. This would be very useful, since deciding on the existence of local roots is very easy. Here is an (easy) example of such a converse.

Proposition 3.5.1 *A number $x \in \mathbb{Q}$ is a square if and only if it is a square in every $\mathbb{Q}_p$, $p \leq \infty$.*

---

[11]Here, of course "∞-adic" means "real." In general, this section will constantly refer to all the absolute values taken together, and thus will constantly use the convention that the usual absolute value corresponds to the "prime" ∞, so that we will write $\mathbb{Q}_\infty = \mathbb{R}$ for the real numbers.

PROOF: This is really very easy: for any $x \in \mathbb{Q}_p$, we have

$$x = \pm \prod_{p < \infty} p^{v_p(x)}.$$

If x is a square at infinity, it is positive. If it is a square at a prime p, then $v_p(x)$ is even. It follows (just write out the prime factorization) that such an x is a square. $\square$

This very basic idea seems to go back to Hensel, but it was first clearly stated by Hasse: *putting together local information at all $p \le \infty$ should give global information*. Exactly in what sense this is true (if it is) depends on each specific problem, but there are many situations in which this principle plays a central role.

A very interesting example of this sort of method is the theory of diophantine equations, in which we are given an equation for which we want to find solutions in $\mathbb{Q}$, or at least to decide if any exist. This is in general an extremely difficult (and absolutely fascinating) subject, but in some cases the question can be decided by the local-global game. Consider, for example, the equation

$$X^2 + Y^2 + Z^2 = 0.$$

One sees at once that the only solution is the trivial one $X = Y = Z = 0$, because this is the only solution in $\mathbb{R}$ (and any other solution in $\mathbb{Q}$ would also be a solution in $\mathbb{R}$). Similarly, it doesn't take too much doing to see that the equation

$$X^2 + Y^2 - Z^2$$

does have a solution in $\mathbb{Q}$, and therefore in all of the $\mathbb{Q}_p$.

What one would hope for in this context is that one would have a perfect correspondence between "global" properties and "local" properties that hold "everywhere." In this example, it is clear that if a global solution (i.e., one in $\mathbb{Q}$) exists, then local solutions exist for all primes (of course, since the solution in $\mathbb{Q}$ belongs to all the $\mathbb{Q}_p$). One would also like the converse to be true, i.e., that the *lack* of a global solution could always be detected locally. To put it in other words, one would like it if the existence of a local solution for every p would guarantee the existence of a global solution. This is far from clear, however, because the local solutions in each $\mathbb{Q}_p$ live in different fields, and there seems to be no compelling reason why they should "glue together" somehow to provide a solution over $\mathbb{Q}$.

For equations like the ones above (of degree 2, homogeneous), a few experiments begin to convince us that the hope is indeed plausible, because for every equation that does not have solutions one can quickly find a prime so that the equation has no solutions in $\mathbb{Q}_p$:

i) $X^2 + Y^2 + Z^2 = 0$ has no nontrivial solutions in $\mathbb{R}$;

ii) $3X^2 + 2Y^2 - Z^2 = 0$ has no nontrivial solutions in $\mathbb{Q}_3$ (check!);

iii) $X^2 - 3Y^2 = 0$ has no nontrivial solutions in $\mathbb{Q}_7$ (check!).

This suggests the following bold statement:

Local-Global Principle: *The existence or non-existence of solutions in $\mathbb{Q}$ (global solutions) of a diophantine equation can be detected by studying, for each $p \leq \infty$, the solutions of the equation in $\mathbb{Q}_p$ (local solutions).*

Of course, as stated, this is too vague to be a "theorem," but the local-global principle has proved to be a valuable guide for the study of diophantine problems. What it has suggested is a "plan of attack" on any given equation (or type of equation): first think locally, then try to put together the local information to obtain global information.

The most naïve version of the principle would be the one we suggested above: the statement that an equation has solutions in $\mathbb{Q}$ if and only if it has solutions in all the $\mathbb{Q}_p$. This sounds wonderful, since it says that "solvable locally everywhere" is the same as "solvable globally." Unfortunately, it is false:

Problem 121 Show that the equation

$$(X^2 - 2)(X^2 - 17)(X^2 - 34) = 0$$

has a root in $\mathbb{Q}_p$ for all $p \leq \infty$, but has no roots in $\mathbb{Q}$.

Problem 122 (This is quite hard.) Show that $X^4 - 17 = 2Y^2$ is solvable locally everywhere, but is not solvable in $\mathbb{Q}$. (The existence of local solutions is easily checked; the non-existence of rational solutions is the hard part.)

One might try to salvage the principle in various ways, for example:

Problem 123 Decide whether it is true that a polynomial in one variable with co-efficients in $\mathbb{Z}$ is irreducible in $\mathbb{Q}[X]$ if and only if it is irreducible in $\mathbb{Q}_p[X]$ for every $p \leq \infty$. (Recall that a polynomial is irreducible if it does not factor into a product of polynomials of lower degree.)

Finally, here is an example where the principle is gloriously successful:

Theorem 3.5.2 (Hasse-Minkowski) *Let*

$$F(X_1, X_2, \ldots X_n) \in \mathbb{Q}[X_1, X_2, \ldots X_n]$$

be a quadratic form (that is, a homogeneous polynomial of degree 2 in n variables). The equation

$$F(X_1, X_2, \ldots X_n) = 0$$

has non-trivial solutions in $\mathbb{Q}$ if and only if it has non-trivial solutions in $\mathbb{Q}_p$ for each $p \leq \infty$.

The proof is just a little out of our reach in this book, since it requires more thorough study of quadratic forms and their properties than we are prepared to spend time on. A very good account of the proof can be found in [Ser73], where it is the culmination of the first half of the book. One should notice that this theorem completely solves the problem of deciding whether a quadratic form has non-trivial zeros, since the local question can be decided rather easily in each case. In fact, for each prime p, an appropriate version of Hensel's Lemma shows that there is a finite procedure for deciding whether the equation is solvable in $\mathbb{Q}_p$ (so that a computer could do it). It is a little worrying that there are infinitely many primes to consider, but it turns out that the whole problem can be sufficiently broken down so that one gets a finite procedure for checking for local solutions at *all* primes, so that (given the Hasse-Minkowski theorem) the whole problem gets reduced to a finite procedure.

In lieu of a proof of the Hasse-Minkowski theorem, it might be fun to work out in detail an example of its application. So let a, b, and c be rational numbers, and consider the equation

$$aX^2 + bY^2 + cZ^2 = 0.$$

We want to use the Hasse-Minkowski theorem to settle completely when it is that such an equation has non-trivial rational solutions ("non-trivial" just means "other than $X = Y = Z = 0$"). To start off, if any of a, b, and c is equal to zero, there certainly is a solution (with one variable non-zero, and the other two equal to zero). Next, it is clear that we can clear denominators, and assume that a, b, and c are integers. We can also assume that they have no common factors (which we could cancel). Finally, we can assume that a, b, and c are square-free (i.e., they have no factors which are squares), by absorbing any square factor into one of the unknowns.

Problem 124 Suppose that $a = a'n^2$. Check that any rational solution (x, y, z) of $aX^2 + bY^2 + cZ^2 = 0$ corresponds to a rational solution (nx, y, z) of $a'X^2 + bY^2 + cZ^2 = 0$. Explain why this means that we can assume that a, b, and c are square-free.

Problem 125 We have already observed that we may assume that a, b, and c have no common factors. Show that in fact we can go farther, and assume that no two of these three numbers have any common factors. In other words, we may assume that the product abc is square-free.

Very well, we are set up now as follows: we have an equation

$$aX^2 + bY^2 + cZ^2 = 0$$

where a, b, and c are pairwise relatively prime integers with no square factors. What Hasse-Minkowski tells us is that we can decide whether this equation has non-trivial rational solutions by looking at each $\mathbb{Q}_p$ in turn. So let's:

1. Suppose $p = \infty$, so that $\mathbb{Q}_p = \mathbb{R}$. It's easy to see that the equation will have a non-trivial solution exactly when a, b, and c are not all positive or all negative. (If you have any doubts, work it out!)

2. Suppose p is an odd prime that does not divide any of the coefficients. The first step towards a solution in $\mathbb{Q}_p$ is to study the solutions modulo p.

Proposition 3.5.3 *Let p be an odd prime, and let a, b, c be pairwise relatively prime integers not divisible by p. Then there exist integers x_0, y_0, and z_0, not all divisible by p, such that*

$$ax_0^2 + by_0^2 + cz_0^2 \equiv 0 \pmod{p}.$$

PROOF: This is a special case of a famous theorem due to Chevalley and Warning. It could be proved more directly, but we give a proof that works in the general case, which makes it somehow the "right" proof.

As x, y, and z run over the integers between 0 and $p-1$ (which, since we are working modulo p, is all we need to worry about), there are p^3 different triples (x, y, z). Let's try to count how many of these are solutions of

$$aX^2 + bY^2 + cZ^2 \equiv 0 \pmod{p}.$$

For that, we use a dastardly trick: notice that

$$(ax^2 + by^2 + cz^2)^{p-1} \equiv \begin{cases} 1 \pmod{p} & \text{if } (x, y, z) \text{ is not a solution} \\ 0 \pmod{p} & \text{if } (x, y, z) \text{ is a solution} \end{cases}$$

This is because, by Fermat's Little Theorem, we have $n^{p-1} \equiv 1 \pmod{p}$ whenever $n \not\equiv 0 \pmod{p}$. This means that if we let N be the total number of *non*-solutions, then

$$N \equiv \sum_{(x,y,z)} (ax^2 + by^2 + cz^2)^{p-1} \pmod{p},$$

where each of x, y, and z ranges through the numbers from 0 to $p-1$. Now, when we expand these powers, we are going to get an equation representing N as a sum of a bunch of sums of the form

$$\sum_{(x,y,z)} \alpha x^{2i} y^{2j} z^{2k}$$

with $2i + 2j + 2k = 2(p-1)$ and $\alpha \in \mathbb{Z}$. We claim that each one of these sums is zero modulo p. To see this, note that we must certainly have that one of $2i$, $2j$, and $2k$ is less than $p-1$ (if they were all $\geq p-1$, then the sum would be $\geq 3(p-1)$, which it isn't). Say $2i < p-1$ (the argument is the same in the other cases). Then we can rewrite our sum as

$$\sum_{(y,z)} \left(\alpha y^{2j} z^{2k} \sum_x x^{2i} \right).$$

Now we invoke a little lemma:

Lemma 3.5.4 *Let n be an integer, $0 \leq n < p - 1$. Then*

$$\sum_{x=0}^{p-1} x^n \equiv 0 \pmod{p}.$$

Assuming the lemma for now (the proof will come later), we see that the inner sum in the last formula is always congruent to zero modulo p. It follows that $N \equiv 0 \pmod{p}$. In other words, the number of triples that are *not* solutions is divisible by p. Since the total number of triples is p^3, we also get that the number of triples that are solutions is divisible by p.

But we already know one solution: $x = y = z = 0$! In other words, the number of triples which are solutions is at the same time divisible by p and at least 1. That means there must be more than one solution, which means there must be a solution (x, y, z) where not all three components are divisible by p, which is what we claimed.   □

To be completely happy, we just need to prove the lemma, which[12] we'll let the reader have some fun with.

Problem 126 Prove the lemma. (Hint: remember that the integers modulo p form a field, with all sorts of nice properties. Note: if $n = 0$, the sum seems to refer to 0^0; read this as simply a synonym for 1.) What is the sum congruent to for other exponents n?

What we know, then, after the proposition, is that when $p \nmid 2abc$ there always are "good" solutions (i.e., solutions that are "non-trivial mod p") of the congruence

$$aX^2 + bY^2 + cZ^2 \equiv 0 \mod p.$$

Once we know that, it's easy to settle the question in $\mathbb{Q}_p$: let (x_0, y_0, z_0) be a "solution mod p" as in the proposition; we know x_0, y_0, and z_0 are not all divisible by p; suppose $p \nmid x_0$ (otherwise, permute the names). Look at the equation

$$aX^2 + by_0^2 + cz_0^2 = 0$$

(in other words, replace the variable Y by the integer y_0, and similarly Z by z_0). This is now a polynomial in one variable, and we know that x_0 is a solution modulo p. Given our assumptions, Hensel's lemma now tells us that there is an $x \in \mathbb{Z}_p$ which is a root of this equation. But then we've done it: (x, y_0, z_0) is a non-trivial solution in $\mathbb{Q}_p$ of the original equation. The upshot:

Corollary 3.5.5 *If p is an odd prime that does not divide abc, then the equation*

$$aX^2 + bY^2 + cZ^2 = 0$$

has a non-trivial solution in $\mathbb{Q}_p$.

---

[12]You saw this coming, no?

Problem 127 Work out the details of the application of Hensel's lemma which we breezed by above.

Problem 128 At which points in the above argument did we use the assumption that $p \nmid abc$?

That handles almost all the primes, but we still have to look at what happens when $p = 2$ and when p divides one of the coefficients (we agreed above that we can assume that no one prime divides two of the coefficients).

3. Suppose $p = 2$, and a, b, and c are all odd. In this case, we will need some special condition to guarantee that there are solutions in $\mathbb{Q}_2$. Suppose a solution (x, y, z) with x, y, $z \in \mathbb{Q}_2$ exists. We can clearly assume that $\max\{|x|_2, |y|_2, |z|_2\} = 1$, i.e, that x, y, and z are 2-adic integers which are not all in $2\mathbb{Z}_2$. (Given a triple that works, multiply by a power of 2 to get this.)

Reducing mod $2\mathbb{Z}_2$, and remembering that the coefficients are all odd, we see that exactly two of x, y, and z will be 2-adic units, and the other will be divisible by 2. Suppose that y and z are units. Now, the square of a 2-adic unit will belong to $1 + 4\mathbb{Z}_2$, while the square of an element in $2\mathbb{Z}_2$ will belong to $4\mathbb{Z}_2$. So, looking mod $4\mathbb{Z}_2$, we get that

$$b + c \equiv 0 \pmod 4.$$

If instead of x being the non-unit, either y or z is, then we get a similar condition involving two other coefficients of the equation.

In other words, if $p = 2$, $2 \nmid abc$, and there is a solution in $\mathbb{Q}_2$, then the sum of two of the coefficients of the equation must be divisible by 4.

It turns out that this is also sufficient:

Problem 129 Suppose a, b, and c are all odd, and the sum of two of them is divisible by 4. Show that the equation

$$aX^2 + bY^2 + cZ^2 = 0$$

has a non-trivial solution in $\mathbb{Q}_2$. (Hint: You'll need an argument using some form of Hensel's lemma, and it'll have to be one that can handle the derivative being divisible by 2.)

4. Suppose $p = 2$, and one of the coefficients is even. We'll leave this and the next one to the reader:

Problem 130 Suppose $p = 2$, and one of a, b, and c is even. Show that if there exists a non-trivial solution of $aX^2 + bY^2 + cZ^2 = 0$ in $\mathbb{Q}_2$, then either the sum of two coefficients or the sum of all three coefficients will be divisible by 8. Show that this condition is also sufficient to guarantee that a non-trivial solution exists.

5. Suppose $p \neq 2$ and a is divisible by p.

Problem 131 Suppose $p \neq 2$ and a is divisible by p. Show that if there exists a non-trivial solution of $aX^2 + bY^2 + cZ^2 = 0$ in $\mathbb{Q}_p$, then there must exist an integer $r \in \mathbb{Z}$ such that
$$b + r^2 c \equiv 0 \pmod{p}.$$
(Another way of putting this is: $-b/c$ is a quadratic residue modulo p.) Show that this condition is also sufficient.

Putting all of this information together, we now have conditions that guarantee, for each p, that there are solutions in $\mathbb{Q}_p$. Using Hasse-Minkowski, we get:

Proposition 3.5.6 *Let a, b, and c be pairwise relatively prime square-free integers. The equation*
$$aX^2 + bY^2 + cZ^2 = 0$$
has non-trivial solutions in $\mathbb{Q}$ if and only if the following conditions are satisfied:

 i) a, b, *and c are not all positive or all negative.*

 ii) *for each odd prime dividing a, there exists an integer $r \in \mathbb{Z}$ such that $b + r^2 c \equiv 0 \pmod{p}$, and similarly for the odd primes dividing b and c.*

 iii) *if a, b, and c are all odd, then there are two of them whose sum is divisible by 4.*

 iv) *if a is even, then either $b + c$ or $a + b + c$ is divisible by 8 (and similarly if one of the others is even).*

A direct proof of this special case of the theorem can be found in chapters 3–5 of [Cas91]. The strategy of the proof is to use conditions *(ii)*, *(iii)*, and *(iv)* and Minkowski's "geometry of numbers," to show that one can find a solution (x, y, z) that satisfies the inequality
$$|a|x^2 + |b|y^2 + |c|z^2 < 4|abc|.$$

(Here $|\ | = |\ |_\infty$ is the "usual" absolute value.) This equation defines an ellipsoid in $\mathbb{R}^3$, and the number of triples (x, y, z) of integers satisfying this condition is finite, so that we can easily run through all of them (on a computer, probably) and find a solution. In other words, Cassels' argument in [Cas91] goes further than merely giving an existence result: it actually gives us the means to find the solution.

Problem 132 The reader who was very attentive to the wording of that last paragraph may have noticed one other feature of Cassels' proof that is worth remarking on: condition *(i)* is never used in the proof. This is rather surprising. For example, it means that if we know that the equation has a solution in $\mathbb{Q}_p$ for every prime p, then it has a solution in $\mathbb{R}$. Or, in more elementary and more dramatic terms, it says that three integers a, b, and c satisfying conditions (ii), (iii), and (iv) cannot all have the same sign. Would you have guessed that something like that was true?

Can you speculate about what might be going on here? (Comment: these are deep waters, but it's always worth the effort to think a little about things like this.)

For equations of degree higher than two, it is unlikely that anything as strong as the Hasse-Minkowski Theorem can be true. In fact, in many cases one has counterexamples that show that one may have local solutions everywhere and still have no global solutions. Still, even in situations where this strong form of the local-global principle is false, the basic idea that getting local information everywhere should give global information often remains useful. For example, in the case of cubic equations, it is *not* true that the existence of local solutions everywhere guarantees the existence of global solutions; nevertheless, there are still strong connections (or at least one suspects so). For example, there is a conjecture, due to Birch and Swinnerton-Dyer, that says, when looked at from this angle, that the quantity of global solutions can be determined in terms of local information. The conjecture is widely believed to be true, and offers one example of how the local-global principle remains one of the fundamental ideas of modern number theory.

4 Elementary Analysis in $\mathbb{Q}_p$

In the field of p-adic numbers we have an object that in many ways is analogous to the field of real numbers: it is a field with an absolute value, and it is complete with respect to the metric given by that absolute value. In fact, the similarities go deeper: $\mathbb{R}$ and the various $\mathbb{Q}_p$ are completions of $\mathbb{Q}$, hence contain $\mathbb{Q}$ as a dense subset; they are all locally compact; none of them are algebraically closed.

These similarities all suggest that much of what is usually done in $\mathbb{R}$ can be extended to $\mathbb{Q}_p$. In particular, the basic structures of the calculus should all extend. The goal of this chapter is to examine what form these basic ideas take in the p-adic context. The central theme will be the theory of infinite series, which we will use to construct a number of different functions on $\mathbb{Q}_p$ which imitate the classical transcendental functions.

The reader will probably remark on the fact that our "elementary analysis" focuses on power series, touching only lightly on the derivatives and completely ignoring the integrals that played such a large role in everyone's Calculus classes. As far as derivatives are concerned, the main reason for this is simply that derivatives are much less interesting in a p-adic context than they are in real analysis. In particular, the fact that the mean value theorem does not hold means that simply working with differentiable functions will usually not be good enough. Functions defined by power series are nicer.

Integration is a different story entirely. It is certainly possible to construct a good p-adic theory of integration. This turns out, however, to be rather subtle, and we have chosen not to pursue it. Students who are interested may find a beginning in Chapter II of [Kob84], where a kind of p-adic integration is used to attack interpolation problems.

Before we go on, we should also note that while there are many similarities between $\mathbb{R}$ and the $\mathbb{Q}_p$, there are also rather large differences; noticing them at this point will prepare us for the changes to come later. To begin with, $\mathbb{R}$ is an *ordered* field: there is a well-defined notion of "bigger than" that is nicely compatible with the operations. This is certainly not true for the $\mathbb{Q}_p$. Secondly, $\mathbb{R}$ is archimedean (more precisely, the absolute value on $\mathbb{R}$ is), while the $\mathbb{Q}_p$ are all non-archimedean. This means, in particular, that $\mathbb{R}$ is *connected* as a metric space, while $\mathbb{Q}_p$, as we saw above, is *totally disconnected*. This means, for example, that there is no clear notion of an interval in $\mathbb{Q}_p$, or any analogue of the notion of a curve. It is these contrasts that will cause most of the differences between real and p-adic analysis.

4.1   Sequences and Series

We begin by studying the basic convergence properties of sequences and series. The most important fact has already been noted: $\mathbb{Q}_p$ is a complete field, so that every Cauchy sequence converges. Furthermore, notice that all of the axioms that hold for the absolute value in $\mathbb{R}$ still hold in $\mathbb{Q}_p$ (being non-archimedean is an *extra* property). Hence, most of the basic theorems still hold in the p-adic context, *with the same proofs!* We will leave it to the reader to look over the basic theory in her real analysis text, and emphasize rather the points where the non-archimedean property introduces serious differences from the real case. Perhaps the most important such difference is the fact, also noted above, that in a non-archimedean context it is easier to test for the Cauchy property.

Lemma 4.1.1 *A sequence (a_n) in $\mathbb{Q}_p$ is a Cauchy sequence, and therefore convergent, if and only if it satisfies*

$$\lim_{n \to \infty} |a_{n+1} - a_n| = 0.$$

PROOF: This is Lemma 3.2.2, which was stated for $\mathbb{Q}$ and the p-adic absolute value, but whose proof clearly only uses that the absolute value is non-archimedean, and hence works just as well for sequences in $\mathbb{Q}_p$.   □

The theory of sequences and their convergence properties is pretty much identical to the theory over $\mathbb{R}$, except for this Lemma. Here are a few examples:

Problem 133 Decide if the following sequences converge in $\mathbb{Q}_p$, and find the limit of those that do:

- $a_n = n!$
- $a_n = n$
- $a_n = 1/n$
- $a_n = p^n$
- $a_n = (1 + p)^{p^n}$

Problem 134 Let a_n be a convergent sequence in $\mathbb{Q}_p$. Show that either $\lim |a_n| = 0$ or there exists an integer M such that $|a_n| = |a_M|$ for every $n \geq M$. In words, the sequence of absolute values of a convergent sequence either tends to zero or becomes constant for large enough n.

As for sequences, so for series: the classical theory still holds. For example, the following is still true:

Problem 135 Let $a_n \in \mathbb{Q}_p$. Show that absolute convergence implies convergence, i.e., that if the series of absolute values $\sum |a_n|$ converges (in $\mathbb{R}$), then the series $\sum a_n$ converges in $\mathbb{Q}_p$.

This is an important and useful result in real analysis. In the p-adic context, however, Lemma 4.1.1 gives us something much better:

Corollary 4.1.2 *An infinite series* $\displaystyle\sum_{n=0}^{\infty} a_n$ *with* $a_n \in \mathbb{Q}_p$ *is convergent if and only if*

$$\lim_{n \to \infty} a_n = 0,$$

in which case we also have

$$\left| \sum_{n=0}^{\infty} a_n \right| \leq \max_n |a_n|.$$

PROOF: A series converges when the sequence of partial sums converges. Now, the n-th term a_n is exactly the difference between the n-th and the $(n-1)$-st partial sums; if it tends to zero, it follows from the lemma that the sequence of partial sums is a Cauchy sequence, hence is convergent.

Finally, the estimate for the sum is a straight extension of the non-archimedean inequality, and we leave its verification to the reader. □

Problem 136 Check the inequality for the absolute value of the sum of a convergent series.

Problem 137 The corollary flies in the face of many admonitions in calculus class: in $\mathbb{R}$, the fact that the general term tends to zero is *not* a sufficient condition for convergence. In other words, the corollary is *false* in $\mathbb{R}$. Give an example of a series in $\mathbb{R}$ whose general term tends to zero, but which does not converge. Give another.

The upshot is that it is much easier to decide on the convergence of an infinite series in the p-adic context than over $\mathbb{R}$. This has the effect of making the theory of series in $\mathbb{Q}_p$ generally a lot simpler than the classical theory. We'll study in detail one example of this: a theorem[1] about double series and reversing the order of summation. We want to consider a "double sequence" b_{ij} of p-adic numbers and ask about the two series we get by summing either first in i, then in j or the other way around. For this to make sense, we need that the b_{ij} tend to zero when we fix one index and let the other go to infinity (otherwise the series won't converge). We'll say that

$$\lim_{i \to \infty} b_{ij} = 0 \qquad \text{uniformly in } j$$

if given any positive number ε we can find an integer N *which does not depend on* j such that

$$i \geq N \implies |b_{ij}| < \varepsilon.$$

---

[1] This is a variant of a theorem given in [Cas86]; I learned it from Keith Conrad.

In other words, for each j the sequence b_{ij} tends to zero when $i \to \infty$, and the convergence is "at the same rate" for all j. The first thing we need is a lemma:

Lemma 4.1.3 *Let $b_{ij} \in \mathbb{Q}_p$, and suppose that*

i) for every i, $\lim_{j \to \infty} b_{ij} = 0$, and

ii) $\lim_{i \to \infty} b_{ij} = 0$ uniformly in j.

Then given any $\varepsilon > 0$ there exists an integer N depending only on ε such that

$$\max(i,\, j) \geq N \implies |b_{ij}| < \varepsilon.$$

PROOF: Given ϵ, the second condition says that we can choose N_0, depending on ε but not on j, such that $|b_{ij}| < \varepsilon$ if $i \geq N_0$. The first condition is weaker: it says that for each i we can find $N_1(i)$ (the notation emphasizes that it *does* depend on i) such that if $j \geq N_1(i)$ we have $|b_{ij}| < \varepsilon$. Now take

$$N = N(\varepsilon) = \max(N_0, N_1(0), N_1(1), \ldots, N_1(N_0 - 1)).$$

This N does the trick: if $\max(i,\, j) \geq N$, then either $i \geq N_0$, and we know $|b_{ij}| < \varepsilon$ regardless of what j is, or $i < N_0$ and $j \geq N$, in which case i must be equal to one of $0, 1, \ldots, N_0 - 1$ and j will be bigger than the appropriate N_1, giving $|b_{ij}| < \varepsilon$ again.   □

The crucial point is that the fact that $b_{ij} \to 0$ uniformly in j allows us to restrict to only a finite number of cases in which we have to use the other condition. Now we can go on to prove our theorem on double series.

Proposition 4.1.4 *Let $b_{ij} \in \mathbb{Q}_p$, and suppose that*

i) for every i, $\lim_{j \to \infty} b_{ij} = 0$, and

ii) $\lim_{i \to \infty} b_{ij} = 0$ uniformly in j.

Then both series

$$\sum_{i=0}^{\infty} \left(\sum_{j=0}^{\infty} b_{ij} \right) \qquad and \qquad \sum_{j=0}^{\infty} \left(\sum_{i=0}^{\infty} b_{ij} \right)$$

converge, and their sums are equal.

PROOF: From the lemma, we know that given ε we can choose N such that if $\max(i,\, j) \geq N$ then $|b_{ij}| < \varepsilon$. In particular, b_{ij} tends to zero for every i when $j \to \infty$ and vice-versa, which means that the internal sums

$$\sum_{j=0}^{\infty} b_{ij} \qquad and \qquad \sum_{i=0}^{\infty} b_{ij}$$

converge (the first for all i, and the second for all j). In addition, for $i \geq$ N we have, by Corollary 4.1.2,

$$\left| \sum_{j=0}^{\infty} b_{ij} \right| \leq \max_{j} |b_{ij}| < \varepsilon;$$

similarly, for $j \geq$ N we have

$$\left| \sum_{i=0}^{\infty} b_{ij} \right| < \varepsilon.$$

In particular, we see that

$$\lim_{i \to \infty} \sum_{j=0}^{\infty} b_{ij} = 0 \qquad \text{and} \qquad \lim_{j \to \infty} \sum_{i=0}^{\infty} b_{ij} = 0,$$

so that both double series converge.

It remains to check that the sums are equal. For that, we continue to use N and ε chosen as above, so that $|b_{ij}| < \varepsilon$ when either i or j is $\geq$ N, and we use over and over the fact that in a non-archimedean field a bound on each term in a sum gives a bound on the sum itself; this is just the ultrametric inequality $|x + y| \leq \max(|x|, |y|)$, as generalized to series in Corollary 4.1.2.

Begin by noticing that

$$\left| \sum_{i=0}^{\infty} \left(\sum_{j=0}^{\infty} b_{ij} \right) - \sum_{i=0}^{N} \left(\sum_{j=0}^{N} b_{ij} \right) \right| =$$

$$= \left| \sum_{i=0}^{N} \left(\sum_{j=N+1}^{\infty} b_{ij} \right) + \sum_{i=N+1}^{\infty} \left(\sum_{j=0}^{\infty} b_{ij} \right) \right|.$$

Now, if $j \geq$ N$+1$, we have $|b_{ij}| < \varepsilon$ for every i; by the ultrametric inequality,

it follows that $\left| \sum_{j=N+1}^{\infty} b_{ij} \right| < \varepsilon$ for every i, and then (use the ultrametric

inequality again!)

$$\left| \sum_{i=0}^{N} \left(\sum_{j=N+1}^{\infty} b_{ij} \right) \right| < \varepsilon$$

Similarly, we get an estimate for the other summand:

$$\left| \sum_{i=N+1}^{\infty} \left(\sum_{j=0}^{\infty} b_{ij} \right) \right| < \varepsilon,$$

and one more application of the ultrametric inequality allows us to conclude that

$$\left| \sum_{i=0}^{\infty} \left(\sum_{j=0}^{\infty} b_{ij} \right) - \sum_{i=0}^{N} \left(\sum_{j=0}^{N} b_{ij} \right) \right| < \varepsilon.$$

Of course, reversing i and j we get a similar inequality for the other double sum. Finally, since clearly one can reverse the order of summation in the finite sum, we can use the ultrametric inequality once again to conclude that

$$\left| \sum_{i=0}^{\infty} \left(\sum_{j=0}^{\infty} b_{ij} \right) - \sum_{j=0}^{\infty} \left(\sum_{i=0}^{\infty} b_{ij} \right) \right| < \varepsilon.$$

But since ε was arbitrary it follows that the two sums must be equal.   □

What this result says is that if the b_{ij} tend to zero in a sufficiently uniform way, then their sum can be taken in any order. This result will prove quite useful in our later applications.

Problem 138 How important is the ultrametric inequality in the proof? What is the best result along these lines in real analysis?

The next two problems show that series can be added and multiplied in the "natural" way.

Problem 139 Show that if $a = \sum a_n$ and $b = \sum b_n$ are convergent, and we set

$$c_n = a_n + b_n,$$

then the series $\sum c_n$ is convergent and has sum $a + b$.

Problem 140 Show that if $a = \sum a_n$ and $b = \sum b_n$ are convergent, and we set

$$c_n = \sum_{i=0}^{n} a_i b_{n-i},$$

then the series $\sum c_n$ is convergent and has sum ab.

4.2   Functions, Continuity, Derivatives

The basic ideas about functions and continuity remain unchanged when we go to the p-adics, since after all they depend only on the metric structure. There are no intervals to work with (in fact, no non-trivial connected sets at all), so usually our functions will be defined in (open or closed) balls. Recall that we write $B(a,r)$ for the open ball with center a and radius r and $\overline{B}(a,r)$ for the closed ball with center a and radius r. The definition of continuity is exactly the same as before:

Definition 4.2.1 *Let $U \subset \mathbb{Q}_p$ be an open set. A function $f : U \to \mathbb{Q}_p$ is said to be* continuous *at $a \in U$ if for every $\varepsilon > 0$ there exists a $\delta > 0$ such that, for every $x \in U$,*

$$|x - a| < \delta \Longrightarrow |f(x) - f(a)| < \varepsilon.$$

The basic results about continuity are true in all metric spaces, and hence are true here too. For example, if U is compact (and remember, in $\mathbb{Q}_p$ it's perfectly possible for a set to be both open *and* compact) and f is continuous at every point of U, then f is uniformly continuous.

Problem 141 Is there a p-adic analogue of the intermediate value theorem?

Derivatives are a bit more interesting, if only because it'll turn out that they don't work as well as in the classical case. It certainly makes perfect sense to define derivatives of functions $f : \mathbb{Q}_p \longrightarrow \mathbb{Q}_p$ in the usual way:

Definition 4.2.2 *Let $U \subset \mathbb{Q}_p$ be an open set, and let $f : U \longrightarrow \mathbb{Q}_p$ be a function. We say f is* differentiable *at $x \in U$ if the limit*

$$f'(x) = \lim_{h \to 0} \frac{f(x + h) - f(x)}{h}$$

exists. If $f'(x)$ exists for every x in U we say f is differentiable in U, and we write $f' : U \longrightarrow \mathbb{Q}_p$ for the function $x \mapsto f'(x)$.

To some extent, the derivative works as expected. For example, we can show that differentiable functions are continuous, in exactly the same way as we do it over $\mathbb{R}$ or $\mathbb{C}$. Along the same vein:

Problem 142 Let $n \in \mathbb{Z}$. What is the derivative of the function $\mathbb{Q}_p \longrightarrow \mathbb{Q}_p$ given by $x \mapsto x^n$?

It is natural to wonder why the derivative seems to play such a minor role in p-adic analysis. One of the reasons for this has to do with the mean value theorem, which is the linchpin of the elementary theory of differentiable functions. If we try to look for a p-adic version of the theorem, we run into the immediate (not-too-serious) difficulty of deciding what it would say. After all, the classical theorem says that given a and b in the domain of a continuously differentiable function, there exists a number ξ *between* a and b such that

$$f(b) - f(a) = f'(\xi)(b - a).$$

The problem, of course, is that "between" doesn't seem to mean anything in the p-adic context, since $\mathbb{Q}_p$ is not an ordered field.

But this initial difficulty is easily resolved. In $\mathbb{R}$, we can redefine "between" by saying that ξ is between a and b if we have

$$\xi = at + b(1 - t) \qquad \text{for } 0 \le t \le 1$$

(draw a picture!). In fact, if we want a mean value theorem that works in the field of complex numbers, we will have to do something like this anyway.

So here is an attempt at a minimal p-adic version of the mean value theorem:

What a p-adic mean value theorem might say: *If a function $f(X)$ is differentiable with continuous derivative on $\mathbb{Q}_p$, then, for any two numbers a and b in $\mathbb{Q}_p$ there exists an element $\xi \in \mathbb{Q}_p$ of the form*

$$\xi = at + b(1 - t) \qquad \text{for some } t, \ |t| \leq 1$$

for which we have
$$f(b) - f(a) = f'(\xi)(b - a).$$

Unfortunately, things aren't that simple.

Proposition 4.2.3 *The "p-adic mean value theorem" we just stated is false.*

PROOF: Take $f(x) = x^p - x$, $a = 0$, $b = 1$. Then $f'(x) = px^{p-1} - 1$ and $f(a) = f(b) = 0$. What the proposed theorem would say, then, would be that there exists a ξ of the above form such that $p\xi^{p-1} - 1 = 0$. But any $\xi = at + b(1 - t) = (1 - t)$ with $t \in \mathbb{Z}_p$ (which is what $|t| \leq 1$ says) must itself belong to $\mathbb{Z}_p$. But then $p\xi^{p-1} - 1$ is clearly a unit in $\mathbb{Z}_p$ (it belongs to $1 + p\mathbb{Z}_p$), and therefore cannot be zero. $\qquad\square$

This makes things much less nice than in the classical case. For example, the next problem asks the reader to show that our most basic intuitions about derivatives fail.

Problem 143 Construct a function $f : \mathbb{Q}_p \longrightarrow \mathbb{Q}_p$ which is differentiable, has zero derivative everywhere, but is *not locally constant*. Such functions are sometimes called "almost constant."

Problem 144 Show that the chain rule is still true in $\mathbb{Q}_p$. Use this fact to show that if f has zero derivative everywhere and g is any continuously differentiable function, then both $f \circ g$ and $g \circ f$ have zero derivative everywhere. Explain why this means that there are a great many "almost constant" functions.

In particular, it follows that two functions which have the same derivative do *not* need to differ by a constant. This shows that knowing that a function is differentiable isn't as useful in the p-adic context as it is classically. This is one of the reasons why we concentrate on functions defined by power series instead.

Of course, the function we used in the example above is a polynomial, so nothing will rescue the "p-adic mean value theorem" we tried to formulate. Nevertheless, it is possible to prove an analogue of the mean value theorem for functions defined by power series, provided one restricts to the case when

$|b - a|$ is small enough. See A. Robert's account of this in [Rob95]. For more information on differentiation in a p-adic context, see K. Mahler's book [Mah73].

4.3   Power Series

In real analysis, power series

$$\sum_{n=0}^{\infty} a_n (X - \alpha)^n$$

offer a convenient way of representing functions, and in particular can be used to *define* several important functions, such as the exponential and trigonometric functions. As one might expect, the p-adic theory turns out to be quite similar to the classical version, except that some of the tricky points become a lot simpler to handle. On the other hand, the non-archimedean property does introduce a few surprises. The biggest of these surprises is the fact that the relation between the formal composition of power series and the composition of the functions they define becomes *more* complicated in the p-adic context than it is in the classical situation. Because this is such an unexpected development, we spend quite a bit of time on it.

The next few sections explore the main ideas about power series and functions defined by power series, focusing, in the end, on the p-adic versions of the logarithm and the exponential. The main influences on our treatment are [Cas86], [Has80], and a set of unpublished notes by Keith Conrad. We have stated most of our results for power series in X, but of course they remain true for power series in $(X - \alpha)$, for the usual reasons.

Consider, then, a power series

$$f(X) = \sum_{n=0}^{\infty} a_n X^n.$$

Given $x \in \mathbb{Q}_p$, we want to consider $f(x)$, which is[2] the series $\sum a_n x^n$; we already know that this converges if and only if $|a_n x^n| \to 0$. As in the classical case, the set of all such x (which we call the *region of convergence*) is a disk.

---

[2]In this section and the following ones, we adopt the convention that X represents an indeterminate, while x usually represents a p-adic number. Hence, $f(X)$ in this statement is to be thought of as the formal power series itself, while $f(x)$ is the numerical series we obtain by substituting x for X. It makes no sense to discuss convergence of $f(X)$: the series is just *there*. It does make sense to discuss the convergence of $f(x)$; whether it converges or not will depend on x. When it does converge, we will write $f(x)$ for *both* the numerical series and its value; this should normally not cause any confusion.

Proposition 4.3.1 *Let* $f(X) = \displaystyle\sum_{n=0}^{\infty} a_n X^n$, *and define*

$$\rho = \frac{1}{\limsup \sqrt[n]{|a_n|}},$$

where we use the usual conventions when the limit is zero or infinity, so that $0 \le \rho \le \infty$.

  i) If $\rho = 0$, *then* $f(x)$ *converges only when* $x = 0$.

  ii) If $\rho = \infty$, *then* $f(x)$ *converges for every* $x \in \mathbb{Q}_p$.

  iii) If $0 < \rho < \infty$ *and* $\lim_{n \to \infty} |a_n| \rho^n = 0$, *then* $f(x)$ *converges if and only if* $|x| \le \rho$.

  iv) If $0 < \rho < \infty$ *and* $|a_n| \rho^n$ *does not tend to zero as* n *goes to infinity, then* $f(x)$ *converges if and only if* $|x| < \rho$.

PROOF: We already know that the region of convergence is

$$\left\{ x \in \mathbb{Q}_p : \lim_{n \to \infty} |a_n x^n| = 0 \right\},$$

so that the point of the theorem is to translate this into more precise information. First of all, it is clear that $f(0)$ converges. Next, if $|x| < \rho$, then it is easy to see (if nothing else, it follows from the theory of powers series over $\mathbb{R}$) that

$$\sum |a_n| |x|^n,$$

which is a power series in $\mathbb{R}$, converges, which implies that the p-adic series does too. Similarly, if $|x| > \rho$, it is easy to see that $|a_n| |x|^n$ cannot tend to zero when n tends to infinity: the definition of ρ implies that for infinitely many n's, $|a_n|$ is close to $1/\rho^n$, and, since $|x| > \rho$, $(|x|/\rho)^n$ gets arbitrarily large as n grows. Finally, the statements about what happens when $|x| = \rho$ are immediate from Corollary 4.1.2. □

As in the archimedean case, the number ρ is called the *radius of convergence* of the series. Notice that, in contrast to what is true in the classical[3] case, what happens at the points on the "boundary" of the region of convergence (i.e., the points with $|x| = \rho$) is rather simple: either the series is convergent at *all* such points or at none of them. (On the other hand, recall that the points such that $|x| = \rho$ are not really the boundary of the open disk!)

---

[3]Over $\mathbb{R}$, the region of convergence may include none, both, or only one of the endpoints of the interval $-\rho < x < \rho$. Over $\mathbb{C}$, it is worse: $|x| < \rho$ is a disk, and the set of points on the boundary for which the series converges can be pretty complicated.

Problem 145 Find the region of convergence of the following p-adic power series:

i) $\sum p^n X^n$

ii) $\sum p^{-n} X^n$

iii) $\sum n! X^n$

One of the nice things about starting from formal power series is the fact that several of the operations we want to do with power series make sense at the formal level. Let's look at these formal operations, and then ask the important question: how do the formal properties translate to properties of the functions defined by the power series?

We start with the easiest operations: given two formal power series $f(X)$ and $g(X)$, we can consider their sum and their product. If

$$f(X) = \sum_{n=0}^{\infty} a_n X^n \qquad \text{and} \qquad g(X) = \sum_{n=0}^{\infty} b_n X^n,$$

then we define

$$(f+g)(X) = \sum_{n=0}^{\infty} (a_n + b_n) X^n$$

and

$$(fg)(X) = \sum_{n=0}^{\infty} \left(\sum_{k=0}^{n} a_k b_{n-k} \right) X^n$$

Notice that if we want to think of these definitions as the result of actually adding or multiplying the series, they imply a lot of reordering and recombining of terms!

As we have defined it, this is a formal operation only. Of course, we'd like to know that it actually works when we plug in numbers for X. It's not too hard to see everything works as expected:

Proposition 4.3.2 *Let $f(X)$ and $g(X)$ be formal power series, and suppose $x \in \mathbb{Q}_p$. If $f(x)$ and $g(x)$ both converge, then:*

i) $(f+g)(x)$ converges and is equal to $f(x) + g(x)$, and

ii) $(fg)(x)$ converges and is equal to $f(x)g(x)$.

PROOF: Basically, all that's needed is an appeal to the results you proved for numerical series in problems 139 and 140. □

Problem 146 Fill in the details of the proof.

Having had such success with adding and multiplying series, we can be more ambitious, and consider the composition of formal series. Suppose we have two formal series

$$f(X) = \sum_{n=0}^{\infty} a_n X^n \qquad \text{and} \qquad g(X) = \sum_{n=0}^{\infty} b_n X^n,$$

and that $b_0 = 0$ (another way of saying that would be to say $g(0) = 0$). We want to check that it makes sense to define the composition $h(X) = f(g(X))$ of the two series. This should be

$$h(X) = a_0 + a_1 g(X) + a_2 g(X)^2 + \cdots + a_n g(X)^n + \ldots$$

That looks like an awful mess, but in fact we can (working formally, of course[4]) reorganize it into a well-behaved power series. The idea is this: since $g(X)$ has no independent term, $g(X)^2$ starts with a term of degree 2, $g(X)^3$ starts with a term of degree 3, and so on. So, when we try to work out what the coefficients of $h(X) = f(g(X)) = \sum c_n X^n$ should be, each coefficient only requires a finite amount of work.

- The zeroth coefficient is just $c_0 = a_0$.

- The first coefficient only requires that we look at the first two terms $a_0 + a_1 g(X)$, and therefore $c_1 = a_1 b_1$.

- The second coefficient requires that we look at the first three terms

$$a_0 + a_1 g(X) + a_2 g(X)^2 = a_0 + a_1(b_1 X + b_2 X^2 + \ldots)$$
$$+ a_2(b_1^2 X^2 + \ldots),$$

  and so $c_2 = a_1 b_2 + a_2 b_1^2$.

- For the third coefficient, look at

$$a_0 + a_1 g(X) + a_2 g(X)^2 + a_3 g(X)^3 = a_0 + a_1(b_1 X + b_2 X^2 + b_3 X^3 \ldots)$$
$$+ a_2(b_1^2 X^2 + 2b_1 b_2 X^3 + \ldots)$$
$$+ a_3(b_1^3 X^3 + \ldots)$$

  so that $c_3 = a_1 b_3 + 2a_2 b_1 b_2 + a_3 b_1^3$.

- And so on! One can clearly find c_n for every n.

---

[4]We could also think like this: if we use the X-adic topology, the completion of the ring $\mathbb{Q}_p[X]$ of polynomials with coefficients in $\mathbb{Q}_p$ is exactly the ring of power series with coefficients in $\mathbb{Q}_p$. In the X-adic topology, the power series

$$a_0 + a_1 g(X) + a_2 g(X)^2 + \cdots + a_n g(X)^n + \ldots$$

converges, and its sum is the power series we're calling $h(X)$.

Problem 147 Can you find a general formula for the c_n?

So we know that given two formal power series $f(X)$ and $g(X)$ with $g(0) = 0$, we have a formal power series $h(X) = f(g(X))$ which is their formal composition. Now we need to ask questions about convergence, and those are not as easy to answer. The point is that plugging a number x into the power series $h(X)$ might give a different answer from what one gets by first plugging x into $g(X)$ and then plugging the result into $f(X)$. One might suspect that there are problems simply by contemplating the amount of rearranging that's going on in our definition of the composite series $h(X)$. In fact, it turns out we need to be very careful. Here's the theorem:

Theorem 4.3.3 *Let* $f(X) = \sum a_n X^n$ *and* $g(X) = \sum b_n X^n$ *be formal power series with* $g(0) = 0$, *and let* $h(X) = f(g(X))$ *be their formal composition. Suppose that*

i) $g(x)$ *converges,*

ii) $f(g(x))$ *converges (this means: plugging the number to which* $g(x)$ *converges into* $f(X)$ *gives a convergent series),*

iii) *for every* n, *we have* $|b_n x^n| \leq |g(x)|$ *(in other words, no term of the series converging to* $g(x)$ *is bigger than the sum).*

Then $h(x)$ *also converges, and* $f(g(x)) = h(x)$.

PROOF: (Following [Has80, chapter 17].) We have

$$f(X) = \sum_{n=0}^{\infty} a_n X^n \qquad \text{and} \qquad g(X) = \sum_{n=1}^{\infty} b_n X^n.$$

Let

$$g(X)^m = \sum_{n=m}^{\infty} d_{m,n} X^n.$$

It's not hard to work out the $d_{m,n}$ by using the formula for the product of formal power series: $d_{m,n} = 0$ if $n < m$, and, for any $n \geq m$,

$$d_{m,n} = \sum_{i_1 + i_2 + \cdots + i_m = n} b_{i_1} b_{i_2} \ldots b_{i_m}$$

(that formula looks uglier than it really is: basically, take all the products of m-tuples of b_i's whose indices add up to n). This allows us to write $h(X) = f(g(X))$ explicitly:

$$h(X) = a_0 + \sum_{n=1}^{\infty} \left(\sum_{m=1}^{n} a_m d_{m,n} \right) X^n.$$

Now let's start thinking about convergence. First of all, since $g(x)$ converges, we can use Proposition 4.3.2 to conclude that the formal series $g(X)^m$ converges when we plug in $X = x$, and in fact converges to $g(x)^m$; in other words,

$$g(x)^m = \sum_{n=m}^{\infty} d_{m,n} x^n.$$

More interesting is the fact that the special assumption we made about $g(X)$ is still true for the series $g(X)^m$: for every n, we have

$$|d_{m,n} x^n| \le |g(x)^m|.$$

To see this, note first that $|g(x)^m| = |g(x)|^m$. Now look at the general term $|d_{m,n} x^n|$. If $n < m$, then $|d_{m,n} x^n| = 0$ and there is nothing to prove. On the other hand, if $n \ge m$, then the ultrametric inequality gives

$$|d_{m,n} x^n| \le \max |b_{i_1} x^{i_1}| \cdot |b_{i_2} x^{i_2}| \ldots |b_{i_m} x^{i_m}|,$$

where the maximum is once again taken over all m-tuples $(i_1, i_2, \ldots, i_m)$ such that $i_1 + i_2 + \cdots + i_m = n$. But we know, from the hypothesis on $g(X)$, that $|b_{i_j} x^{i_j}| \le |g(x)|$ for every i_j; multiplying all these inequalities gives $|d_{m,n} x^n| \le |g(x)^m|$, which is the inequality we want.

So now we know that $g(x)$ converges, that powers of $g(x)$ converge, and that both the series for $g(x)$ and for $g(x)^m$ satisfy the extra condition that no term is larger than the final sum. We also know, from our assumptions, that $f(g(x))$ converges, that is, that $a_m(g(x))^m$ tends to zero as m grows. We have

$$f(g(x)) = a_0 + \sum_{m=1}^{\infty} a_m g(x)^m = a_0 + \sum_{m=1}^{\infty} a_m \left(\sum_{n=m}^{\infty} d_{m,n} x^n \right)$$

$$= a_0 + \sum_{m=1}^{\infty} \sum_{n=m}^{\infty} a_m d_{m,n} x^n$$

(where the order of the summations is crucial, of course), and, on the other hand,

$$h(x) = a_0 + \sum_{n=1}^{\infty} \left(\sum_{m=1}^{n} a_m d_{m,n} \right) x^n = a_0 + \sum_{n=1}^{\infty} \sum_{m=1}^{n} a_m d_{m,n} x^n.$$

These series are obtained from each other by reversing the order of the summation, so what we need to do is check that this is legal and that both series will have the same sum. That's what Proposition 4.1.4 is for!

To apply the proposition, we need to show that the general term $a_m d_{m,n} x^n$ tends to zero sufficiently uniformly. So let's study that general term. The crucial thing is to notice that we can use the fact that $g(x)^m$ is larger than any term of the series to get a uniform bound:

$$|a_m d_{m,n} x^n| \leq |a_m g(x)^m|,$$

where the important thing is that the right-hand side is independent of n. Given ε, we can, since $a_m g(x)^m \to 0$, choose N such that $|a_m g(x)^m| < \varepsilon$ if $m \geq N$. This shows part of what we want:

$$\lim_{m \to \infty} a_m d_{m,n} x^n = 0, \qquad \text{uniformly in } n.$$

On the other hand, for each m we know that the series

$$g(x)^m = \sum_{n=0}^{\infty} d_{m,n} x^n$$

converges, and it follows (after multiplying by a_m) that, for every m,

$$\lim_{n \to \infty} a_m d_{m,n} x^n = 0.$$

That's what we need to be able to apply Proposition 4.1.4 and conclude both that the series for $h(x)$ converges and that its sum is equal to $f(g(x))$, which is what we needed to prove. □

The extra assumption on $g(X)$ is essential! In other words, an equality of formal power series that involves composition does not need to imply equality of the functions unless we can check this extra condition. We will discuss an important example of this, involving the logarithm and the exponential functions, later in this chapter.

Problem 148 If you can't wait, consider the following example in $\mathbb{Q}_2$. Let

$$f(X) = 1 + X + \frac{X^2}{2!} + \cdots + \frac{X^n}{n!} + \cdots$$

be the usual formal series for the exponential, let

$$g(X) = 2X^2 - 2X,$$

and let

$$h(X) = f(g(X)).$$

We will show later that $f(x)$ converges for every $x \in 4\mathbb{Z}_2$ and diverges otherwise. Since $g(X)$ is a polynomial, $g(x)$ converges for every x. In particular, suppose we take $x = 1$. Then $g(1) = 0$ and so $f(g(1)) = 1$.

  i) Check that the first two conditions in the theorem are satisfied, but the third is not.

 ii) Let $h(X) = \sum a_n X^n$. Show that if $n \geq 2$ then $v_2(a_n) \geq 1 + n/4$, and conclude that $h(x)$ converges for all $x \in \mathbb{Z}_2$. (This part is rather hard.)

iii) By computing out the first few terms of $h(X)$ and using the estimate for the valuation of the a_n, show that $h(1) \equiv 3 \pmod 4$.

iv) Conclude that $h(1) \neq f(g(1))$.

It is interesting to remark that in this case classical analysis is actually easier: if the radius of convergence of $f(X)$ is ρ and $|g(x)| < \rho$, then $h(x)$ converges and we have $f(g(x)) = h(x)$. See, for example, Proposition 5.1 in Section 2 of Chapter 1 of [Car95].

Problem 149 One other operation with power series which we didn't mention is differentiation. Given a power series $f(X) = \sum a_n X^n$, we define its formal derivative to be $f'(X) = \sum n a_n X^{n-1}$. Show that this has the usual properties of a derivative:

i) $(f + g)'(X) = f'(X) + g'(X)$

ii) $(fg)'(X) = f'(X)g(X) + f(X)g'(X)$

iii) If $h(X) = f(g(X))$, then $h'(X) = f'(g(X))g'(X)$

Notice that these are equalities of formal series!

4.4   Functions Defined by Power Series

We will use power series to define functions. In other words, given a power series $f(X)$ we will think of it as defining a function whose domain is the set of x for which $f(x)$ converges. Just as in the classical case, the functions that are defined by power series have nice properties. The simplest one is continuity.

Lemma 4.4.1 *Let* $f(X) = \sum a_n X^n$ *be a power series with coefficients in* $\mathbb{Q}_p$, *and let* $\mathcal{D} \subset \mathbb{Q}_p$ *be its region of convergence, i.e., the set of* $x \in \mathbb{Q}_p$ *for which* $f(x)$ *converges. The function*

$$f : \mathcal{D} \to \mathbb{Q}_p$$

defined by $x \mapsto f(x)$ *is continuous on* $\mathcal{D}$.

PROOF: Identical to the proof over $\mathbb{R}$.   □

Problem 150 Prove the lemma. In $\mathbb{R}$, continuity at the endpoints of the interval of convergence is a problem. Make sure that your p-adic proof handles those points as well.

As in the classical case, we can change the center of the series expansion, i.e., to re-write our function as a power series in $(X - \alpha)$ for any α in the region of convergence. In the classical case, the resulting series can (usually does) have a different region of convergence than the original series, and this fact is one of the ways to obtain "analytic continuations." Surprisingly, this never happens in the p-adic case.

Proposition 4.4.2 *Let $f(X) = \sum a_n X^n$ be a power series with coefficients in $\mathbb{Q}_p$, and let $\alpha \in \mathbb{Q}_p$ be a point for which $f(\alpha)$ converges. For each $m \geq 0$, define*

$$b_m = \sum_{n \geq m} \binom{n}{m} a_n \alpha^{n-m},$$

and consider the power series

$$g(X) = \sum_{m=0}^{\infty} b_m (X - \alpha)^m.$$

i) The series defining b_m converges for every m, so that the b_m are well-defined.

ii) The power series $f(X)$ and $g(X)$ have the same region of convergence, that is, $f(\lambda)$ converges if and only if $g(\lambda)$ converges.

iii) For any λ in the region of convergence, we have $g(\lambda) = f(\lambda)$.

PROOF: Claim (i) is easy to see: since α belongs to the region of convergence for $f(X)$, we get (for fixed m)

$$\left| \binom{n}{m} a_n \alpha^{n-m} \right| \leq |a_n \alpha^{n-m}| = |\alpha|^{-m} \cdot |a_n \alpha^n| \to 0,$$

which gives the desired convergence by Lemma 4.1.2.

To show (ii) and (iii), take any λ in the region of convergence of $f(X)$, and compute

$$f(\lambda) = \sum_n a_n (\lambda - \alpha + \alpha)^n = \sum_n \sum_{m \leq n} \binom{n}{m} a_n \alpha^{n-m} (\lambda - \alpha)^m.$$

The last sum looks a lot like a partial sum for $g(\lambda)$, except that it needs to be re-ordered. For that, we use Proposition 4.1.4. To check that the condition is satisfied, set

$$\beta_{nm} = \begin{cases} \binom{n}{m} a_n \alpha^{n-m} (\lambda - \alpha)^m & \text{if } m \leq n \\ \\ 0 & \text{if } m > n \end{cases}$$

We need to check that the sequence β_{nm} satisfies the conditions in Proposition 4.1.4. Note first that

$$|\beta_{nm}| = \left| \binom{n}{m} a_n \alpha^{n-m} (\lambda - \alpha)^m \right| \leq |a_n \alpha^{n-m} (\lambda - \alpha)^m|,$$

so that the problem is bounding this last expression. To do that, recall that the region of convergence is a (closed or open) disk of some radius ρ; since both λ and α are in the region of convergence, there exists a radius ρ_1 such that

- the closed disk of radius ρ_1 is contained in the region of convergence, and

- we have both $|\lambda| \leq \rho_1$ and $|\alpha| \leq \rho_1$.

(This dodge takes care of the question of whether the region of convergence is the open or the closed disk: if it is the closed disk, we can take $\rho_1 = \rho$; if the open, take ρ_1 to be the larger of the absolute values of α and λ, so that $\rho_1 < \rho$. The second choice actually works in both cases.) Then we have

- $|\alpha|^{n-m} \leq \rho_1^{n-m}$ by construction, and

- $|\lambda - \alpha|^m \leq \max\{|\lambda|, |\alpha|\}^m \leq \rho_1^m$ by the non-archimedean property.

Going back to the terms we want to estimate, we get

$$|\beta_{nm}| \leq |a_n \alpha^{n-m}(\lambda - \alpha)^m| \leq |a_n|\rho_1^n,$$

which is independent of m and tends to zero as $n \to \infty$. This means that given any $\varepsilon > 0$ there exists an N for which $|\beta_{nm}| < \varepsilon$ if $n \geq$ N and any m. This shows that β_{nm} tends to zero uniformly in m. The other condition is easy: if $m > n$, we have $\beta_{nm} = 0$, hence it's certainly true that for every n we have $\beta_{nm} \to 0$ when $m \to \infty$. Thus, the conditions in Proposition 4.1.4 are satisfied, and we can reverse the order of summation.

Changing the order of summation in the expression for $f(\lambda)$ gives the expression for $g(\lambda)$, so that applying Proposition 4.1.4 allows us to conclude that $g(\lambda)$ converges and is equal to $f(\lambda)$. This shows that g converges whenever f does, and in that case their values are equal. To conclude, notice that we can switch the roles of g and f in the argument, which shows that in fact the regions of convergence are identical.   □

Problem 151 Before you relax from that long proof, are you sure that the smoke-and-mirrors phrase "switch the roles of g and f" is really justified? Does anything further need to be checked?

Problem 152 Prove the following relative of the proposition: Let $f(X)$ be a power series such that $f(x)$ converges for $|x| < \rho$, and suppose $|a| = 1$ and $|b| < \rho$. Then the function $g(x) = f(ax + b)$ is given by a power series $g(X)$ which converges for $|x| < \rho$.

As in the classical theory, functions which can be expressed as power series in a closed disk $\overline{B}(a, r)$ are called *analytic* on $\overline{B}(a, r)$. Functions of this kind in general have very nice properties, and this is also true in $\mathbb{Q}_p$. Unfortunately, the theory is not so nice as, for example, the theory over the complex numbers. One of the crucial reasons is the theorem we have just proved: we cannot get an "analytic continuation" for a function by choosing another center and expanding in a power series. Doing so in $\mathbb{Q}_p$ produces a power series with exactly the same region of convergence, which therefore does not allow us to "continue" the function to a larger domain.

Also unpleasant is the fact that many functions are "locally analytic" for trivial reasons having to do with the fact that $\mathbb{Q}_p$, since it is non-archimedean, is totally disconnected. In fact, consider the function given by

$$f(x) = \begin{cases} 1 & \text{if } x \in \mathbb{Z}_p \\ 0 & \text{if } x \notin \mathbb{Z}_p \end{cases}$$

Since both $\mathbb{Z}_p$ and its complement are open sets, around any point one can find a ball in which $f(x)$ is constant, and hence can be written as a (constant) power series! One would not want to think of such a function as "analytic." Hence, while the set of analytic functions on a closed ball behaves well, it isn't clear how to move from that "local" theory to a "global" notion of analytic function.

It turns out that one can get around such difficulties, and come up with a good concept of "analytic" functions and of "analytic continuation." Unfortunately, this requires quite a sophisticated approach. The resulting theory is developed in what is called *Rigid Analytic Geometry*; its foundations are due to John Tate, and it has become a very important branch of modern number theory. For an introduction to this rather difficult subject, the reader might look at [BGR84].

We will stick to simpler things. First of all, if a function is given by a power series it completely determines that power series. As in the classical case, this can be shown by using derivatives (see below), but we prove something stronger. Let's say a sequence $\{x_m\}$ converging to a limit L is *stationary* if there exists an n such that $x_m = L$ for all $m \geq n$.

Proposition 4.4.3 *Let $f(X)$ and $g(X)$ be formal power series, and suppose there is a non-stationary sequence $x_m \in \mathbb{Q}_p$ converging to zero in $\mathbb{Q}_p$ and such that $f(x_m) = g(x_m)$ for every m. Then $f(X) = g(X)$ (i.e., $f(X)$ and $g(X)$ have the same coefficients).*

PROOF: (This is identical to the classical proof.) Replacing the sequence $\{x_m\}$ by a subsequence if necessary, we can assume $x_m \neq 0$ for all m. If we consider the difference $h(X) = f(X) - g(X) = \sum a_n X^n$, then we have $h(x_m) = 0$ for every m, and we want to show that $a_n = 0$ for every n. Suppose not; then let r be the least index for which $a_r \neq 0$, so that

$$\begin{aligned} h(X) &= a_r X^r + a_{r+1} X^{r+1} + a_{r+2} X^{r+2} \ldots \\ &= X^r(a_r + a_{r+1}X + a_{r+2}X^2 + \ldots) \\ &= X^r h_1(X), \end{aligned}$$

where $h_1(0) = a_r \neq 0$. Since h_1 is a function defined by a power series, it is continuous, so $h_1(x_m) \to a_r$ as $m \to \infty$ (remember that our assumption is that $x_m \to 0$); in particular, $h_1(x_m)$ is non-zero for large enough m. It follows that $h(x_m) = x_m^r h_1(x_m)$ is non-zero for large enough m, which is a contradiction. $\qquad\square$

Problem 153 Suppose $f(X)$ and $g(X)$ are formal power series, and suppose that x_m is a non-constant sequence in $\mathbb{Q}_p$ converging to a point x such that both $f(x)$ and $g(x)$ converge. Show that if $f(x_m) = g(x_m)$ for every m, then $f(X) = g(X)$.

In the problem 149, we considered a "formal derivative" operation on formal power series. If a function is defined by a power series, we want its derivative to correspond to the formal derivative of the power series, and it does:

Proposition 4.4.4 Let $f(X) = \sum a_n X^n$ be a power series, with non-zero radius of convergence, and let $f'(X)$ be its formal derivative. Let $x \in \mathbb{Q}_p$. If $f(x)$ converges, then so does $f'(x)$, and we have

$$f'(x) = \lim_{h \to 0} \frac{f(x+h) - f(x)}{h}.$$

PROOF: (Following [Has80].) Note, first, that there are indeed elements $h \to 0$ for which $f(x + h)$ converges, since the region of convergence is a (closed or open) ball centered at the origin. In fact, let ρ be the radius of convergence. If $x = 0$, any h with $|h| < \rho$ works; if $x \neq 0$, then any h with $|h| < |x|$ works. (Remember that if $|h| < |x|$, then $|x + h| = |x|$.) In particular, the limit that appears in the proposition does make sense.

Suppose, then, that $f(x)$ converges, which is equivalent to saying that $a_n x^n \to 0$. If $x = 0$ then it is clear that $f'(x)$ converges. If $x \neq 0$, notice that

$$|na_n x^{n-1}| \leq |a_n x^{n-1}| = \frac{1}{|x|}|a_n x^n| \to 0,$$

and again we see[5] that $f'(x)$ converges.

Recall that either $f(X)$ converges in the closed ball $\overline{B}(0, \rho)$ or in the open ball $B(0, \rho)$. In the first case, set $\rho_1 = \rho$. In the second case, choose ρ_1 such that $|x| \leq \rho_1 < \rho$.

Since we only care about h close to zero, we may assume, if $x \neq 0$, that $|h| < |x| \leq \rho_1$. Otherwise, $x = 0$ and we can simply assume $|h| \leq \rho_1$. Now,

$$f(x + h) = \sum_{n=0}^{\infty} a_n(x + h)^n = \sum_{n=0}^{\infty} a_n \sum_{m=0}^{n} \binom{n}{m} x^{n-m} h^m.$$

Subtracting $f(x)$ and dividing by h, we get

$$\frac{f(x + h) - f(x)}{h} = \sum_{n=1}^{\infty} \sum_{m=1}^{n} a_n \binom{n}{m} x^{n-m} h^{m-1}.$$

Now, since we have $|x| \leq \rho_1$ and $|h| \leq \rho_1$, we have

$$\left| a_n \binom{n}{m} x^{n-m} h^{m-1} \right| \leq |a_n| \rho_1^{n-1},$$

---

[5]This is one of those places where Corollary 4.1.2 really simplifies our life!

and since $\rho_1 < \rho$ we have $|a_n|\rho_1^n \to 0$. This shows that the series converges uniformly[6]in h. By a standard theorem (check your real analysis textbook!), this means we can take the limit term-by-term. In this case, that amounts to setting $h = 0$, which gives

$$f'(x) = \sum_{n=1}^{\infty} na_n x^{n-1},$$

which is what we want.   □

Problem 154 Let $f(X) = a_n X^n$ be a formal power series, and suppose $f(x)$ converges. Show that for every k the k-th derivative $f^{(k)}(x)$ exists, and is given by

$$f^{(k)}(x) = k! \sum_{n \geq k} \binom{n}{k} a_n (x - a)^{n-k};$$

in particular, we have

$$a_k = \frac{f^{(k)}(a)}{k!}.$$

Problem 155 Can you think of any reason why one would want to write the derivative as above, with $k!$ factored out?

To demonstrate that we've actually proved quite a bit, here's an easy consequence of our results. As we pointed out above, it is possible for two p-adic functions to have the same derivative without it being the case that their difference is constant. That doesn't happen for functions defined by power series.

Corollary 4.4.5 *Suppose $f(X)$ and $g(X)$ are power series, and suppose that both series converge for $|x| < \rho$. If $f'(x) = g'(x)$ for all $|x| < \rho$, then there exists a constant $c \in \mathbb{Q}_p$ such that $f(X) = g(X) + c$ as power series. In particular, $f(X)$ and $g(X)$ have the same disk of convergence, and we have $f(x) = g(x) + c$ for all x in the disk of convergence.*

PROOF: Let $f(X) = \sum a_n X^n$, $g(X) = \sum b_n X^n$, and let $f'(X)$ and $g'(X)$ be the formal derivatives. By 4.4.4 and 4.4.3, we can conclude that $a_n = b_n$ for all $n \geq 1$, and the conclusion follows.   □

The next theorem we want to look at is a fundamental result about the zeros of functions defined by power series.

---

[6]This is another one of those concepts from analysis. Basically, a series $\sum a_n(h)$ converges to a sum $s(h)$ uniformly in h if for every ε we can find an N *independent of h* such that if $m > N$ then

$$\left| s(h) - \sum_{n=0}^{m} a_n(h) \right| < \varepsilon.$$

In our case, we have an estimate for $a_n(h)$ that doesn't depend on h, hence the convergence is uniform.

Theorem 4.4.6 (Strassman) *Let*

$$f(X) = \sum_{n=0}^{\infty} a_n X^n = a_0 + a_1 X + a_2 X^2 + \cdots$$

be a non-zero power series with coefficients in $\mathbb{Q}_p$, and suppose that we have $\lim_{n \to \infty} a_n = 0$, so that $f(x)$ converges for all $x \in \mathbb{Z}_p$. Let N be the integer defined by the two conditions

$$|a_N| = \max_n |a_n| \qquad and \qquad |a_n| < |a_N| \quad for \ n > N.$$

Then the function $f : \mathbb{Z}_p \longrightarrow \mathbb{Q}_p$ defined by $x \mapsto f(x)$ has at most N zeros.

(The existence of N follows from the fact that the coefficients a_n tend to zero: there is a largest absolute value, and N is the index of the last coefficient for which the maximum is attained.) Strassman's theorem is usually proved using a high-powered result known as the p-adic Weierstrass preparation theorem. We will look at that way of doing things in Chapter 6, but for now we forgo that approach and give the direct proof found in [Cas86].

PROOF: We use induction on N.

a) If N $= 0$, we must have $|a_0| > |a_n|$ for all $n \geq 1$, and what we want to prove is that in that case there are no zeros: $f(x) \neq 0$ for all $x \in \mathbb{Z}_p$. Indeed, if we had $f(x) = 0$, then

$$0 = f(x) = a_0 + a_1 x + a_2 x^2 + \cdots,$$

from which it would follow that

$$
\begin{aligned}
|a_0| &= |a_1 x + a_2 x^2 + \cdots| \\
&\leq \max_{n \geq 1} |a_n x^n| \\
&\leq \max_{n \geq 1} |a_n|.
\end{aligned}
$$

But this contradicts the assumption that $|a_0| > |a_n|$ for all $n \geq 1$.

b) To handle the induction step, we use an idea from the algebra of polynomials: a zero implies a factorization. Suppose that

$$|a_N| = \max_n |a_n| \qquad and \qquad |a_n| < |a_N| \quad for \ n > N,$$

and suppose that $f(\alpha) = 0$ for some $\alpha \in \mathbb{Z}_p$. Choose any $x \in \mathbb{Z}_p$. Then we have

$$
\begin{aligned}
f(x) &= f(x) - f(\alpha) = \sum_{n \geq 1} a_n (x^n - \alpha^n) \\
&= (x - \alpha) \sum_{n \geq 1} \sum_{j=0}^{n-1} a_n x^j \alpha^{n-1-j}.
\end{aligned}
$$

By Proposition 4.1.4, we can re-order the series as a power series in x, which gives

$$f(x) = (x - \alpha) \sum_{j=0}^{\infty} b_j x^j = (x - \alpha)g(x),$$

where $g(X)$ is the power series with coefficients

$$b_j = \sum_{k=0}^{\infty} a_{j+1+k}\alpha^k.$$

It is easy to see that $b_j \to 0$ as $j \to \infty$. In fact, we have

$$|b_j| \leq \max_{k \geq 0} |a_{j+1+k}| \leq |a_N|$$

for every j, and

$$|b_{N-1}| = |a_N + a_{N+1}\alpha + a_{N+2}\alpha^2 + \cdots| = |a_N|,$$

and finally, if $j \geq N$,

$$|b_j| \leq \max_{k \geq 0} |a_{j+k+1}| \leq \max_{j \geq N+1} |a_j| < |a_N|.$$

This shows that the magic number in Strassman's theorem when applied to $g(X)$ is $N - 1$. By induction, we can assume that $g(X)$ has at most $N - 1$ zeros in $\mathbb{Z}_p$, which implies that $f(X)$ has at most N zeros (those of $g(X)$, plus α). This proves the theorem. □

Problem 156 Check that the application of Proposition 4.1.4 in the proof of Strassman's theorem is valid.

Strassman's theorem is only the first of several important theorems[7] about zeros of functions on $\mathbb{Q}_p$ defined by power series. Even so, it is a very powerful theorem. Here are some consequences.

Corollary 4.4.7 Let $f(X) = \sum a_n X^n$ be a non-zero power series which converges on $\mathbb{Z}_p$, and let $\alpha_1, \ldots, \alpha_m$ be the roots of $f(X)$ in $\mathbb{Z}_p$. Then we can find a power series $g(X)$ which converges on $\mathbb{Z}_p$ but has no zeros in $\mathbb{Z}_p$, for which

$$f(X) = (X - \alpha_1) \cdots (X - \alpha_m)g(X).$$

PROOF: Clear from the proof of the theorem and Proposition 4.4.3. □

Since $\mathbb{Z}_p$ is just the closed unit ball in $\mathbb{Q}_p$, we can extend the result to other disks by simple scaling.

---

[7]For example, the p-adic Weierstrass Preparation Theorem and the theory of Newton polygons, which allows very detailed control of the zeros. See Chapter 6 for more details.

Corollary 4.4.8 *Let $f(X) = \sum a_n X^n$ be a non-zero power series which converges on $p^m \mathbb{Z}_p$, for some $m \in \mathbb{Z}$. Then $f(X)$ has a finite number of zeros in $p^m \mathbb{Z}_p$.*

PROOF: Define $g(X) = f(p^m X) = \sum a_n p^{mn} X^n$. Since $f(X)$ converges in $p^m \mathbb{Z}_p$, $g(x)$ converges for $x \in \mathbb{Z}_p$, and applying the theorem to $g(X)$ gives the finiteness.   □

Problem 157 Strassman's Theorem actually gives a bound for the number of roots in $\mathbb{Z}_p$. What is a bound for the number of roots in $p^m \mathbb{Z}_p$?

Problem 158 Say all you can about the zeros of the functions defined by the power series in problem 145.

This result allows us to prove a variant of Proposition 4.4.3:

Corollary 4.4.9 *Let $f(X) = \sum a_n X^n$ and $g(X) = \sum b_n X^n$ be two p-adic power series which converge in a disk $p^m \mathbb{Z}_p$. If there exist infinitely many numbers $\alpha \in p^m \mathbb{Z}_p$ such that $f(\alpha) = g(\alpha)$, then $a_n = b_n$ for all $n \geq 0$.*

PROOF: Apply the previous corollary to $f(X) - g(X)$.   □

Notice that since $p^m \mathbb{Z}_p$ is compact, the existence of infinitely many α as above implies the existence of a convergent sequence of such α, so that this result could also be proved directly from Proposition 4.4.3.

One consequence of this is something of a surprise:

Corollary 4.4.10 *Let $f(X) = \sum a_n X^n$ be a p-adic power series which converges in some disk $p^m \mathbb{Z}_p$. If the function $p^m \mathbb{Z}_p \longrightarrow \mathbb{Q}_p$ defined by $f(X)$ is periodic, that is, if there exists $\pi \in p^m \mathbb{Z}_p$ such that $f(x + \pi) = f(x)$ for all $x \in p^m \mathbb{Z}_p$, then $f(X)$ is constant.*

PROOF: The series $f(X) - f(0)$ has zeros at $n\pi$ for all $n \in \mathbb{Z}$. Since $\pi \in p^m \mathbb{Z}_p$ implies $n\pi \in p^m \mathbb{Z}_p$, this gives infinitely many zeros, and hence the series $f(X) - f(0)$ must be identically zero, i.e., $f(X)$ must be constant.   □

This offers an intriguing contrast to the classical case, where the sine and cosine functions are both periodic and "entire," i.e., they can each be expressed as a power series that converges everywhere. The crucial difference is that in the classical case it never happens that all the multiples of the period are in the same bounded interval, while in our case the non-archimedean property guarantees just that.

While periodicity is very different in the classical and the p-adic situations, the zeros of an entire function are distributed similarly in both cases:

Corollary 4.4.11 *Let $f(X) = \sum a_n X^n$ be a p-adic power series, and suppose that $f(X)$ is entire, i.e., that $f(x)$ converges for every $x \in \mathbb{Q}_p$. Then $f(X)$ has at most a denumerable set of zeros. Furthermore, if the set of zeros is not finite then the zeros form a sequence α_n with $|\alpha_n| \to \infty$.*

PROOF: This is clear, because the number of zeros in each bounded disk $p^m \mathbb{Z}_p$ is finite. $\qquad\square$

It is natural (and tempting) to conjecture from these results that there should be a representation of any entire function as an infinite product over the zeros; something like

$$f(X) = h(X) \prod (1 - \alpha^{-1} X),$$

where α ranges over the zeros of $f(X)$ and $h(X)$ is an entire function with no zeros. (Why it's best to write the expansion in terms of the inverses of the roots may be a little mysterious now; we will go back to this in Chapter 6.) It is easy to see that such a representation does exist, but it will not be very interesting unless we are ready to go to the algebraic closure of $\mathbb{Q}_p$, since even polynomials may fail to have roots in $\mathbb{Q}_p$. When we have the necessary machinery set up for working over the algebraic closure, we will be able to obtain a very precise description of entire functions in this spirit.

This brings out a rather embarrassing point: in the case of $\mathbb{R}$, the algebraic closure is an old friend, the field of complex numbers. By contrast, we really know very little about the algebraic closure of $\mathbb{Q}_p$. In fact, we do not even know whether the p-adic absolute value on $\mathbb{Q}_p$ can be extended to the algebraic closure. It turns out that this extension is indeed possible (we will discuss this a little later), but that the algebraic closure *is not complete* with respect to this absolute value. (This is very different from the classical case, where $\mathbb{C}$ is just as complete as $\mathbb{R}$.) The obvious thing to do is to go through the completion process again. The resulting field, usually called $\mathbb{C}_p$, is both complete and algebraically closed, and is the p-adic analog of the complex numbers. From many points of view, the field $\mathbb{C}_p$ is the "correct" context in which to do p-adic analysis, and we will go through the process of constructing it and studying the results in chapters 5 and 6 of this book. For now, we want to stay at a more intuitive level, and hence will continue working in $\mathbb{Q}_p$. What we will do, however, is be careful to construct our arguments in such a way that they will be easy to generalize to other fields. This will save us a lot of work later on.

4.5   Some Elementary Functions

In this section, our goal is to use power series to define p-adic functions which are analogous to classical functions. We begin with p-adic versions of the exponential and the logarithm functions. In contrast to the archimedean case, it is the logarithm that has the better convergence properties.

We begin with the usual power series for the logarithm:

$$\mathbf{f}(X) = \log(1 + X) = \sum_{n=1}^{\infty} (-1)^{n+1} \frac{X^n}{n} = X - \frac{X^2}{2} + \frac{X^3}{3} + \cdots$$

(We use **log**—rather than log—to emphasize that we are considering the formal power series, and not the function, which after all we have not yet defined in the p-adic context.) Since the coefficients of this power series are rational numbers, it makes sense to think of the series as a power series in $\mathbb{Q}_p$ (for any prime p). The first step towards understanding it is, of course, to compute its radius of convergence. Before we jump into the limit calculation, however, we should note another classical vs. p-adic contrast. In the classical case, all the integers in the denominators help the convergence, because they tend to make the terms of the series smaller. In the p-adic case, this is exactly reversed: integers in the denominator either do not change the absolute value (when they are not divisible by p) or make it *bigger* (when they are). What saves convergence in the case of this series is that "in general" n is not too divisible by p.

To compute ρ, let $\mathbf{f}(X) = \sum_{n=1}^{\infty}(-1)^{n+1}\dfrac{X^n}{n}$, so that

$$|a_n| = \left|\frac{1}{n}\right| = p^{v_p(n)}.$$

From this, we get

$$\sqrt[n]{|a_n|} = p^{v_p(n)/n} \to 1$$

as $n \to \infty$. (Check!) Hence, $\rho = 1$. This doesn't decide for us whether the convergence happens on the open or closed ball of radius 1. To decide, we need to look at what happens when $|x| = 1$. But it is clear that in that case $|a_n x^n| = |a_n| = |1/n|$ does not tend to zero. So we get

Lemma 4.5.1 *The series*

$$\mathbf{f}(X) = \sum_{n=1}^{\infty}(-1)^{n+1}\frac{X^n}{n} = X - \frac{X^2}{2} + \frac{X^3}{3} - \frac{X^4}{4} + \cdots$$

converges for $|x| < 1$ (and diverges otherwise).

Problem 159 Check that

$$\lim_{n \to \infty} p^{v_p(n)/n} = 1.$$

(The main idea is to estimate $v_p(n)$ as a function of n.)

The conclusion is that $\mathbf{f}(X)$ defines a function on the open ball $B(0, 1)$ of radius 1 and center 0. This suggests that we should define the logarithm in the obvious way, so that $\mathbf{f}(x) = \log(1 + x)$.

Definition 4.5.2 *Let $B = B(1, 1) = \{x \in \mathbb{Z}_p : |x - 1| < 1\} = 1 + p\mathbb{Z}_p$. We define the p-adic logarithm of $x \in B$ as*

$$\log_p(x) = \mathbf{log}(1 + (x - 1)) = \sum_{n=1}^{\infty}(-1)^{n+1}\frac{(x - 1)^n}{n}.$$

Of course, if we want this function to deserve to be called a logarithm, we had better check that it satisfies the functional equation that characterizes logarithms.

Proposition 4.5.3 *Suppose* $a, b \in 1 + p\mathbb{Z}_p$. *Then*

$$\log_p(ab) = \log_p(a) + \log_p(b).$$

PROOF: In the literature, this is often proved by noting that there is an underlying identity of power series. The problem with this is that verifying condition (iii) of Theorem 4.3.3 is somewhat problematic. So instead we give a direct proof that mimics the classical proof. For any $x \in p\mathbb{Z}_p$, let

$$f(x) = \log_p(1 + x) = \sum_{n=1}^{\infty} (-1)^{n+1} \frac{x^n}{n}.$$

Then, by our results on derivatives of functions defined by power series, we have

$$f'(x) = \sum_{n=0}^{\infty} (-1)^n x^n = \frac{1}{1 + x}.$$

Now fix $y \in p\mathbb{Z}_p$ and define

$$g(x) = \log_p((1 + x)(1 + y)) = f(y + (1 + y)x).$$

By the result in problem 152, this is a power series that converges for $|x| < 1$. Now use the chain rule to compute the derivative of g:

$$g'(x) = (1 + y)f'(y + (1 + y)x) = \frac{(1 + y)}{1 + y + (1 + y)x} = \frac{1}{1 + x} = f'(x).$$

Since both $f(x)$ and $g(x)$ are defined by power series that converge for $|x| < 1$, it follows by Corollary 4.4.5 that $g(x) = f(x) + c$. Plugging in $x = 0$ shows that $c = g(0) = f(y)$. Hence we've shown that $g(x) = f(x) + f(y)$; translating back to logarithms, this says

$$\log_p((1 + x)(1 + y)) = \log_p(1 + x) + \log_p(1 + y),$$

and we are done. $\qquad\qquad\qquad\qquad\qquad\qquad\qquad\qquad\qquad\qquad\qquad$ □

Problem 160 Show that if $p = 2$ then $-1 \in B$, so that it make sense to compute $\log_p(-1)$. Show that $\log_p(-1) = 0$. Compare with the example in section 3 of Chapter 1. Can you estimate the highest power of 2 that divides the n-th partial sum?

In the previous chapter, we used Hensel's Lemma to determine for which m there exist m-th roots of unity in $\mathbb{Q}_p$. Our method restricted us to the case where $p \nmid m$, so we left open the possibility of the existence of p^n-th roots of unity in $\mathbb{Q}_p$. It turns out, as we said then, that these do not exist, except for the trivial case when $p = 2$ and $n = 1$. The next three problems use the p-adic logarithm to prove this claim. The idea is that if x is a root of unity and $x \in 1 + p\mathbb{Z}_p$, then we must have $\log_p(x) = 0$, so that studying the zeros of the logarithm will gives us a handle on the roots of unity.

Problem 161 Use Strassman's Theorem to show that for $p \neq 2$ we have $\log_p(x) = 0$ if and only if $x = 1$. If $p = 2$, show that $\log_p(x) = 0$ if and only if $x = \pm 1$. (Hint: one can't use Strassman's Theorem directly, because the series does not converge in $\mathbb{Z}_p$, but rather in $p\mathbb{Z}_p$. But that is easily handled with a change of variables.)

Problem 162 Let $p \neq 2$. Show that if $x \in 1 + p\mathbb{Z}_p$ and $x^p = 1$, then $x = 1$. Conclude that there are no p-th roots of unity (and hence no p^n-th roots of unity) in $\mathbb{Q}_p$.

Problem 163 Let $p = 2$. Show that if $x \in 1 + 2\mathbb{Z}_2$ and $x^4 = 1$, then $x = \pm 1$. Conclude that there are no fourth roots of unity in $\mathbb{Q}_2$. (There are, of course, the square roots of unity ± 1.)

Since knowing the roots of unity in $\mathbb{Q}_p$ turns out to be very useful, we summarize all that we know about them:

- for $p = 2$, the only roots of unity in $\mathbb{Q}_p$ are ± 1

- for $p \neq 2$, $\mathbb{Q}_p$ contains all of the $(p - 1)$-st roots of unity, and no others.

(Recall that the existence of the $(p - 1)$-st roots of unity was proved as an application of Hensel's Lemma in the last chapter.)

Having obtained a logarithm, exponentials cannot be far behind. In the classical case, the series

$$\exp(X) = \sum_{n=0}^{\infty} \frac{X^n}{n!} = 1 + X + \frac{X^2}{2} + \frac{X^3}{6} + \cdots$$

converges for all $x \in \mathbb{R}$, because the coefficients $1/n!$ tend very quickly to zero with respect to the real absolute value. In the p-adic context, of course, this changes drastically, because $n!$ tends to zero, so that $1/n!$ becomes arbitrarily large as n grows. This means that we cannot expect to have a large radius of convergence. To determine what that radius will be, we have to work out exactly how fast the coefficients $1/n!$ grow, i.e., we have to work out the how divisible $n!$ is by p.

Lemma 4.5.4 *Let p be a prime. Then*

$$v_p(n!) = \sum_{i=0}^{\infty} \left\lfloor \frac{n}{p^i} \right\rfloor < \frac{n}{p-1},$$

where $\lfloor \cdot \rfloor$ is the greatest integer function. In particular

$$|n!|_p > p^{-n/(p-1)}$$

PROOF: The formula

$$v_p(n!) = \sum_{i=1}^{\infty} \left\lfloor \frac{n}{p^i} \right\rfloor$$

is well known and easy to prove. We leave it as the next problem. The inequality then follows, because $\lfloor x \rfloor \leq x$, so that

$$v_p(n!) = \sum_{i=1}^{\infty} \left\lfloor \frac{n}{p^i} \right\rfloor \leq \sum_{i=1}^{\infty} \frac{n}{p^i} = \frac{n}{p-1}$$

by the usual formula for geometric series. $\square$

Problem 164 Prove that

$$v_p(n!) = \sum_{i=1}^{\infty} \left\lfloor \frac{n}{p^i} \right\rfloor$$

Here is another version of the same formula, which sometimes is useful:

Problem 165 Let n be a positive integer, and let $n = a_0 + a_1 p + a_2 p^2 + \cdots + a_k p^k$ be its expansion in base p. Let $s = a_0 + a_1 + \cdots + a_k$ be the sum of the digits in the expansion. Show that

$$v_p(n!) = \frac{n-s}{p-1}.$$

(Hint: work out the difference between n/p^i and its integral part in terms of the base p expansion.)

Now we use these estimates to work out the convergence of the exponential.

Lemma 4.5.5 *Let*

$$\mathbf{g}(X) = \sum_{n=0}^{\infty} \frac{X^n}{n!} = 1 + X + \frac{X^2}{2!} + \frac{X^3}{3!} + \cdots$$

Then $\mathbf{g}(x)$ *converges if and only if* $|x| < p^{-1/(p-1)}$.

PROOF: Since

$$|a_n| = |1/n!| = p^{v_p(n!)} < p^{n/(p-1)}$$

by our first estimate, we get

$$\rho \geq p^{-1/(p-1)}.$$

Thus, the series certainly converges for $|x| < p^{-1/(p-1)}$.

On the other hand, let $|x| = p^{-1/(p-1)}$ and let $n = p^m$ be a power of p. In this case, we have

$$v_p(n!) = v_p(p^m!) = 1 + p + \cdots + p^{m-1} = \frac{p^m - 1}{p - 1}$$

(notice that this is a special case of the result in problem 165). Then, since $v_p(x) = 1/(p-1)$,

$$v_p\left(\frac{x^n}{n!}\right) = v_p\left(\frac{x^{p^m}}{p^m!}\right) = \frac{p^m}{p-1} - \frac{p^m-1}{p-1} = \frac{1}{p-1}.$$

This does not depend on m, hence $x^n/n!$ cannot tend to zero, and the series doesn't converge. Since we know that the region of convergence is a disk, this proves the lemma.   □

REMARK: There is something a little strange about the inequality in the lemma. If $p \neq 2$ and $x \in \mathbb{Z}_p$, then the absolute value of x can either be greater than or equal to 1 (which is bigger than $p^{-1/(p-1)}$) or less than or equal to p^{-1} (which is smaller): there are no values "in the middle." Thus,

$$|x| < p^{-1/(p-1)} \iff |x| \leq p^{-1} \iff x \in p\mathbb{Z}_p \iff |x| < 1,$$

so that the disk in the lemma is just the open disk of radius one. This seems to suggest that all our care in working out the precise radius of convergence is wasted. This is not really the case. The point is that our estimates will work in any field *containing* $\mathbb{Q}_p$ (with an absolute value extending the one on $\mathbb{Q}_p$), and in such fields there indeed may be elements with

$$p^{-1/(p-1)} \leq |x| < 1.$$

This will be particularly important when one wants to work in the field $\mathbb{C}_p$ which we mentioned above.

In any case, it is worth noting that as long as we stay in $\mathbb{Q}_p$, we have:

- if $p \neq 2$, $\mathbf{g}(x) = \exp(x)$ converges for $x \in p\mathbb{Z}_p$,

- if $p = 2$, $\mathbf{g}(x) = \exp(x)$ converges for $x \in 4\mathbb{Z}_2$,

since $-1/(2-1) = -1$.

Now we can define the p-adic exponential.

Definition 4.5.6 *Let* $D = B(0, p^{-1/(p-1)}) = \{x \in \mathbb{Z}_p : |x| < p^{-1/(p-1)}\}$. *The p-adic exponential is the function* $\exp_p : D \longrightarrow \mathbb{Q}_p$ *defined by*

$$\exp_p(x) = \sum_{n=0}^{\infty} \frac{x^n}{n!}.$$

Just as in the case of the logarithm, the formal property of the exponential is preserved.

Proposition 4.5.7 *If $x, y \in D$ we have $x + y \in D$ and*

$$\exp_p(x+y) = \exp_p(x)\exp_p(y).$$

PROOF: This is essentially a formal manipulation of power series:

$$\exp_p(x+y) = \sum_{n=0}^{\infty} \frac{(x+y)^n}{n!} = \sum_{n=0}^{\infty} \frac{1}{n!} \sum_{k=0}^{n} \binom{n}{k} x^{n-k} y^k$$

$$= \sum_{n=0}^{\infty} \sum_{k=0}^{n} \frac{1}{n!} \frac{n!}{(n-k)!k!} x^{n-k} y^k$$

$$= \sum_{n=0}^{\infty} \sum_{k=0}^{n} \frac{x^{n-k}}{(n-k)!} \frac{y^k}{k!}$$

$$= \left(\sum_{m=0}^{\infty} \frac{x^m}{m!} \right) \left(\sum_{k=0}^{\infty} \frac{y^k}{k!} \right)$$

$$= \exp_p(x) \exp_p(y),$$

as claimed. □

Problem 166 Are there convergence issues to check in the proof?

This shows that, apart from the smallish radius of convergence, we have obtained something that is a lot like the classical exponential.

There is of course one more formal property we would like to be true also in the p-adic context: the fact that the logarithm and the exponential are inverses, i.e., the relation

$$\exp(\log(1+X)) = 1 + X$$

and its inverse. This is a formal equality of power series, so that we only need to check that the conditions in Theorem 4.3.3 hold.

Proposition 4.5.8 Let $x \in \mathbb{Z}_p$, $|x| < p^{-1/(p-1)}$. Then we have

$$|\exp_p(x) - 1| < 1$$

so that $\exp_p(x)$ is in the domain of $\log_p$, and

$$\log_p(\exp_p(x)) = x.$$

Conversely, if $|x| < p^{-1/(p-1)}$ we have

$$|\log_p(1+x)| < p^{-1/(p-1)}$$

so that $\log_p(1+x)$ is in the domain of $\exp_p$, and

$$\exp_p(\log_p(1+x)) = 1 + x.$$

PROOF: We need to check the estimates to know that all the series converge, and we also need to check condition (iii) from Theorem 4.3.3. Note first that both identities are clearly true when $x = 0$, so that we can assume $x \neq 0$.

To compute $\log_p(\exp_p(x))$, we are actually plugging $\exp_p(x) - 1$ into the series $\log(1 + X)$, so that is the quantity we need to estimate. We start from

$$\left|\frac{x^n}{n!}\right| = |x|^n \cdot p^{v_p(n!)} < |x|^n p^{n/(p-1)},$$

which we get from Lemma 4.5.4. Since $|x| < p^{-1/(p-1)}$, this is less than 1, and it follows that

$$|\exp_p(x) - 1| = \left|\sum_{n=1}^{\infty} \frac{x^n}{n!}\right| < 1,$$

as claimed.

But in fact we can do better by using the result in problem 165; suppose $n \geq 2$; to make the computation easier, let's use the valuation v_p instead of absolute values. Since $v_p(x) > 1/(p-1)$, we get

$$v_p\left(\frac{x^{n-1}}{n!}\right) = (n-1)v_p(x) - v_p(n!) > \frac{n-1}{p-1} - \frac{n-s}{p-1} = \frac{s-1}{p-1} \geq 0,$$

where, as in problem 165, s is the sum of the digits in the expansion of n in base p (so that $s \geq 1$). It follows that

$$\left|\frac{x^{n-1}}{n!}\right| < 1,$$

and so

$$\left|\frac{x^n}{n!}\right| < |x|.$$

But this implies that $|\exp_p(x) - 1| = |x| > |x^n/n!|$ for all $n \geq 2$, so that condition (iii) in Theorem 4.3.3 is satisfied. (Notice that this also shows that $|\exp_p(x) - 1| < p^{-1/(p-1)}$, a stronger inequality than claimed in the theorem.) Applying Theorem 4.3.3, we can conclude from the formal equality of power series that if $|x| < p^{-1/(p-1)}$ we have

$$\log_p(\exp_p(x)) = x.$$

Now let's consider the composition in the opposite order. This time we're plugging $\log_p(1 + x)$ into $\exp(X)$, so we need to estimate the series for the logarithm. Suppose, then, that $|x| < p^{-1/(p-1)}$, or, in valuation language, that $v_p(x) > 1/(p-1)$.

If $n > 1$, we get

$$v_p\left(\frac{(-1)^{n+1}x^n}{n}\right) - v_p(x) = (n-1)v_p(x) - v_p(n)$$

$$> \frac{n-1}{p-1} - v_p(n)$$

$$= (n-1)\left(\frac{1}{p-1} - \frac{v_p(n)}{n-1}\right).$$

If we can show that the last expression is never negative, it will follow both that the estimate we claimed above holds and that the extra condition is satisfied. So let $n = p^v n'$ with n' not divisible by p. Then

$$\frac{v_p(n)}{n-1} = \frac{v}{p^v n' - 1} \le \frac{v}{p^v - 1} = \frac{1}{p-1}\frac{v}{p^{v-1} + \cdots + p + 1} \le \frac{1}{p-1}.$$

Putting all the inequalities together gives that for $n > 1$ we have

$$v_p\left(\frac{(-1)^{n+1}x^n}{n}\right) - v_p(x) > 0,$$

or, going back to absolute values,

$$\left|\frac{(-1)^{n+1}x^n}{n}\right| < |x|.$$

Now we appeal to Corollary 4.1.2; it tells us that

$$|\log_p(x)| = |x| < p^{-1/(p-1)},$$

which shows both that $\log_p(x)$ is in the domain of the exponential and that condition (iii) in Theorem 4.3.3 is satisfied. Hence the formal equality of power series implies what we want:

$$\exp_p(\log_p(1+x)) = 1 + x.$$

This finishes the proof.   □

Problem 167 The one step that might need checking is the the inequality

$$\frac{v}{p^{v-1} + \cdots + p + 1} \le 1.$$

Can you prove it?

The hypotheses of the theorem are indeed necessary, for two reasons. The first, and less crucial one, is that if $|x| < 1$ but $|x| \ge p^{-1/(p-1)}$, it can very well be that $\log_p(1+x)$ does not belong to the domain of the exponential.

(Can you find an example with $p = 2$?) But much more serious is the fact that it can happen that we have $|x| < 1$, $|x| \geq p^{-1/(p-1)}$, and also

$$|\log_p(1 + x)| < p^{-1/(p-1)},$$

so that all the series involved converge, but

$$\exp_p(\log_p(1 + x)) \neq 1 + x.$$

This is due to the fact that the extra condition in Theorem 4.3.3 really does matter. To see this concretely, consider what happens[8] when we take $p = 2$ and $x = -2$: in that case, $1 + x = 1 - 2 = -1$, so that

$$\log_p(1 + x) = \log_p(-1) = 0.$$

Then, when we plug into the series for the exponential, we get

$$\exp_p(\log_p(-1)) = \exp(0) = 1 \neq -1.$$

In other words, the p-adic exponential and logarithm are inverses only within the restricted domains specified in the proposition.

Problem 168 Why doesn't Theorem 4.3.3 apply in this situation?

Problem 169 Use power series to define p-adic analogues of the sine and cosine functions, and determine their regions of convergence. Show that if $p \equiv 1 \pmod 4$ then there exists $i \in \mathbb{Q}_p$ such that $i^2 = -1$, and the classical relation

$$\exp_p(ix) = \cos_p(x) + i \sin_p(x)$$

holds for any x in the common region of convergence. The classical trigonometric functions are periodic; are the p-adic versions periodic?

As an application of the p-adic logarithm and exponential, we can study the group of p-adic units $\mathbb{Z}_p^\times$ a little more carefully. We've already shown, using Hensel's lemma, that $\mathbb{Z}_p^\times$ contains the $(p-1)$-st roots of unity. Now we want to know what "the rest of $\mathbb{Z}_p^\times$" looks like. The idea is to look carefully at the domains and images of the logarithm and exponential functions.

To simplify the notation, let's introduce a parameter q as follows:

- if p is an odd prime, then $q = p$;

- if $p = 2$, then $q = 4$.

The point is that then the p-adic exponential $\exp_p(x)$ will be defined for $x \in q\mathbb{Z}_p$, and $\log_p(x)$ will be defined for $x \in 1 + p\mathbb{Z}_p$. Notice also that $\mathbb{Q}_p$ contains the $(p-1)$-st roots of unity when p is odd, and contains the square roots of unity when $p = 2$. If we use Euler's φ function, defined by $\varphi(n) =$ the number of integers between 1 and n which are relatively prime to n, then the number of roots of unity in $\mathbb{Q}_p$ is always $\varphi(q)$, since $\varphi(p) = p - 1$ for any prime, and $\varphi(4) = 2$.

---

[8]This is the example we referred to above when we discussed Theorem 4.3.3.

Problem 170 Another way to define the φ function is to say that $\varphi(n)$ is the number of elements in $(\mathbb{Z}/n\mathbb{Z})^\times$, i.e., the number of invertible elements in the ring $\mathbb{Z}/n\mathbb{Z}$. Check that the two definitions are equivalent.

Let's define two subsets of $\mathbb{Z}_p^\times$:

$$U = \{x \in \mathbb{Z}_p^\times : |x - 1| < 1\} = 1 + p\mathbb{Z}_p$$

$$U_1 = \{x \in \mathbb{Z}_p^\times : |x - 1| < p^{-1/(p-1)}\} = 1 + q\mathbb{Z}_p.$$

Notice that

- $U_1 \subset U \subset \mathbb{Z}_p^\times$,

- $U = U_1$ except if $p = 2$,

- if $p = 2$, then $U = \mathbb{Z}_p^\times$, and

- U and U_1 are *subgroups* of $\mathbb{Z}_p^\times$.

The elements of U are often called the "1-units" (which comes from "the units which are congruent to 1 (mod $p\mathbb{Z}_p$)").

Problem 171 Check that U and U_1 are indeed subgroups of $\mathbb{Z}_p^\times$.

We can now determine the structure of $\mathbb{Z}_p^\times$ quite precisely:

Proposition 4.5.9 *Let* U *and* U_1 *be as above, and let*

$$W = \{x \in \mathbb{Z}_p : |x| < p^{-1/(p-1)}\} = q\mathbb{Z}_p,$$

considered as an additive group.

i) The p-adic logarithm $\log_p$ *defines a homomorphism of groups*

$$\log_p : U \longrightarrow \mathbb{Z}_p^+,$$

whose image is contained in the valuation ideal $\wp = p\mathbb{Z}_p$.

ii) The p-adic logarithm $\log_p$ *defines an isomorphism of groups*

$$\log_p : U_1 \overset{\sim}{\longrightarrow} W,$$

with inverse $\exp_p$. *In particular,* $U_1 \cong W \cong \mathbb{Z}_p^+$ *is torsion-free.*

PROOF: This is a straight translation of the discussion above into the language of groups. Recall that a group is torsion-free if there exist no elements $x \neq 1$ such that $x^m = 1$ for some m. The last statement follows from the fact that the *additive* group $\mathbb{Z}_p^+$ is torsion-free. □

Corollary 4.5.10 *For any prime p, we have an isomorphism $\mathbb{Z}_p^\times \cong V \times U_1$, where $U_1 \cong \mathbb{Z}_p^+$ is a torsion-free pro-p-group and V is the torsion part of $\mathbb{Z}_p^\times$. Furthermore:*

i) V is the set of roots of unity in $\mathbb{Q}_p$, which is a subgroup of $\mathbb{Z}_p^\times$, and

ii) $V \cong (\mathbb{Z}/q\mathbb{Z})^\times$, so that V is a cyclic group of order $\varphi(q)$;

PROOF: It is easy to see that there is an exact sequence

$$1 \longrightarrow U_1 \longrightarrow \mathbb{Z}_p^\times \overset{\pi}{\longrightarrow} (\mathbb{Z}/q\mathbb{Z})^\times \longrightarrow 0.$$

(Remember that this means that the kernel of each homomorphism in the sequence is equal to the image of the previous one, so that this is basically just the *definition* of U_1.) In fancy language, what we want to prove is that the exact sequence "splits," but we will just give a direct proof.

We already know (from a combination of Hensel's Lemma and Strassman's Theorem) that $\mathbb{Z}_p^\times$ contains a group V of roots of unity. It is a cyclic group of order $p-1$ when p is odd, and of order 2 when $p=2$, so in any case it case $\varphi(q)$ elements, i.e., just as many elements as $(\mathbb{Z}/q\mathbb{Z})^\times$ does. We know that any two of these elements are distinct modulo q (Hensel's Lemma for odd p, check directly for $p=2$—see the next problem). Suppose ζ_1 and ζ_2 have the same image under the map π. Then $\zeta_1 \zeta_2^{-1} \in U_1$, so that $\zeta_1 \zeta_2^{-1} = 1 + qx$ for some $x \in \mathbb{Z}_p$, so that $\zeta_1 \equiv \zeta_2 \pmod{q}$, which we know can't happen unless $\zeta_1 = \zeta_2$. In other words, π induces an isomorphism between V and $(\mathbb{Z}/q\mathbb{Z})^\times$, and the other assertions in the theorem follow easily. □

Problem 172 Prove that two different roots of unity (of order prime to p) cannot be congruent modulo $q\mathbb{Z}_p$. (This means: their difference cannot belong to $q\mathbb{Z}_p$. There are a whole lot of ways to do this.)

Problem 173 Fill in whatever is missing in the proof of the Corollary.

One thing that follows from this result is that for any odd prime p there exists an inclusion

$$\omega : \mathbb{F}_p^\times \cong V \hookrightarrow \mathbb{Z}_p^\times,$$

where $\mathbb{F}_p$ is the field with p elements. We can extend ω to $\mathbb{F}_p$ by setting $\omega(0) = 0$. The function ω is called *the Teichmüller character*, and it appears quite frequently in many different guises. If we compose it with the "reduction modulo p" map from $\mathbb{Z}$ to $\mathbb{F}_p$,

$$\mathbb{Z} \overset{(\bmod\ p)}{\longrightarrow} \mathbb{F}_p \overset{\omega}{\longrightarrow} \mathbb{Z}_p,$$

we get a Dirichlet character[9] with values in $\mathbb{Z}_p$, which is also usually called the Teichmüller character and denoted by ω. To complete the confusion, one

---

[9]Basically, a multiplicative function on $\mathbb{Z}$.

often also uses ω to denote the projection from $\mathbb{Z}_p^\times$ onto its direct factor V, so that every $x \in \mathbb{Z}_p^\times$ is written uniquely as

$$x = \omega(x) \cdot x_1$$

with $x_1 \in 1 + q\mathbb{Z}_p$. This makes sense, because if we extend this projection to all of $\mathbb{Z}_p$ by mapping non-units to 0, and then restrict back to $\mathbb{Z}$, we get the Dirichlet character ω. The apparently confusing notation turns out, then, not to be so bad, because all the different maps denoted by ω are closely related.

Problem 174 When $p = 2$, some of the above needs to be modified. What changes are needed?

To introduce one more bit of notation, one often uses $\langle x \rangle$ to denote the projection of x on $U_1 = 1 + q\mathbb{Z}_p$, so that the direct product decomposition looks like

$$x = \omega(x)\langle x \rangle.$$

The next problem gives a different way of obtaining ω.

Problem 175 Show that if $p \neq 2$ and $x \in \mathbb{Z}_p^\times$, then we have

$$\omega(x) = \lim_{n \to \infty} x^{p^n}.$$

(Hint: one idea is to start with the expression of x as a product of $\omega(x)$ and $\langle x \rangle$.)

We want to conclude our exploration of the p-adic elementary functions by considering the binomial series. In $\mathbb{R}$, we know that we function $(1 + X)^\alpha$ can be expanded as a power series which converges for $|x| < 1$:

$$(1 + X)^\alpha = \mathbf{B}(\alpha, X) = \sum_{n=0}^\infty \binom{\alpha}{n} X^n,$$

where

$$\binom{\alpha}{n} = \frac{\alpha(\alpha - 1)\ldots(\alpha - n + 1)}{n!}.$$

We want to use this series to define the p-adic version of this function. (Of course, as is the case over $\mathbb{R}$, this is only interesting when α is not an integer, but it will work in that case also.) In the p-adic context, the convergence properties of the series will depend on the choice of the p-adic number α. We only consider the case when $\alpha \in \mathbb{Z}_p$ is a p-adic *integer*. The case when $\alpha \in \mathbb{Q}_p$ but is not in $\mathbb{Z}_p$ is actually easier, and we leave it as an exercise for the reader.

So take a p-adic integer α, and consider the binomial series

$$(1 + X)^\alpha = \mathbf{B}(\alpha, X) = \sum_{n=0}^\infty \binom{\alpha}{n} X^n.$$

The first thing is to check that the coefficients are be p-adic integers.

Lemma 4.5.11 *If $\alpha \in \mathbb{Z}_p$ and $n \geq 0$, then $\binom{\alpha}{n} \in \mathbb{Z}_p$.*

PROOF: For each n, consider the polynomial

$$P_n(X) = \frac{X(X-1)\dots(X-n+1)}{n!} \in \mathbb{Q}[X].$$

Just as any polynomial does, $P_n(X)$ defines a continuous function from $\mathbb{Q}_p$ to $\mathbb{Q}_p$. Now, we know that the binomial coefficient $\binom{m}{n}$ of two *positive integers* $m, n \in \mathbb{Z}_+$ is in $\mathbb{Z}$. Hence, for $\alpha \in \mathbb{Z}_+$, we have

$$P_n(\alpha) = \binom{\alpha}{n} \in \mathbb{Z}.$$

In other words, the continuous function P_n maps the set $\mathbb{Z}_+$ of positive integers to $\mathbb{Z}$. By continuity, it must map the closure of $\mathbb{Z}_+$ in $\mathbb{Z}_p$ to the closure of $\mathbb{Z}$. But remember that any element in $\mathbb{Z}_p$ is the limit of a sequence of positive integers (the partial sums of its p-adic expansion). Hence the closure of $\mathbb{Z}_+$ is all of $\mathbb{Z}_p$, and we conclude that P_n maps $\mathbb{Z}_p$ to $\mathbb{Z}_p$, which is what we want to prove.   □

Corollary 4.5.12 *If $\alpha \in \mathbb{Z}_p$ and $|x| < 1$, the series*

$$\mathbf{B}(\alpha, x) = \sum_{n=0}^{\infty} \binom{\alpha}{n} x^n$$

converges.

PROOF: Clear.   □

Problem 176 The Corollary makes no claim that the radius of convergence is in fact equal to 1, nor that the series diverges when $|x| = 1$. What are the facts?

As for the logarithm and exponential, it follows from an equality of formal power series that for $\alpha = a/b \in \mathbb{Z}_{(p)}$ and $|x| < 1$ we have

$$\left(\mathbf{B}\left(\frac{a}{b}, x\right) \right)^b = (1+x)^a,$$

so that it makes sense to write

$$\mathbf{B}\left(\frac{a}{b}, x\right) = (1+x)^{a/b}.$$

This suggests that we should *define*, for any $\alpha \in \mathbb{Z}_p$ and any $x \in p\mathbb{Z}_p$,

$$(1+x)^\alpha := \mathbf{B}(\alpha, x).$$

One should be careful, however, to distinguish the p-adic function $\mathbf{B}(a/b, x)$ from its real analogue, *even when x is rational and $1 + x$ is a b-th power in* $\mathbb{Q}$. The following neat example is taken from [Kob84].

EXAMPLE: (Following Koblitz) Let $p = 7$, $\alpha = 1/2$, and $x = 7/9$, so that $x \in 7\mathbb{Z}_7$ and $1 + x = 16/9$ is a rational square. In $\mathbb{R}$, we have

$$(1 + x)^{1/2} = \frac{4}{3}.$$

In $\mathbb{Q}_7$, on the other hand, we have $|x| = 1/7$, so that, for $n \geq 1$,

$$\left| \binom{1/2}{n} x^n \right| \leq |x|^n = \frac{1}{7^n} < 1.$$

This implies that

$$(1 + x)^{1/2} = 1 + \sum_{n \geq 1} \binom{1/2}{n} x^n \in 1 + 7\mathbb{Z}_7,$$

or, in terms of absolute values, that

$$\left| (1 + x)^{1/2} - 1 \right| < 1.$$

But

$$\left| \frac{4}{3} - 1 \right| = \left| \frac{1}{3} \right| = 1,$$

so that we cannot have $\mathbf{B}(\frac{1}{2}, \frac{7}{9}) = \frac{4}{3}$. In fact, what happens is that in $\mathbb{Q}_7$ we have

$$(1 + 7/9)^{1/2} = \mathbf{B}\left(\frac{1}{2}, \frac{7}{9} \right) = -\frac{4}{3} = 1 - \frac{7}{3} \in 1 + 7\mathbb{Z}_7.$$

The point is that the same series $\sum a_n$ with $a_n \in \mathbb{Q}$ can converge in both $\mathbb{R}$ and some $\mathbb{Q}_p$, but have different limits (even different rational limits), since the topologies are completely different.

In any case, we *will* write $(1 + x)^\alpha$ instead of $\mathbf{B}(\alpha, x)$, and let the context decide in which field we are working. The point is to keep in mind that the meaning of the symbol depends on the underlying field.

Problem 177 Study the convergence properties of the binomial series when α is not a p-adic integer.

Problem 178 Show that the value of $\mathbf{B}(\alpha, x)$ *does not* depend on the field we are working in when $x \in \mathbb{Q}$ and $\alpha \in \mathbb{Z}$ is an integer.

The next exercise is taken from [Kob84]. It attempts to decide exactly when $(1 + x)^{1/2}$ is equal to the positive square root.

Problem 179 (Koblitz) Choose $x \in \mathbb{Q}$ such that $1 + x$ is a square in $\mathbb{Q}$; say $\sqrt{1 + x} = a/b$ with a and b positive and relatively prime. Let S be the set of primes (including the infinite prime, if applicable) for which the binomial series $\mathbf{B}(1/2, x)$ converges in $\mathbb{Q}_p$. (The limit will have to be a square root of $1 + x$, hence will equal either a/b or $-a/b$.) Prove that:

i) If p is an odd prime, then $p \in S$ if and only if $p|(a+b)$ or $p|(a-b)$, and in $\mathbb{Q}_p$ we will have $\mathbf{B}(1/2,x) = -a/b$ in the first case, $\mathbf{B}(1/2,x) = a/b$ in the second.

ii) We will have $2 \in S$ if and only if a and b are both odd; the limit in $\mathbb{Q}_2$ will be a/b if $a \equiv b \pmod 4$, and $-a/b$ if $a \equiv -b \pmod 4$.

iii) We will have $\infty \in S$ if and only if $0 < a/b < \sqrt{2}$, and the sum in $\mathbb{R}$ will always be a/b.

iv) There is no x for which the set S is empty, and S will have only one element if and only if $x \in \{8, \frac{16}{9}, 3, \frac{5}{4}\}$.

v) Except for the x mentioned in the previous item, there always exist primes $p, q \in S$ such that the sum in $\mathbb{Q}_p$ is different from the sum in $\mathbb{Q}_q$.

For other interesting results along these lines, tracking what happens when we look at the same series in various different $\mathbb{Q}_p$, see the article [BS96].

4.6   Interpolation

The idea of interpolating a known function to obtain a related p-adic function has become very important in number theory, where the standard targets for this method have been the zeta and L-functions. The point of this section is to give a first example of this. Our example is very simple, and it illustrates only some of the many ideas that have arisen in the literature. We refer the reader to the standard references.[10]

In the previous section, we considered the binomial series, and used it to define a p-adic function $x \mapsto x^\alpha$ for $x \in 1 + p\mathbb{Z}_p$ and $\alpha \in \mathbb{Z}_p$. What we would like to do in this section is to invert the situation, and think of x^α *as a function of α*. We would like to interpret this as an interpolation problem, in the following way.

Suppose $n \in \mathbb{Z}_p$ is any p-adic integer, and α is an integer. Then it certainly makes sense to compute n^α. Thus, we can consider the function

$$f(\alpha) = n^\alpha,$$

which is well-defined for $\alpha \in \mathbb{Z}$. What we would like to do is to extend this function to the widest possible range of p-adic values of α. Since $\mathbb{Z}$ is dense in $\mathbb{Z}_p$, such an extension, if continuous, is unique, because two continuous functions that coincide on a dense subset are identical. Indeed, one can even work with smaller subsets of $\mathbb{Z}$: the set of positive integers, or the set of negative integers, or any other set of integers which is dense in $\mathbb{Z}_p$. The problem of finding such an extension is called the problem of finding a p-adic interpolation of the function $f(\alpha) = n^\alpha$.

The first thing to say about p-adic interpolation is that in a certain sense the whole thing is trivial. This is because we know perfectly well when it is

---

[10]The idea of working out this example is due to Koblitz, who goes through a similar discussion in [Kob84].

that a function defined on a dense subset of $\mathbb{Z}_p$ has a continuous extension
to all of $\mathbb{Z}_p$. The crucial notion here is *uniform continuity*, and we recall
what that means. If $f(x)$ is a function defined on (a subset of) a field $\Bbbk$ with
an absolute value, we say that f is uniformly continuous if it satisfies the
following condition:

> Given any real number $\epsilon > 0$, one can find a real number $\delta > 0$
> such that for any $x, y \in \Bbbk$
>
> $$|x - y| < \delta \implies |f(x) - f(y)| < \epsilon$$

The point, of course, is that mere continuity guarantees that for each fixed
choice of x one can find a δ that works, but uniform continuity requires that
the *same* δ work for every x. The reason this is relevant to the interpolation
problem is a well-known theorem which we leave as an exercise:

Problem 180 Show that any continuous function defined on a compact set is automatically uniformly continuous and bounded.

Problem 181 Can you give an example of a function $\mathbb{Z} \longrightarrow \mathbb{Z}_p$ which is continuous
but not uniformly continuous? (This may be a little hard.)

Now suppose our $f(\alpha)$ could indeed be extended to $\mathbb{Z}_p$. Then, since $\mathbb{Z}_p$ is
compact, the extension would have to be bounded and uniformly continuous.
Hence (restricting back), so would $f(\alpha)$. It turns out that in fact these two
conditions are sufficient.

Proposition 4.6.1 *Let S be a dense subset of $\mathbb{Z}_p$, and let $f : S \longrightarrow \mathbb{Q}_p$ be
a function. Then there exists a continuous extension $\tilde{f} : \mathbb{Z}_p \longrightarrow \mathbb{Q}_p$ of f to
$\mathbb{Z}_p$ if and only if f is bounded and uniformly continuous. If it exists, this
extension in unique.*

PROOF: We know that the condition is necessary, and that the extension
is unique if it exists, by the discussion above. The difficulty is to prove the
sufficiency, i.e., to show that uniform continuity and boundedness are enough
to guarantee the existence of the extension.

The key is the continuity. If $x \in \mathbb{Z}_p$, there exists a sequence

$$\alpha_1, \alpha_2, \ldots, \alpha_k, \ldots$$

of elements of S which tends to x (because S is dense). If $\tilde{f}$ exists, then we
will have

$$\tilde{f}(x) = \lim_{k \to \infty} \tilde{f}(\alpha_k) = \lim_{k \to \infty} f(\alpha_k).$$

This shows the way to proceed.

First of all, since the sequence α_k tends to x, it is a Cauchy sequence, so that

$$\lim |\alpha_{k+1} - \alpha_k| = 0.$$

Since f is uniformly continuous and bounded, it follows that

$$\lim |f(\alpha_{k+1}) - f(\alpha_k)|) = 0$$

(check!), so that the $f(\alpha_k)$ form a Cauchy sequence, hence have a limit in $\mathbb{Q}_p$. Now we can *define* $\tilde{f}$ by the condition we know it has to satisfy:

$$\tilde{f}(x) := \lim_{k \to \infty} f(\alpha_k)$$

for any sequence α_k converging to x. This gives the extension. $\qquad \square$

There are a whole bunch of things to check, and the reader should:

Problem 182 Check that the image of a Cauchy sequence α_k by a bounded and uniformly continuous function f is again a Cauchy sequence.

Problem 183 Check that the function $\tilde{f}$ defined above does not depend on the choice of the sequence α_k.

Problem 184 Check that the function $\tilde{f}$ defined above is indeed a continuous function on $\mathbb{Z}_p$.

One less obvious fact is that one can replace $\mathbb{Z}_p$ in the proposition by any compact subset of $\mathbb{Q}_p$:

Problem 185 Check that the proposition remains true if we replace $\mathbb{Z}_p$ by any compact subset of $\mathbb{Q}_p$, such as $\mathbb{Z}_p^\times$, $1 + p\mathbb{Z}_p$, or $p^m \mathbb{Z}_p$. (Hint: the point is that only the compactness was used.)

This result may seem to completely settle the issue, but that is far from being the case, for several important reasons. For one thing, one often wants to know more about $\tilde{f}$ than its bare existence. For example, can it be written as a power series? Does it extend to a set larger than $\mathbb{Z}_p$? Can we give a good method to compute (better: to approximate) it? Another point is that we can exploit the "if and and only if" in the proposition: if what we want to prove is the uniform continuity, then finding an interpolation will prove just that! Finally, thinking in terms of interpolation often gives us useful new ideas, as we shall see below when we get to the nitty-gritty of our example.

Before we go on to the example, however, it may be useful to unwind what uniform continuity really means in our case. We will take $f(\alpha)$ to be a function defined on a dense subset S of $\mathbb{Z}_p$, with values in $\mathbb{Q}_p$. Then being "close" in S amounts to being congruent modulo a high power of p, and being close in $\mathbb{Q}_p$ is the same. Hence, f will be uniformly continuous if it satisfies the following congruence condition:

Given $m \in \mathbb{Z}$, there exists $N \in \mathbb{Z}$ such that

$$\alpha \equiv \beta \pmod{p^N} \implies f(\alpha) \equiv f(\beta) \pmod{p^m}.$$

Thus, uniform continuity has a simple translation in terms of congruence properties. This turns out to be quite important.

Now we return to the exponential function $\alpha \mapsto n^\alpha$. This is defined, at first, for $\alpha \in \mathbb{Z}$ and $n \in \mathbb{Z}_p$, and we would like to extend it to all $\alpha \in \mathbb{Z}_p$ The answer, as it happens, depends quite seriously on n.

First of all, suppose n is a 1-unit, that is, $n \in 1 + p\mathbb{Z}_p$. Then we can use the binomial series to get our interpolation:

Corollary 4.6.2 *For any $n \in 1 + p\mathbb{Z}_p$ there exists a continuous function $f_n : \mathbb{Z}_p \longrightarrow \mathbb{Q}_p$ such that for any $\alpha \in \mathbb{Z}$ we have $f_n(\alpha) = n^\alpha$.*

PROOF: We can just define $f_n(\alpha) = \mathbf{B}(\alpha, n - 1)$, which converges because we are assuming $n \in 1 + p\mathbb{Z}_p$. Checking continuity, however, is not all that easy (remember that we want continuity in α, rather than in n, so it is not just a matter of saying that power series are continuous functions). We leave the verification to the reader as a challenging problem.   □

Problem 186 Show that $\mathbf{B}(\alpha, x)$ is continuous as a function of α.

One might also try to go the direct route, and show that if $n \in 1 + p\mathbb{Z}_p$ then $\alpha \mapsto n^\alpha$ is bounded and uniformly continuous. Boundedness is easy: any integral power of n will be in $\mathbb{Z}_p$ (and even in $1 + p\mathbb{Z}_p$), because n is a unit (there are negative powers in this game too!). As for uniform continuity, that is also not hard to show; notice, first that

$$(1 + pk)^{p^m} \equiv 1 \pmod{p^{m+1}},$$

so that if $\beta = \alpha + ip^m$ we get

$$n^\beta = n^\alpha \cdot (n^{p^m})^i \equiv n^\alpha \pmod{p^{m+1}},$$

which is what we want. This establishes the *existence* of f_n. Proving that $f_n(\alpha) = \mathbf{B}(\alpha, n - 1)$ requires showing that the latter is continuous.

This does the trick for $n \in 1 + p\mathbb{Z}_p$. We would like, however, to consider more general p-adic integers. Unfortunately, that turns out to be quite tricky. To begin with, suppose p divides n. Then, as the integer α becomes bigger, n^α becomes p-adically closer and closer to zero. This messes everything up. For example, take $n = p$, and look at the sequence $\alpha_k = 1 + p^k$. Then

$$\lim_{k \to \infty} \alpha_k = 1, \qquad \text{but} \qquad \lim_{k \to \infty} p^{\alpha_k} = 0 \neq p^1,$$

so that the map $\alpha \mapsto p^\alpha$ is not even continuous.

We might have a better chance if we tried to work only with p-adic units. When $p = 2$, this gives nothing new, since $\mathbb{Z}_2^\times = 1 + 2\mathbb{Z}_2$. For odd primes p, however, we know that $1 + p\mathbb{Z}_p$ is a subgroup of index $p - 1$ in $\mathbb{Z}_p^\times$, so that going from $n \in 1 + p\mathbb{Z}_p$ to $n \in \mathbb{Z}_p^\times$ would be progress. Even this, however, turns out to be a little tricky, basically because of the presence of the roots of unity.

Let $p \neq 2$; for $n \in \mathbb{Z}_p^\times$, we will try to interpolate the function $\alpha \mapsto n^\alpha$. Since we have already done this for $n \in 1 + p\mathbb{Z}_p$, the easiest way to do this is to use the known relation between $\mathbb{Z}_p^\times$ and its subgroup $1 + p\mathbb{Z}_p$. Recall that we showed that there is a direct product decomposition

$$\mathbb{Z}_p^\times = V \times U_1 \cong \mathbb{F}_p^\times \times (1 + p\mathbb{Z}_p),$$

and that for $x \in \mathbb{Z}_p^\times$ this decomposition gives $x = \omega(x)\langle x \rangle$ with $\omega(x) \in V$ and $\langle x \rangle \in 1 + p\mathbb{Z}_p$. Then, for any integer α, we have

$$n^\alpha = \omega(n)^\alpha \langle n \rangle^\alpha.$$

The first thing to note is that $\omega(n)$ is a $(p-1)$-st root of unity, and hence if $\alpha \equiv \alpha_0 \pmod{p-1}$ we can re-write the formula as

$$n^\alpha = \omega(n)^{\alpha_0} \langle n \rangle^\alpha.$$

Now, since $\langle n \rangle \in 1 + p\mathbb{Z}_p$, we already know how to interpolate its part of the function, i.e., we know how to interpolate the function $\alpha \mapsto \langle n \rangle^\alpha$. But this is almost enough to solve the problem, since we've reduced everything to this known interpolation together with the choice of α_0. In fact, the best way to think of this is to do a complete turnaround, and change the function to be interpolated!

Rather than considering the function $\alpha \mapsto n^\alpha$ for all integers α, consider it only for those integers congruent to a fixed α_0 modulo $(p-1)$. There are of course $p - 1$ different functions of this kind, each corresponding to a choice of α_0. The kicker, of course, is that the set of integers α which are congruent to a fixed α_0 is itself *dense* in $\mathbb{Z}_p$, so that it makes sense to ask for an interpolation from this set to all of $\mathbb{Z}_p$. And this, by the discussion above, is easily done: consider the p-adic function $f_{\alpha_0} : \mathbb{Z}_p \longrightarrow \mathbb{Z}_p$ given by

$$f_{\alpha_0}(\alpha) = \omega(n)^{\alpha_0} \langle n \rangle^\alpha.$$

This, first of all, makes sense, by the discussion above, since we do know how to compute the α-th power of the 1-unit $\langle n \rangle$. Next, it does coincide with the function $\alpha \mapsto n^\alpha$ whenever α is an integer satisfying $\alpha \equiv \alpha_0 \pmod{p-1}$. So it does give a (somewhat skewed) solution to our interpolation problem, which we state as a theorem:

Proposition 4.6.3 *Let $n \in \mathbb{Z}_p^\times$ and $\alpha_0 \in \{0, 1, \ldots, p-1\}$, and let*

$$A_{\alpha_0} = \{\alpha \in \mathbb{Z} : p \nmid \alpha \text{ and } \alpha \equiv \alpha_0 \pmod{p-1}\} \subset \mathbb{Z}.$$

Then

$$f_{\alpha_0}(\alpha) = \omega(n)^{\alpha_0} \langle n \rangle^\alpha$$

defines a function $f_{\alpha_0} : \mathbb{Z}_p \longrightarrow \mathbb{Z}_p$ *such that*

$$f_{\alpha_0}(\alpha) = n^\alpha \qquad \text{whenever} \quad \alpha \in A_{\alpha_0}.$$

Notice that all the different f_{α_0} coincide if $n \in 1 + p\mathbb{Z}_p$, so that this is a genuine extension of our first interpolation. What happens, though, if we compute f_{α_0} on the wrong sort of $\alpha \in \mathbb{Z}$? Well, we get something like

$$\begin{aligned}
f_{\alpha_0}(\alpha) &= \omega(n)^{\alpha_0} \langle n \rangle^\alpha \\
&= \omega(n)^{\alpha_0 - \alpha} \omega(n)^\alpha \langle n \rangle^\alpha \\
&= \omega(n)^{\alpha_0 - \alpha} n^\alpha
\end{aligned}$$

In words, f_{α_0} actually interpolates a function that is slightly different from our original function: rather than giving n^α, it gives a "twisted" version which ends up being equal to a root of unity times n^α. For the special α's that belong to A_{α_0}, the root of unity disappears, and we get our original function. So we're close, but we haven't really done exactly what we set out to do. This is in fact as good a result as one might hope for, as the example in the next problem shows.

Problem 187 Show that the function $\mathbb{Z} \longrightarrow \mathbb{Z}$ given by $\alpha \mapsto (-1)^\alpha$ can only be interpolated to a function $\mathbb{Z}_p \longrightarrow \mathbb{Z}_p$ when $p = 2$ (in which case -1 is a 1-unit). For $p = 3$, the Proposition above claims that there exist *two* functions f_0 and f_1 which "together" give an interpolation. Describe the two functions f_0 and f_1. (Hint: they are not very interesting).

Some readers may find this situation a bit unsatisfactory: rather than one interpolating function, we have ended up with a whole bunch, each of which gives an interpolation for a restriction of the original function to a smaller set. One way of jazzing this up a bit is the following. The collection of all the f_{α_0} together define a function

$$\mathcal{F} : \mathbb{Z}_p \times \mathbb{Z}/(p-1)\mathbb{Z} \longrightarrow \mathbb{Z}_p$$

given by $\mathcal{F}(\alpha, \alpha_0) = f_{\alpha_0}(\alpha)$. Now, one has the "diagonal inclusion"

$$\mathbb{Z} \hookrightarrow \mathbb{Z}_p \times \mathbb{Z}/(p-1)\mathbb{Z},$$

given by $\alpha \mapsto (\alpha, \alpha)$. (The first α to be thought of as an element of $\mathbb{Z}_p$, the second as an integer modulo $p - 1$.) In other words, if $\alpha \in \mathbb{Z}$, its image under the inclusion is the pair (α, α_0), where $\alpha_0 \equiv \alpha \pmod{p-1}$. Thus, if we restrict $\mathcal{F}$ to the image of $\mathbb{Z}$ we get

$$\alpha \mapsto \mathcal{F}(\alpha, \alpha_0) = f_{\alpha_0}(\alpha) = \omega(n)^{\alpha_0} \langle n \rangle^\alpha = n^\alpha.$$

This means that we *can* think of $\mathcal{F}$ as giving an interpolation of the function $\alpha \mapsto n^\alpha$, provided we think of $\mathbb{Z}$ as included in this larger set.

Interpolation problems of this kind are very important in the applications of p-adic analysis to number theory, and several of the features of our toy example persist in the more interesting ones. First, one often has to "remove the p-part." In our case, this was accomplished by restricting the base n to be a p-adic unit. Second, the interpolation often requires us to consider "twisted" versions of the original function. In our case, these were the several f_{α_0} functions, and restricting one of the f_{α_0} to $\mathbb{Z}$ does not give the function $\alpha \mapsto n^\alpha$, but rather the function

$$\alpha \mapsto \omega(n)^{\alpha_0 - \alpha} n^\alpha.$$

This kind of modification, when something is multiplied by a root of unity, if often referred to as "twisting." The upshot: one cannot interpolate the function $\alpha \mapsto n^\alpha$, but one can interpolate appropriate twists of that function. This phenomenon is actually quite common.

An obvious question should be mentioned here: what is the point? Why should one want to interpolate "classical" functions in this fashion?

The question is hard to answer in elementary terms, without delving into the complexities of the specific interpolation problems that mathematicians have been interested in. But we can give some idea of what is going on by saying that many classical functions have interesting "special values," that is, their values at certain magical points have a special significance. For example, the values of the Riemann zeta function $\zeta(s)$ at positive even integers involve the Bernoulli numbers, which hide within themselves quite a lot of information about the arithmetic of cyclotomic fields. (Its pole at $s = 1$ also carries this kind of information).

Now suppose one can interpolate these special values with a p-adic function. This gives us a p-adic function which shares with its classical analogue the same (or similar, if a twist creeps in) special values. Well, this means that one can get information on those values by looking at either function... and the p-adic function is often easier to handle. This yields a basic strategy that has been applied over and over in modern number theory, with very interesting results. The reader may want to browse through the articles in [CT91] to get some idea of what the goals of this particular enterprise are. To begin to study the enterprise itself, one might start with the treatment in [Kob84].

5 Vector Spaces and Field Extensions

Up to now, we have kept our attention focused on the field $\mathbb{Q}$ and its p-adic completions. We have already felt, however, the need to consider other fields (for example, when we dealt with the zeros of a function defined by a power series). In fact, just as we have emphasized the natural analogy between the p-adic fields $\mathbb{Q}_p$ and the field $\mathbb{R}$ of real numbers, it is a very natural thing to do to look for an extension of $\mathbb{Q}_p$ that is analogous to the complex numbers. In other words, we would like to look for ways to extend $\mathbb{Q}_p$ in order to obtain a field that is not only complete (so that we can do analysis), but also algebraically closed (so that all polynomials have roots). This turns out to be more subtle (and therefore more interesting) than one might expect. It turns out, first of all, that to get an algebraically closed field one must make a very large extension of $\mathbb{Q}_p$. This extension turns out not to be complete any more, so there is no other recourse but to go through the completion process again, and this finally yields the field we wanted. This is very different from the classical case, where going from $\mathbb{R}$ to an algebraically closed field is just a small step (just add i), and the resulting field (the complex numbers) is already complete. The goal of this chapter is to tell the p-adic version of this story in its entirety.

In order to get there, we begin by considering vector spaces over $\mathbb{Q}_p$ and the norms one might define on them. This is a step in the right direction, since any field containing $\mathbb{Q}_p$ will also be a $\mathbb{Q}_p$-vector space. We then go on to considering the fields themselves. This will necessarily involve some knowledge of abstract algebra; as usual, we have tried to make the facts we use explicit, in order to make it easier to look up the material we need in the standard texts. We start with finite field extensions, and only after we have understood them well do we try to go on to an algebraic closure.

The reader should note that we have taken one of two possible points of view in addressing our subject. We will be investigating extensions of the p-adic fields $\mathbb{Q}_p$. It would be just as interesting to consider extensions of $\mathbb{Q}$ itself, and to attempt to construct a theory of absolute values on such fields. This leads to an interesting theory, which we have decided not to address at all (because it requires more knowledge of Galois theory than we wish to assume, and because it properly belongs in an introduction to algebraic number theory). This means that we must of necessity fail to mention certain topics, such as the extension to bigger fields of the product formula, of Ostrowski's theorem, or of the local-global principle. Instead, we take a "strictly local" perspective: we are living in the p-adic world from the start. There is a good discussion of the global (or semi-local) aspect in [Cas86] and in many introductions to algebraic number theory.

Once we have absolute values on extensions of $\mathbb{Q}_p$, we will be in a position to extend to such fields much of what was done in chapters 3 and 4. Rather than do so in full detail, we will often be content with "this clearly extends;" the reader for whom the "clearly" is not clear should go back and check.[1] We will also need to prove a few results about these fields that will allow us to understand what goes on when one puts them all together to get an algebraic closure.

5.1 Normed Vector Spaces over Complete Valued Fields

The algebraic part of the theory of vector spaces over $\mathbb{Q}_p$ is, of course, identical to the theory of vector spaces over any other field. This is simply because that part of the theory does not depend on the specific field at all: it only requires the knowledge of the basic field properties. Therefore, we won't bother to discuss the basics about vector spaces, subspaces, bases, dimension, and so on.

What we would like to focus on, then, is the point where the vector spaces acquire a metric. This is usually done by putting a *norm* on the vector space. For example, in the classical case, we can metrize $\mathbb{R}^2$ using the norm

$$\|(x, y)\| = \sqrt{x^2 + y^2},$$

and similarly for all the $\mathbb{R}^n$. Of course, there isn't just one choice of norm. For example, the following two choices of norms on $\mathbb{R}^2$ are also popular:

$$\|(x, y)\|_1 = |x| + |y|$$

and

$$\|(x, y)\|_{\text{sup}} = \max\{|x|, |y|\}$$

(the subscript 1 here has nothing to do with the subscripts on the p-adic norms; there should be no serious confusion involved).

We want to build up an analogous theory for norms on vector spaces over $\mathbb{Q}_p$. We begin, as we did in Chapter 2, by considering a general theory of normed vector spaces over valued fields, because it is no more difficult than doing things over $\mathbb{Q}_p$. As we did then, we will restrict to $\mathbb{Q}_p$ whenever that makes things easier.

We begin with a field $\Bbbk$, which we assume has an absolute value $|\ |$ on it. (We do not make any assumption about whether the absolute value is archimedean, but we *do* assume it is non-trivial, because otherwise things

---

[1]Instructors should note that this may mean that more time than usual may need to be spent on this chapter!

are pretty silly.) In order to get an interesting theory, we assume that $\mathbb{k}$ is *complete* with respect to its absolute value. We will also assume for simplicity that $\mathbb{k}$ is of *characteristic zero*,[2] so that it contains $\mathbb{Q}$. The reader should keep both $\mathbb{R}$ and $\mathbb{Q}_p$ in mind as examples.

Let V be a vector space over $\mathbb{k}$. At first we make no further assumptions on V, but later we will want to concentrate on the case where V is finite-dimensional.

Definition 5.1.1 *Let $\mathbb{k}$ be a complete valued field of characteristic zero with an absolute value $|\ |$. A norm on a $\mathbb{k}$-vector space V is a function*

$$\|\ \| : V \longrightarrow \mathbb{R}_+$$

satisfying the following conditions:

  i) $\|\mathbf{v}\| = 0$ *if and only if* $\mathbf{v} = 0$,

  ii) for any two vectors $\mathbf{v}, \mathbf{w} \in V$, *we have* $\|\mathbf{v} + \mathbf{w}\| \le \|\mathbf{v}\| + \|\mathbf{w}\|$,

  iii) for any $\mathbf{v} \in V$ *and any* $\lambda \in \mathbb{k}$, *we have* $\|\lambda\mathbf{v}\| = |\lambda|\,\|\mathbf{v}\|$.

A vector space V which has a norm $\|\ \|$ is called a normed vector space over $\mathbb{k}$.

In other words, a norm is just a way to measure the size of vectors, and the conditions merely require that it behave as we would expect such a notion of length to behave. One is tempted, of course, to introduce the notion of non-archimedean norms, but it is less clear that it is a good idea. For example, consider the norm on $V = \mathbb{Q}_p \times \mathbb{Q}_p$ given by

$$\|(x,y)\| = \sqrt{|x|_p^2 + |y|_p^2}.$$

One easily checks that this is indeed a norm, but that it does not satisfy the naïve analogue of the non-archimedean inequality (by which we mean that something like

$$\|(x + x', y + y')\| \le \max\{\|(x,y)\|, \|(x',y')\|\},$$

does not hold). But this norm is still "non-archimedean" in the sense that given two vectors it may not be possible to find an integer multiple of one which is bigger than the other (check this!). In fact, this suggests that normed vector spaces over non-archimedean complete fields are automatically "non-archimedean" in any reasonable sense, so that there is nothing to define.

---

[2]The reader will recall, I hope, that the characteristic of a field is the smallest number of ones that need to be added together to get zero, when this is possible, and is zero when it is not possible. For example, the characteristic of $\mathbb{Q}$ is zero, and the characteristic of $\mathbb{F}_p$ is p. It is an easy exercise to prove that if the characteristic is non-zero, then it must be a prime number.

Given a norm, we can easily define a metric (i.e., a way of measuring distance) on V, by saying that the distance between two vectors is (what else?) the size of their difference:

Definition 5.1.2 *Let V be a normed vector space with norm $\| \ \|$. We define a metric on V by putting, for any $\mathbf{v}, \mathbf{w} \in V$,*

$$d(\mathbf{v}, \mathbf{w}) = \|\mathbf{v} - \mathbf{w}\|.$$

Problem 188 Show that the metric thus defined is indeed a metric, that is, it has the properties listed in Problem 42.

Once we have a metric, we have, as in Chapter 2, a topology, so that we can talk about open and closed balls, open sets, and convergence. (We urge the reader who is hesitant about this to re-read the appropriate section of Chapter 2.)

Problem 189 Let V be a normed vector space. The point of this problem is to check that the metric $d(x, y)$ (or, equivalently, the norm it is derived from) relates well to the operations in V:

i) Fix $\mathbf{v}_0, \mathbf{w}_0 \in V$. Show that for any $\varepsilon > 0$ there exists a $\delta > 0$ such that, whenever $d(\mathbf{v}, \mathbf{v}_0) < \delta$ and $d(\mathbf{w}, \mathbf{w}_0) < \delta$, we have $d(\mathbf{v} + \mathbf{w}, \mathbf{v}_0 + \mathbf{w}_0) < \varepsilon$. In other words, addition is a continuous function.

ii) Fix $\mathbf{v}_0 \in V$ and $\lambda_0 \in \mathbf{k}$. Show that for any $\varepsilon > 0$ there exists a $\delta > 0$ such that, whenever $d(\mathbf{v}, \mathbf{v}_0) < \delta$ (distance in V) and $d(\lambda, \lambda_0) < \delta$ (distance in $\mathbf{k}$), we have $d(\lambda \mathbf{v}, \lambda_0 \mathbf{v}_0) < \varepsilon$. In other words, multiplication of a vector by an element of $\mathbf{k}$ is a continuous function.

This shows that the metric $d(\mathbf{v}, \mathbf{w})$ makes V a *topological vector space* over the topological field $\mathbf{k}$. (Compare Problem 43.)

Let's consider some examples. For these, we assume V is finite-dimensional, and we fix a basis $\{\mathbf{v}_1, \mathbf{v}_2, \ldots, \mathbf{v}_n\}$. Any vector in V can then be written (uniquely) in the form $\mathbf{v} = a_1 \mathbf{v}_1 + a_2 \mathbf{v}_2 + \cdots + a_n \mathbf{v}_n$ with $a_i \in \mathbf{k}$, and we exploit this to obtain norms on V from the absolute value on $\mathbf{k}$:

i) We can define a norm by putting

$$\|a_1 \mathbf{v}_1 + a_2 \mathbf{v}_2 + \cdots + a_n \mathbf{v}_n\|_{\sup} = \max_{1 \leq i \leq n} |a_i|.$$

This is called the sup-norm on V with respect to our choice of basis.

ii) We can also define, for each real number $r \geq 1$, the r-norm

$$\|a_1 \mathbf{v}_1 + a_2 \mathbf{v}_2 + \cdots + a_n \mathbf{v}_n\|_r = \left(|a_1|^r + |a_2|^r + \cdots + |a_n|^r\right)^{1/r}.$$

These are analogous to norms on spaces of functions that are often used in analysis.

Notice that if $k = \mathbb{R}$, $V = \mathbb{R}^2$, $v_1 = (1, 0)$, $v_2 = (0, 1)$, and we take $r = 1$ or $r = 2$, we get the examples mentioned in the introduction to this section. We leave it to the reader to check that these are indeed norms.

Problem 190 Check that the sup-norm and the r-norms are indeed norms.

Problem 191 Let $k = \mathbb{R}$, $V = \mathbb{R}^2$, and use the canonical basis $\{(1, 0), (0, 1)\}$. Sketch the closed ball of radius 1 with respect to (a) the sup-norm, (b) the r-norms for $r = 1, 2, 3$.

Problem 192 Show, with an example, that the norms we have defined depend quite seriously on the choice of basis. (Hint: this is very easy; just use the simplest vector space you can think of.)

Problem 193 Let $V = \mathbb{Q}_p \times \mathbb{Q}_p$, and define $\|(x, y)\| = |x + y|$. Does this define a norm?

As in the case of fields, we need to define a notion of equivalence for norms, just as we defined equivalence of absolute values.

Definition 5.1.3 *We say two norms $\| \ \|_1$ and $\| \ \|_2$ on a k-vector space V are equivalent if there exist positive real numbers C and D such that, for every vector $v \in V$, we have*

$$\|v\|_1 \leq C\|v\|_2 \qquad and \qquad \|v\|_2 \leq D\|v\|_1.$$

To get a good feeling for this notion, the reader is invited to work through a few elementary facts about it:

Problem 194 Show that two norms on V are equivalent if and only if they define the same topology on V (i.e., a set is open with respect to one norm if and only if it is open with respect to the other).

Problem 195 Sometimes it's useful to state the condition for equivalence in another way. Suppose $\| \ \|_1$ and $\| \ \|_2$ are equivalent. Show that any open ball around 0 with respect to norm $\| \ \|_1$ contains an open ball around 0 with respect to $\| \ \|_2$ and is contained in an open ball around 0 with respect to $\| \ \|_2$. Show that this condition is equivalent to the two inequalities.

Problem 196 Show that if two norms are equivalent, then they have the same Cauchy sequences; in other words, a sequence is Cauchy with respect to one of them if and only if it is Cauchy with respect to the other.

One should note that the fact that equivalent norms give the same Cauchy sequences is a priori *stronger* than the fact that they induce the same topology. That the two things end up being the same in our case is directly linked to the fact that our metric comes from a norm on a vector space, and that extra structure yields extra information.

Problem 197 Show that the norms on $\mathbb{R}^2$ that we mentioned above are equivalent. (Hint: your sketches from problem 191 might prove helpful.)

Problem 198 Let $V = \mathbb{Q}_p \times \mathbb{Q}_p$, and define the norms

$$\|(a, b)\|_{\mathsf{sup}} = \max\{|a|, |b|\}$$

and

$$\|(a, b)\|_1 = |a| + |b|.$$

Prove that these norms are equivalent.

Once we have a metric, we can ask about completeness, just as in Chapter 3. Recall that we say V is complete with respect to a norm $\| \; \|$ if any Cauchy sequence in V (with respect to $\| \; \|$) converges. (Note that this depends only on the equivalence class of the norm, which is as we want it.) It is well-known that $\mathbb{R}^2$, for example, is complete with respect to all of the norms mentioned above. Here is another example where one can show completeness for a whole bunch of spaces and norms in one blow:

Proposition 5.1.4 *Let V be a finite-dimensional vector space over a complete valued field $\Bbbk$. Choose a basis $\{v_1, v_2, \ldots, v_m\}$ for V, and let $\| \; \|$ be the sup-norm with respect to this basis. Then V is complete. Specifically, a sequence (w_n) with*

$$w_n = a_{1n}v_1 + a_{2n}v_2 + \cdots + a_{mn}v_m$$

is Cauchy in V if and only if the sequences of basis coefficients (a_{1n}), (a_{2n}), $\ldots (a_{mn})$ are Cauchy sequences in $\Bbbk$, and the limit is obtained by taking the limits of the coefficients:

$$\lim_{n \to \infty} w_n = (\lim_{n \to \infty} a_{1n})v_1 + (\lim_{n \to \infty} a_{2n})v_2 + \cdots + (\lim_{n \to \infty} a_{mn})v_m.$$

PROOF: Since the norm is simply given by the largest of the basis coefficients, saying that $\|w_{n_1} - w_{n_2}\|$ tends to zero just amounts to saying that *all* the differences $a_{in_1} - a_{in_2}$ do. And that is enough to prove everything we've claimed is true.   □

Finally, here are some problems that suggest some avenues for further exploration:

Problem 199 Let V and W be normed vector spaces, and write $\| \; \|_v$ and $\| \; \|_w$ for their norms. Let $f : V \longrightarrow W$ be a linear transformation . Show that the following are equivalent:

i) f is continuous at $0 \in V$;

ii) $\sup\limits_{\|v\|_v \leq 1} \|f(v)\|_w$ is finite;

iii) there exists an M such that we have $\|f(v)\|_w \leq M\|v\|_v$ for all $v \in V$;

iv) f is continuous at all $v \in V$.

Problem 200 (Hard) Let V be the space of sequences (a_n) with $a_n \in \mathbb{Q}_p$ and $\lim a_n = 0$. Define a norm on V by $\|(a_n)\| = \sup_n |a_n|$.

i) Is V complete with respect to this norm?

ii) Consider the subspace $W \subset V$ defined by the condition that $\sum |a_n|$ converges (in $\mathbb{R}$, of course). Is W a closed subspace of V? On W, we have two norms: the norm induced by the norm on V, and the 1-norm given by $\|(a_n)\| = \sum |a_n|$. Are these norms equivalent?

Problem 201 Let V be the space of all polynomials with coefficients in $\mathbb{Q}_p$. Choose a positive real number $c \in \mathbb{R}$ and define, for $f(X) = a_n X^n + \cdots + a_1 X + a_0$,

$$\|f(X)\|_c = \max_{0 \leq i \leq n} |a_i| c^i.$$

i) Show that this is a norm on V.

ii) Is V complete with respect to this norm?

iii) We know how to multiply polynomials. Is it true that the norm we just defined is multiplicative, i.e., that $\|f(X)g(X)\|_c = \|f(X)\|_c \|g(X)\|_c$?

iv) Explain why this norm is interesting.

v) Now suppose we vary c; we get a whole family of norms. Are they equivalent?

5.2   Finite-dimensional Normed Vector Spaces

The problems at the end of the previous section already hint that there is a fundamental difference between finite and infinite-dimensional spaces when it comes to the theory of norms. This is indeed the case, and in this section we prove the fundamental theorem about finite dimensional normed vector spaces over complete fields. What this theorem says is that, after Proposition 5.1.4, we already know all that there is to know about the finite-dimensional case. This is because it turns out that *any* norm on such a vector space is equivalent to the sup-norm (with respect to any given basis); in particular, all the sup-norms are equivalent. The proof of this result, which we give next, is often given only for locally compact complete fields; we give a general proof, following Cassels in [Cas86].

Theorem 5.2.1 *Let V be a finite-dimensional vector space over a complete valued field $\Bbbk$. Then any two norms on V are equivalent. Moreover, V is complete with respect to the metric induced by any norm.*

This is a tricky theorem to prove, so we do this in several parts. Take V to be a finite-dimensional vector space over $\mathbb{k}$ (which have assumed to be complete, of course). Fix a basis $\{v_1, v_2, \ldots, v_n\}$ of V, and let $\|\ \|_0$ be the sup-norm with respect to this basis. Finally, let $\|\ \|_1$ be any other norm on V. We want to prove that $\|\ \|_1$ is equivalent to $\|\ \|_0$, which means that we want to show that there are positive real numbers C and D such that, for every $v \in V$, we have

$$\|v\|_1 \leq C\|v\|_0 \qquad \text{and} \qquad \|v\|_0 \leq D\|v\|_1.$$

The first inequality is not very hard to obtain:

Proposition 5.2.2 *Let*

$$C = n \cdot \max_{1 \leq i \leq n} \|v_i\|_1.$$

Then we have, for any $v \in V$,

$$\|v\|_1 \leq C\|v\|_0.$$

PROOF: Take $v \in V$, and write it in terms of the basis as

$$v = a_1 v_1 + a_2 v_2 + \cdots + a_n v_n.$$

Then $\|v\|_0 = \max |a_i|$. Now just follow the path of least resistance:

$$\|v\|_1 = \|a_1 v_1 + a_2 v_2 + \cdots + a_n v_n\|_1$$

$$\leq \|a_1 v_1\|_1 + \|a_2 v_2\|_1 + \cdots + \|a_n v_n\|_1$$

$$= |a_1|\|v_1\|_1 + |a_2|\|v_2\|_1 + \cdots + |a_n|\|v_n\|_1$$

$$\leq n \max |a_i| \max \|v_i\|_1 = C \max |a_i| = C\|v\|_0,$$

which is exactly what we want.   □

The converse inequality takes a lot more proving. We will do it by induction on the dimension of V.

Proposition 5.2.3 *There exists a positive real number D such that, for every $v \in V$, we have $\|v\|_0 \leq D\|v\|_1$. In particular, V is complete with respect to $\|\ \|_1$.*

PROOF: (Take a deep breath. Here goes.) Notice, first of all, that once the inequality is proved, it follows that $\|\ \|_1$ is equivalent to the sup-norm, and we already know that V is complete with respect to $\|\ \|_0$, so that V will also be complete with respect to $\|\ \|_1$. In other words, once we have proved the first statement of the Proposition, we will have proved the second statement too.

We will prove the inequality by induction on the dimension of V, noting first that it is trivially true for spaces of dimension 1.[3] Thus, we only need to prove the induction step: assume that the proposition is true for spaces of dimension $n - 1$, and show that it is then also true for spaces of dimension n.

Let V, then, be a space of dimension n. As above, we fix a basis

$$\{v_1, v_2, \ldots, v_n\}.$$

We want to show that there exists a number D such that

$$\|w\|_0 \le D\|w\|_1 \qquad \text{for all } w \in V.$$

Well, suppose not. In that case, the quotient $\|w\|_1/\|w\|_0$ must get arbitrarily close to zero as w ranges through the vectors in V (because otherwise we can let $E > 0$ be a number such that the quotient is always bigger than E, and then taking $D = 1/E$ will do the trick). This means that, given any integer m we can find a vector w_m such that

$$\|w_m\|_1 < \frac{1}{m}\|w_m\|_0.$$

We want to argue that the w_m can be chosen in a particular (rather peculiar) way. Note, first, that the sup-norm $\|w_m\|_0$ is equal to the largest of the n basis coefficients. Since there is a finite number of basis vectors and an infinite number of m's, there must be some index i such that there are infinitely many m's for which $\|w_m\|_0$ is equal to the i-th basis coefficient. (Got it? Read it again.) After permuting the basis vectors, we can assume that $i = n$, i.e., that there are infinitely many m's such that $\|w_m\|_0 = $ the n-th basis coefficient. Let $m_1, m_2, \ldots$ be the sequence of those m's, arranged in increasing order; we will now restrict ourselves to the corresponding sequence of w's

$$w_{m_1}, w_{m_2}, w_{m_3}, \ldots, w_{m_k}, \ldots$$

Recall that these satisfy the inequality

$$\|w_{m_k}\|_1 < \frac{1}{m_k}\|w_{m_k}\|_0,$$

and that we've also arranged things so that if we set β_k equal to the n-th basis coefficient of w_{m_k}, then $\|w_{m_k}\|_0 = |\beta_k|$.

Now consider the vectors $\beta_k^{-1}w_{m_k}$. These have two nice properties:

i) their n-th basis coefficient is 1, so that we can write

$$\beta_k^{-1}w_{m_k} = u_k + v_n,$$

with u_k belonging to the subspace $W \subset V$ spanned by the vectors $v_1, v_2, \ldots, v_{n-1}$. (We take this equation as the definition of the u_k.)

---

[3]Prove it!

ii) We have

$$\|\mathbf{u}_k + \mathbf{v}_n\| = \|\beta_k^{-1} \mathbf{w}_{m_k}\|_1 = |\beta_k|^{-1} \|\mathbf{w}_{m_k}\|_1 = \frac{\|\mathbf{w}_{m_k}\|_1}{\|\mathbf{w}_{m_k}\|_0} < \frac{1}{m_k}$$

where m_k is an infinite increasing sequence of integers.

It follows that we have constructed a sequence of vectors $\mathbf{u}_k$, all of which lie in the $(n-1)$-dimensional subspace W, and such that the norms $\|\mathbf{u}_k + \mathbf{v}_n\|_1$ (where, remember, $\mathbf{v}_n$ is the n-th vector in our chosen basis for V) tend to zero as $k \to \infty$.

Now clearly, the $\mathbf{u}_k$ form a Cauchy sequence in W, since

$$\|\mathbf{u}_{k+l} - \mathbf{u}_k\|_1 \le \|\mathbf{u}_{k+l} + \mathbf{v}_n\|_1 + \|\mathbf{u}_k + \mathbf{v}_n\|_1 < \frac{1}{m_{k+l}} + \frac{1}{m_k}.$$

By induction, we know that W is complete, so that there must be a vector $\mathbf{u} \in W$ such that $\mathbf{u}_k \to \mathbf{u}$. But then we must have

$$\|\mathbf{u} + \mathbf{v}_n\|_1 = \lim \|\mathbf{u}_k + \mathbf{v}_n\|_1 = 0,$$

which means that $\mathbf{u} = -\mathbf{v}_n$, which is a contradiction, since $\mathbf{v}_n \notin W$ by the definition of the subspace W. This contradiction shows that D must exist, and therefore proves the theorem.   □

That is quite a long haul, but worth it, since it says that, as long as our vector spaces are finite-dimensional, the theory is essentially quite simple, and we might as well work with the sup-norm all the time.

There is one extra property of finite-dimensional normed spaces that is worth pointing out. This has to do with *local compactness*, which we discussed above when we showed that the p-adic fields $\mathbb{Q}_p$ were locally compact, as are $\mathbb{R}$ and $\mathbb{C}$ (see Section 3.3). In the vector space context, we have the following:

Proposition 5.2.4 *Let* $\mathbb{k}$ *be a locally compact complete valued field, and let* V *be a finite-dimensional (and therefore complete) normed vector space over* $\mathbb{k}$. *Then* V *is locally compact.*

PROOF: To show that V is locally compact, we need to find a neighborhood of the zero vector which is compact. The neighborhood we will choose will be the closed unit ball B around zero, so that

$$B = \{\mathbf{v} \in V : \|\mathbf{v}\| \le 1\}.$$

Using the main theorem, we see that we can take any norm on V (being locally compact is a topological property, and all the norms are equivalent). We choose the sup-norm with respect to some fixed basis $\{\mathbf{v}_1, \mathbf{v}_2, \dots \mathbf{v}_n\}$. Then a vector $\mathbf{v} = a_1 \mathbf{v}_1 + a_2 \mathbf{v}_2 + \cdots + a_n \mathbf{v}_n$ belongs to B if and only if each a_i belongs to the closed unit ball in $\mathbb{k}$. This is promising, since we know that the closed unit ball in $\Bbbk$ is compact (because we are assuming that $\Bbbk$ is locally compact).

In Chapter 3, we saw that to prove that a set is compact it is enough to show that it is complete and that it is totally bounded (which means: for every positive number ε there is a finite covering of the set by balls of radius ε). The first part is done already: B is a closed subset of a complete space, and therefore is complete.

To show that B is totally bounded, we use the fact that the unit ball in $\Bbbk$ is totally bounded. Given an ε, cover the unit ball in $\Bbbk$ with a finite number, say N, of balls of radius ε. Let $c_1, c_2, \ldots c_N$ be the centers of those balls. Then consider the n^N vectors in V each of whose basis coefficients is one of the c_i. Around each of these vectors, take a ball of radius ε. We claim that these balls cover B, that is, that any vector in B belongs to at least one of them.

To see that, take a vector $\mathbf{v} = a_1\mathbf{v}_1 + a_2\mathbf{v}_2 + \cdots + a_n\mathbf{v}_n \in B$. Since this means that the coefficients a_j are in the unit ball in $\Bbbk$, we know that each a_j is within less than ε of one of the centers c_i; call this one c_{i_j}. Then $\mathbf{v}$ belongs to the ball of radius ε (remember, with respect to the sup-norm) around the vector $c_{i_1}\mathbf{v}_1 + c_{i_2}\mathbf{v}_2 + \cdots + c_{i_n}\mathbf{v}_n$, which proves what we wanted. $\qquad\square$

Problem 202 Draw a picture to explain the proof we just gave. It should clarify things immensely.

The converse of this proposition is also true: any locally compact normed vector space over $\Bbbk$ is of necessity finite-dimensional. This is harder to prove, however, so we leave it to the reader to puzzle it out or look it up.

In contrast to the finite-dimensional case, the theory of infinite-dimensional normed vector spaces is quite rich and complex. It is the starting point of the field called "functional analysis," which has a long and distinguished history. Given our point of view, of course, we would mostly be interested in *non-archimedean* functional analysis, which is a much younger, but still very interesting, subject. We refer the interested reader to the references (see [Ami75, BGR84, Mon70, Sch84, vR78]).

5.3   Finite Field Extensions

We now go on to what we are really interested in, which is considering extensions of the field $\mathbb{Q}_p$. These are simply fields K containing $\mathbb{Q}_p$. For example, if 2 is not a square in $\mathbb{Q}_p$, we might want to consider the extension $K = \mathbb{Q}_p(\sqrt{2})$. More generally, we might want to obtain K by adjoining a root of some irreducible polynomial, or even to consider a field like $\mathbb{Q}_p(X)$ (rational functions with coefficients in $\mathbb{Q}_p$). For this section, we will restrict ourselves to *finite* extensions (the definition follows just below).

So let K be a field containing $\mathbb{Q}_p$. This means, among other things, that K is a vector space over $\mathbb{Q}_p$, and we say that K is a *finite* extension of $\mathbb{Q}_p$ if its dimension as a $\mathbb{Q}_p$-vector space is finite. We will write $[K : \mathbb{Q}_p] = \dim_{\mathbb{Q}_p} K$, and call this number the *degree* of K over $\mathbb{Q}_p$. We want to consider absolute values on K, but to keep things interesting we will require that these absolute values extend the p-adic absolute value on $\mathbb{Q}_p$. In other words, we are looking for a function $|\ | : K \longrightarrow \mathbb{R}_+$ which is an absolute value, and hence satisfies the usual properties:

   i) $|x| = 0$ if and only if $x = 0$,

   ii) $|xy| = |x|\,|y|$ for any $x, y \in K$,

   iii) $|x + y| \leq |x| + |y|$ for any $x, y \in K$,

and that also satisfies the extra condition that

   iv) $|\lambda| = |\lambda|_p$ whenever $\lambda \in \mathbb{Q}_p$.

There are several things to note. First, any such function will be a norm on K as a $\mathbb{Q}_p$-vector space (restrict x to $\mathbb{Q}_p$ in the second property, and we have the defining properties of a norm). Second, the absolute value $|\ |$ will have to be non-archimedean, since this depends only on the absolute values of the elements of $\mathbb{Z}$, which are in $\mathbb{Q}_p$ (see Theorem 2.2.2).

We begin by showing that *if* such an absolute value exists, it must have certain properties. Later, we will use these properties to obtain a construction which shows that the extension we are looking for *does* exist.[4]

The first thing is easy:

Proposition 5.3.1 *Let K be a finite extension of $\mathbb{Q}_p$. If there exists an absolute value $|\ |$ on K extending the p-adic absolute value on $\mathbb{Q}_p$, then*

   i) K is complete with respect to $|\ |$, and

   ii) we can take the limit of a sequence in K by taking the limits of the coefficients with respect to any given basis $\{x_1, x_2, \ldots, x_n\}$ of K as a $\mathbb{Q}_p$-vector space.

In particular, the topology on K induced by $|\ |$ is simply the unique topology on K as a normed $\mathbb{Q}_p$-vector space, and therefore is independent of the particular choice of absolute value.

PROOF: Obvious, of course, because all norms on a finite-dimensional vector space are equivalent. The statement about convergence just says that they are equivalent to the sup-norm with respect to any given basis.   □

---

[4]This is another standard mathematician's ruse: study the properties an object must have if it exists, and this may lead to a proof that it does exist. St. Anselm would understand.

Problem 203 Let $p = 5$. Check that 2 is not a square in $\mathbb{Q}_5$. Let $K = \mathbb{Q}_5(\sqrt{2})$. Give an example of a norm on K which is not an absolute value. Can you arrange things in your example so that the norm gives the same as the 5-adic absolute value when computed on elements of $\mathbb{Q}_5$?

A very important fact follows from the Proposition.

Corollary 5.3.2 *There is at most one absolute value on K extending the p-adic absolute value on $\mathbb{Q}_p$.*

PROOF: Suppose $|\ |$ and $\|\ \|$ are two absolute values on K which extend the p-adic absolute value. We first show that they are equivalent[5] (as absolute values), and then we show that they are identical.

To show that $|\ |$ and $\|\ \|$ are equivalent, we need to show that for any $x \in K$, we have

$$|x| < 1 \Longleftrightarrow \|x\| < 1.$$

To see this, remember that $|x| < 1$ if and only if $x^n \to 0$ with respect to the topology defined by $|\ |$, and similarly that $\|x\| < 1$ if and only if $x^n \to 0$ with respect to the topology defined by $\|\ \|$. But we already know that $|\ |$ and $\|\ \|$ are equivalent as norms on the vector space K, and hence define the same topology. Therefore, we have convergence with respect to one absolute value exactly when we have convergence with respect to the other, and this proves our claim. (Notice how seriously the field structure, rather than just the vector space structure, comes into that argument.)

This shows that $|\ |$ and $\|\ \|$ are equivalent absolute values on K; according to Lemma 3.1.2, this means that there is a positive real number α such that we have $|x| = \|x\|^\alpha$ for every $x \in K$. But $|x|$ and $\|x\|$ must be *equal* whenever $x \in \mathbb{Q}_p$, since both absolute values extend the p-adic absolute value; computing both at $x = p$ shows that we must have $\alpha = 1$, i.e., the two absolute values are the same. $\square$

We know, then, that there can be at most one extension of the p-adic absolute value to K, and that K will be complete with respect to that extension. None of this establishes, however, that such an extension does exist.[6] To show the existence of the absolute value, we will need to give a construction.

One consequence of the uniqueness, however, should be noted (and will be used when we construct the absolute value). It is simply this: suppose that we have two extensions K and L, one containing the other, so that, say, $\mathbb{Q}_p \subset L \subset K$, and suppose that we have found absolute values $|\ |_L$ on L and $|\ |_K$ on K, both extending the p-adic absolute value on $\mathbb{Q}_p$. The restriction of $|\ |_K$ to elements of L is an absolute value on L which extends the p-adic

---

[5]The definition is at the beginning of Chapter 3; see especially Lemma 3.1.2.

[6]We do know that there are many vector space norms on K, and we can arrange for these to have the right value on elements of $\mathbb{Q}_p$, but it is not at all clear that any of these norms will be an absolute value, i.e., will work well with the multiplication in K.

absolute value; by uniqueness, it must be the same as $|\ |_L$. In other words, if $x \in L \subset K$, then

$$|x|_L = |x|_K.$$

In words, *the absolute value of x does not depend on the context.* We will use this, when defining $|x|$, in two different ways: at times we will want to work in $\mathbb{Q}_p(x)$, the smallest extension of $\mathbb{Q}_p$ containing x; at other times, we will want to work in a bigger field that may have nicer properties.

In order to be able to give the construction, we need to recall a few facts from the theory of field extensions. We assume that the reader has met these concepts before, and hence only sketch out the basic facts; for more details, see any standard text on abstract algebra.

So let K and F be fields, and assume that $F \subset K$ and that $[K : F]$ is finite; we will say that K/F is a finite field extension. Recall that we are always assuming that our fields have characteristic zero.

Let $\mathbf{C}$ be any algebraically closed field containing F (or, to be fancy, *fix* an inclusion of F into such a field $\mathbf{C}$, and *identify* F with its image under the inclusion.). We will say the field extension K/F is *normal* if all the (necessarily injective) homomorphisms[7] $\sigma : K \hookrightarrow \mathbf{C}$ which induce the identity (or, if we're being fancy, our fixed inclusion) on F have the same image. Another way to say this is to identify K with one of its images, and then say that it is normal if every σ maps K to itself. If K/F is normal, then we can think of σ as an automorphism $K \longrightarrow K$ which induces the identity on F. To make a picture, any such σ fits into a diagram like this:

$$
\begin{array}{ccc}
\mathbf{C} & \longrightarrow & \mathbf{C} \\
\uparrow & & \uparrow \\
K & \overset{\sigma}{\longrightarrow} & K \\
\uparrow & & \uparrow \\
F & = & F
\end{array}
$$

where the vertical arrows are inclusions.

When K/F is normal, it is clear that the choice of $\mathbf{C}$ doesn't much matter, since any σ maps K to itself anyway. We call a map $\sigma : K \longrightarrow K$ which induces the identity on F an *automorphism of the extension K/F*. It is known that when K/F is normal (and of characteristic zero[8]) the automorphisms of K/F form a finite group[9] whose order is equal to the degree $[K : F]$ (this group is called the *Galois group* of the field extension).

The following problems give a few examples.

---

[7]Recall that a field homomorphism is a mapping that (*i*) sends 1 to 1, and (*ii*) works well with the field operations, so that $\sigma(x + y) = \sigma(x) + \sigma(y)$ and $\sigma(xy) = \sigma(x)\sigma(y)$. It is a nice exercise, especially recommended to the reader who is unsure of his footing at this point, to show that such a function must always be injective.

[8]In characteristic p, an extra condition, called "separability," is needed.

[9]It is easy to prove that they form a group, and makes a nice exercise.

Problem 204 Let $F = \mathbb{Q}$ and $K = \mathbb{Q}(\sqrt[3]{2})$. Show that K/F is not a normal extension by taking $\mathbf{C} = \mathbb{C}$, considering K as a subfield of $\mathbb{R}$ (and hence of $\mathbb{C}$) in the obvious way, and noting that any $\sigma : K \longrightarrow \mathbb{C}$ must map $\sqrt[3]{2}$ to a cube root of 2...what are the choices?

Problem 205 Let $F = \mathbb{Q}$ and $K = \mathbb{Q}(i)$, where, as usual, $i^2 = -1$. Show that this extension is normal. In fact, show that any extension that is obtained by adding to F the square root of some element will be normal. (Hint: there are only two possibilities for σ.)

Problem 206 Let $F = \mathbb{Q}$ and $K = \mathbb{Q}(\sqrt[3]{2}, \zeta)$, where

$$\zeta = \frac{-1 + i\sqrt{3}}{2}$$

is a cube root of 1. Show that K/F is a normal extension. Show also that K is the smallest normal extension of $\mathbb{Q}$ containing $\sqrt[3]{2}$.

Normal extensions are very nice, and it is comforting (and useful) to know that the process suggested above in the case of $\mathbb{Q}(\sqrt[3]{2})$ and $\mathbb{Q}(\sqrt[3]{2}, \zeta)$ works in general: given any finite extension K/F, there exists a finite *normal* extension of F containing K. The smallest such is called the *normal closure* of K/F. This will be useful in what follows, because it means that to construct the absolute value of an element $x \in K$ we might as well assume that K is a normal extension (otherwise just replace K by its normal closure, since the absolute value does not depend on context).

The crucial fact that we will need is that there exists a function

$$\mathbf{N}_{K/F} : K \longrightarrow F,$$

which is called the *norm from K to F*. (It is a bit unfortunate that this "norm" has the same name as the vector space "norm," but both terms have been standard for such a long time that there is no chance of ever changing them. Watch out for the context to avoid confusion.) This will be useful because it gives a natural way to "go down" from elements of the bigger field K to elements of F.

The norm function can be defined in several ways, each useful in certain contexts; here are three:

i) Take $\alpha \in K$, think of K as a finite-dimensional F-vector space, and consider the F-linear map from K to K given by multiplication by α. Since this is linear, it corresponds to a matrix. Then we define $\mathbf{N}_{K/F}(\alpha)$ to be the determinant of this matrix.

ii) Take $\alpha \in K$, and consider the subextension $F(\alpha)$, i.e., the smallest field containing both F and α (this is clearly a subfield of K). Set $r = [K : F(\alpha)]$ to be the degree of K as an extension of $F(\alpha)$. Let

$$f(X) = X^n + a_{n-1}X^{n-1} + \cdots + a_1 X + a_0 \in F[X]$$

be the minimal polynomial of α over F, that is, the lowest degree monic polynomial with coefficients in F such that $f(\alpha) = 0$. Then we define $\mathbf{N}_{K/F}(\alpha) = (-1)^{nr} a_0^r$.

iii) Suppose the extension K/F is *normal*. Then we can define $\mathbf{N}_{K/F}(\alpha)$ to be the product of all the $\sigma(\alpha)$, where σ runs through the (finite) set of all the automorphisms of K/F.

Before we discuss why these definitions are equivalent, note some useful facts. First, if $\alpha \in F$ (rather than in the bigger field K), then $\mathbf{N}(\alpha) = \alpha^n$, where $n = [K : F]$ is the degree of the extension. (This is essentially obvious from any of the definitions—check!) Next, norms are multiplicative. This is probably easiest to see from the first definition, since determinants are multiplicative, but it's pretty obvious from the last one, too. (Less so for the middle definition—can you give a direct proof using that version?) In any case, we will need to know that

$$\mathbf{N}_{K/F}(\alpha\beta) = \mathbf{N}_{K/F}(\alpha)\mathbf{N}_{K/F}(\beta)$$

for any α, $\beta \in K$. Notice, by contrast, that the norm of a sum has no clear relation to the norms of the summands.

The equivalence of these definitions is not hard to prove; we suggest that the reader who has not seen it proved work through the next few exercises.

Problem 207 Prove the equivalence of the first two definitions in the case that $K = F(\alpha)$, by considering the basis of K which consists of $\{1, \alpha, \alpha^2, \ldots, \alpha^{n-1}\}$.

Problem 208 Using the first definition, show that if we have three fields $F \subset L \subset K$, then, for any $\alpha \in K$, we have

$$\mathbf{N}_{L/F}(\mathbf{N}_{K/L}(\alpha)) = \mathbf{N}_{K/F}(\alpha).$$

(Suggestion: one can make a basis for L over F by multiplying all elements of a basis of L over K by all elements of a basis of K over F. This will give the matrix corresponding to multiplication by α a kind of "block structure.") Use this to conclude that the first two definitions are equivalent also when K is bigger than $F(\alpha)$. You will need the two facts mentioned above.

Problem 209 Suppose K/F is normal and that $K = F(\alpha)$. Show that the images $\sigma(\alpha)$ as σ runs through the automorphisms of K/F are exactly the roots of the polynomial $f(X)$. (It's easy to see that any $\sigma(\alpha)$ is a root—just compute $\sigma(f(\alpha))$—but it's less clear that for each root there is a unique σ for which $\sigma(\alpha)$ is equal to that root.) Conclude that the second and third definitions are equivalent in this case.

Problem 210 Finish off the proof that all three definitions give the same answer. (One loose end to consider is the case where K/F is normal, but K is not equal to $F(\alpha)$. What then?)

Problem 211 Suppose K/F is *not* normal. Can you give a version of the third definition that makes sense?

After all that theory, we need some concrete examples to keep us afloat. Let's take a really easy one, and put $F = \mathbb{Q}_5$, $K = \mathbb{Q}_5(\sqrt{2})$. Take a generic element $a + b\sqrt{2} \in K$; let's compute its norm using all three definitions:

i) A basis for K over $\mathbb{Q}_5$ is $\{1, \sqrt{2}\}$. The linear map "multiplication by $a + b\sqrt{2}$" maps 1 to $a + b\sqrt{2}$ and $\sqrt{2}$ to $2b + a\sqrt{2}$, so its matrix with respect to our basis is

$$\begin{bmatrix} a & 2b \\ b & a \end{bmatrix},$$

which has determinant $a^2 - 2b^2$. Therefore, $\mathbf{N}_{K/F}(a + b\sqrt{2}) = a^2 - 2b^2$.

ii) We will have $r = 1$ unless $b = 0$, in which case $r = 2$. If $b = 0$, we have $\alpha = a$, whose minimal polynomial is $X - a$, and the norm is then $(-1)^2 a^2 = a^2$. If $b \neq 0$, we must work out the minimal polynomial; it must be of degree two. Since $(a + b\sqrt{2})^2 = a^2 + 2b^2 + 2ab\sqrt{2}$, we will get zero by combining as follows:

$$(a + b\sqrt{2})^2 - 2a(a + b\sqrt{2}) + (a^2 - 2b^2) = 0.$$

(Can you see how that was found?) Hence, the minimal polynomial is

$$X^2 - 2aX + (a^2 - 2b^2),$$

and the norm is $a^2 - 2b^2$. Thus, whether b is zero or not, we have $\mathbf{N}_{K/F}(a + b\sqrt{2}) = a^2 - 2b^2$.

iii) Finally, we have two automorphisms: the identity, and

$$\sigma : a + b\sqrt{2} \mapsto a - b\sqrt{2}.$$

The product of the images of $a + b\sqrt{2}$ is

$$(a + b\sqrt{2})(a - b\sqrt{2}) = a^2 - 2b^2,$$

so that once again we have $\mathbf{N}_{K/F}(a + b\sqrt{2}) = a^2 - 2b^2$.

The general case is nowhere near as easy, of course. Here are a few more relatively simple examples:

Problem 212 Do the same for

i) a general quadratic extension $\mathbb{Q}_p(\sqrt{n})$,

ii) some specific elements of the extension $\mathbb{Q}(\sqrt[3]{2}, \zeta)$ with $\zeta = (-1 + i\sqrt{3})/2$ (notice that this is an extension of degree 6; working with the general element wouldn't be too pleasant).

To see why the norm is going to play a central role, notice the following. Suppose $K/\mathbb{Q}_p$ is a normal extension, and let σ be an automorphism. Let $|\ |$ be an absolute value on K. Then the function $x \mapsto |\sigma(x)|$ is also an absolute value on K (check!), and also gives the p-adic absolute value over $\mathbb{Q}_p$, since σ induces the identity on $\mathbb{Q}_p$. But we have shown that there is *only one* such absolute value! Thus, we must have $|\sigma(x)| = |x|$ for any $x \in K$. Multiplying over all the σ's (and remembering that there are exactly $n = [K : \mathbb{Q}_p]$ of them) we get that

$$\left| \prod_\sigma \sigma(x) \right| = |x|^n.$$

Now, since the product is equal to the norm, this translates to

$$|x|^n = |\mathbf{N}_{K/\mathbb{Q}_p}(x)|,$$

or, taking the root,

$$|x| = \sqrt[n]{|\mathbf{N}_{K/\mathbb{Q}_p}(x)|}.$$

But this last gives a formula which we can compute just from the knowledge of the p-adic absolute value, since the norm is an element of $\mathbb{Q}_p$!

So far, that only works for normal extensions, but note the following:

Lemma 5.3.3 *Let L and K be finite extensions of $\mathbb{Q}_p$ which form a tower: $\mathbb{Q}_p \subset L \subset K$. Let $x \in L$. Set $m = [L : \mathbb{Q}_p]$ and $n = [K : \mathbb{Q}_p]$. Then*

$$\sqrt[m]{|\mathbf{N}_{L/\mathbb{Q}_p}(x)|_p} = \sqrt[n]{|\mathbf{N}_{K/\mathbb{Q}_p}(x)|_p}.$$

PROOF: We have

$$\mathbf{N}_{K/F}(x) = \mathbf{N}_{L/\mathbb{Q}_p}\big(\mathbf{N}_{K/L}(x)\big),$$

and $\mathbf{N}_{K/L}(x) = x^{[K:L]}$. Remembering that $[K : \mathbb{Q}_p] = [K : L][L : \mathbb{Q}_p]$ and plugging everything into the formulas gives the equality.   □

This is very nice, since it says that the value of $\sqrt[n]{|\mathbf{N}_{K/\mathbb{Q}_p}(x)|_p}$ is the same for any field K containing x (as above, $n = [K : \mathbb{Q}_p]$). In particular, this shows that it must be equal to the absolute value of x also when the extension is not normal (pass to the normal closure!). In other words, we have proved the following:

Proposition 5.3.4 *If there is an absolute value on K extending the p-adic absolute value, then it must be given by the formula*

$$|x| = \sqrt[n]{|\mathbf{N}_{K/\mathbb{Q}_p}(x)|_p},$$

where $n = [K : \mathbb{Q}_p]$ is the degree of the extension.

Notice that the fact that the value of our formula does not depend on the choice of field containing x matches exactly the fact that the same is true of the absolute value (if it exists). This is encouraging, since we want to prove that the formula does define an absolute value. We are now, finally, in a position to prove that. The proof we give is taken from [EHH+91].

Theorem 5.3.5 *Let $K/\mathbb{Q}_p$ be a finite extension of degree n. The function $|\,| : K \longrightarrow \mathbb{R}_+$ defined by*

$$|x| = \sqrt[n]{|\mathbf{N}_{K/\mathbb{Q}_p}(x)|_p}$$

is a non-archimedean absolute value on K which extends the p-adic absolute value on $\mathbb{Q}_p$.

PROOF: Several things are immediate. First, $|x| = 0$ will only happen if $\mathbf{N}_{K/\mathbb{Q}_p}(x) = 0$, which (using the first definition of the norm) will only happen if multiplication by x is not invertible; since K is a field, that only happens if $x = 0$. Next, since $\mathbf{N}_{K/\mathbb{Q}_p}(xy) = \mathbf{N}_{K/\mathbb{Q}_p}(x)\mathbf{N}_{K/\mathbb{Q}_p}(y)$, we will certainly have $|xy| = |x|\,|y|$. Finally, if $x \in \mathbb{Q}_p$ then $\mathbf{N}_{K/\mathbb{Q}_p}(x) = x^n$, so that $|x| = \sqrt[n]{|x|_p^n} = |x|_p$.

It remains only to show the non-archimedean inequality, i.e., that

$$|x + y| \leq \max\{|x|, |y|\}$$

for any $x, y \in K$. Dividing through by y, we see that this amounts to showing that for any $x \in K$ we have

$$|x + 1| \leq \max\{|x|, 1\},$$

and this will follow from

$$|x| \leq 1 \Longrightarrow |x - 1| \leq 1.$$

Proof that this is sufficient: To see this, notice that $x + 1 = -(-x - 1)$, so that if this implication is true, then we also have

$$|x| \leq 1 \Longrightarrow |-x| \leq 1 \Longrightarrow |-x - 1| = |x + 1| \leq 1.$$

Now just consider cases: if $|x| \leq 1$, then $\max\{|x|, 1\} = 1$, and the implication proves the inequality; if $|x| > 1$, then we get $|1/x| < 1$, which we are assuming yields $|1 + 1/x| < 1$. So we have

$$\left| \frac{x+1}{x} \right| = \left| 1 + \frac{1}{x} \right| \leq 1,$$

which says $|x + 1| \leq |x|$, which is what we want. So we have shown that it is sufficient to show, for every $x \in K$, that

$$|x| \leq 1 \Longrightarrow |x - 1| \leq 1.$$

So let's prove that this is true. Looking at the definition, we see that $|x| \leq 1$ will happen exactly when $|\mathbf{N}_{K/\mathbb{Q}_p}(x)|_p \leq 1$. Hence, what we need to show is that

$$|\mathbf{N}_{K/\mathbb{Q}_p}(x)|_p \leq 1 \Longrightarrow |\mathbf{N}_{K/\mathbb{Q}_p}(x-1)|_p \leq 1,$$

or, in more algebraic terms, that

$$\mathbf{N}_{K/\mathbb{Q}_p}(x) \in \mathbb{Z}_p \Longrightarrow \mathbf{N}_{K/\mathbb{Q}_p}(x-1) \in \mathbb{Z}_p.$$

We will do this by using the definition of the norm in terms of the minimal polynomial. By the lemma, we may assume that $K = \mathbb{Q}_p(x)$ is the smallest field containing x (and note that we will always have $\mathbb{Q}_p(x) = \mathbb{Q}_p(x-1)$, since any field containing x will also contain $x-1$ and vice-versa). Let

$$f(X) = X^n + a_{n-1}X^{n-1} + \cdots + a_1 X + a_0$$

be the minimal polynomial for x. Then the minimal polynomial for $x-1$ is clearly

$$f(X+1) = X^n + (a_{n-1}+n)X^{n-1}\cdots + (1 + a_{n-1} + \cdots + a_1 + a_0)$$

(because clearly $f(x-1) = 0$ and the degree is right!). Thus, using the second definition for the norm, we have

$$\mathbf{N}_{K/\mathbb{Q}_p}(x) = (-1)^n a_0$$

and

$$\mathbf{N}_{K/\mathbb{Q}_p}(x-1) = (-1)^n(1 + a_{n-1} + \cdots + a_1 + a_0).$$

What we want to prove will follow, then, from the assertion that if

$$f(X) = X^n + a_{n-1}X^{n-1} + \cdots + a_1 X + a_0$$

is an irreducible polynomial and $a_0 \in \mathbb{Z}_p$, then we have

$$1 + a_{n-1} + \cdots + a_1 + a_0 \in \mathbb{Z}_p.$$

In fact, we will prove something that is even better.

Lemma 5.3.6 *If* $f(X) = X^n + a_{n-1}X^{n-1} + \cdots + a_1 X + a_0$ *is a monic irreducible polynomial with coefficients in* $\mathbb{Q}_p$ *and* $a_0 \in \mathbb{Z}_p$, *then all of the coefficients* $a_{n-1}, \ldots a_1, a_0$ *belong to* $\mathbb{Z}_p$.

PROOF OF THE LEMMA: This is the crux of the matter, and we follow the proof given by Neukirch in [EHH+91]. We will use the second form of Hensel's Lemma, proved way back in Chapter 3, to show that if some of the coefficients are not in $\mathbb{Z}_p$ then $f(X)$ will be reducible.

So let $f(X) = X^n + a_{n-1}X^{n-1} + \cdots + a_1 X + a_0$, and assume that $a_0 \in \mathbb{Z}_p$ but some $a_i \notin \mathbb{Z}_p$. Choose m to be the smallest exponent such that $p^m a_i \in \mathbb{Z}_p$

for every i, and "clear denominators" by multiplying the whole polynomial by p^m. Set

$$g(X) = p^m f(X) = b_n X^n + b_{n-1} X^{n-1} + \cdots + b_1 X + b_0,$$

so that $b_i = p^m a_i$. Since $f(X)$ is monic, $b_n = p^m$ is divisible by p; since $a_0 \in \mathbb{Z}_p$ (our main hypothesis), $b_0 = p^m a_0$ is also divisible by p; by our choice of m, all the b_i are in $\mathbb{Z}_p$, and at least one is not divisible by p. Let k be the smallest i such that b_i is not divisible by p. Then we have a factorization

$$g(X) \equiv (b_n X^{n-k} + \cdots + b_k) X^k \pmod{p},$$

and the two factors are clearly relatively prime modulo p. By the second form of Hensel's Lemma, it follows that $g(X) = p^m f(X)$ is reducible, and therefore so is $f(X)$ itself. This proves the lemma, and therefore also the theorem. □

This gives the extension we needed. In other words, given any finite extension K of $\mathbb{Q}_p$, we have shown that there exists a unique absolute value on K which extends the p-adic absolute value on $\mathbb{Q}_p$; we call it, of course, the p-adic absolute value on K. We know that K is complete with respect to this absolute value.

To complete this section, we go on to consider an algebraic closure of $\mathbb{Q}_p$. This is a field $\overline{\mathbb{Q}}_p$ which contains all the roots of all the polynomials with coefficients in $\mathbb{Q}_p$. To construct it, we just take the union of all the finite extensions of $\mathbb{Q}_p$ (and then we prove that this is an algebraically closed field).

We claim that we have already constructed an absolute value on the algebraic closure. The point is this: given any $x \in \overline{\mathbb{Q}}_p$, the extension $\mathbb{Q}_p(x)$ is finite (its degree is the degree of the minimal polynomial of x over $\mathbb{Q}_p$). Since x then lives in the finite extension $\mathbb{Q}_p(x)$, we can define $|x|$ by using the unique extension of the p-adic absolute value to $\mathbb{Q}_p(x)$. But we already know that this absolute value does not depend on the field we take it in; in other words, it just depends on x itself (as the root of some polynomial over $\mathbb{Q}_p$). Thus, it makes sense to say it is the absolute value of the element $x \in \overline{\mathbb{Q}}_p$. This shows that we have actually defined a function

$$| \ | : \overline{\mathbb{Q}}_p \longrightarrow \mathbb{R}_+$$

which extends the p-adic absolute value, and it is easy to see that this function is an absolute value. Our construction, then, shows that there is a unique p-adic absolute value on $\overline{\mathbb{Q}}_p$.

Problem 213 Prove that the function we have defined is an absolute value, i.e., that it satisfies the three conditions listed in the beginning of this section.

It is not clear (in fact, it is not true) that $\overline{\mathbb{Q}}_p$ is complete with respect to this absolute value, because $\overline{\mathbb{Q}}_p$ is an *infinite* extension of $\mathbb{Q}_p$. Proving this will take knowing a lot more about the absolute value on $\overline{\mathbb{Q}}_p$. For now, we will content ourselves to showing that $\overline{\mathbb{Q}}_p$ is indeed an infinite extension of $\mathbb{Q}_p$. To do this, it is enough to show that that there are irreducible polynomials of arbitrarily large degree over $\mathbb{Q}_p$. Since the root of an irreducible polynomial of degree n generates an extension of degree n, this means that $\overline{\mathbb{Q}}_p$ contains extensions of degree n for every n, and hence is not a finite extension. We conclude this section by showing that this is in fact the case. We first need the following lemma:

Lemma 5.3.7 *Suppose that $f(X) \in \mathbb{Z}_p[X]$ factors (in a non-trivial way) in $\mathbb{Q}_p[X]$, so that*

$$f(X) = g(X)h(X)$$

with $g(X), h(X) \in \mathbb{Q}_p[X]$ and non-constant. Then there exist non-constant polynomials $g_0(X), h_0(X) \in \mathbb{Z}_p[X]$ such that $f(X) = g_0(X)h_0(X)$.

PROOF: If $k(X) = a_n X^n + \cdots + a_1 X + a_0 \in \mathbb{Q}_p[X]$ is any polynomial, define

$$w(k(X)) = \min_{0 \le i \le n} v_p(a_i).$$

This is a kind of p-adic valuation on polynomials, since $w(k(X))$ is the largest power of p that divides *all* the coefficients of $k(X)$. It is easy to see that for $a \in \mathbb{Q}_p$ we have $w(a\,k(X)) = v_p(a) + w(k(x))$. Also, it is clear that $k(X) \in \mathbb{Z}_p[X]$ if and only if $w(k(X)) \ge 0$. We will use the "valuation" w to prove the lemma.

Step 1: If the lemma is true for the case when $w(f(X)) = 0$, then it is true in general.

Proof of step 1: Consider the general case, in which all we know is that $w(f(X)) \ge 0$. By the definition of w, there exists some number $a \in \mathbb{Q}_p$ such that $w(f(X)) = -v_p(a)$ (just take a to be the inverse of the coefficient with the smallest valuation); since we know $f(X) \in \mathbb{Z}_p[X]$, we know that $a^{-1} \in \mathbb{Z}_p$. Then it is clear that $w(a\,f(X)) = 0$; set $\tilde{f}(X) = a\,f(X)$ and, say, $\tilde{g}(X) = a\,g(X)$, so that $\tilde{f}(X) = \tilde{g}(X)h(X)$ and $w(\tilde{f}(X)) = 0$.

If we know that the theorem is true in this case, then we can decompose $\tilde{f}(X)$ as a product of two polynomials in $\mathbb{Z}_p[X]$, say, $\tilde{f}(X) = G_0(X) H_0(X)$. Then we have

$$f(X) = a^{-1}\tilde{f}(X) = a^{-1}G_0(X) H_0(X),$$

and, since we know $a^{-1} \in \mathbb{Z}_p$, this decomposition is of the kind we want: just absorb the a^{-1} into one of the factors by putting $g_0(X) = a^{-1}G_0(X)$ and $h_0(X) = H_0(X)$, and we get the decomposition we want:

$$f(X) = g_0(X)\,h_0(X).$$

This proves step 1. In other words, *we may assume, without loss, that $w_0(f(X)) = 0$*, i.e., that at least one coefficient of $f(X)$ is a p-adic unit.

Step 2: The lemma is indeed true when $w(f(X)) = 0$.

Proof of step 2: Assume, then, that $w(f(X)) = 0$. Using the same reasoning as above, we can find $b \in \mathbb{Q}_p$ such that $w(b\,g(X)) = 0$ and $c \in \mathbb{Q}_p$ such that $w(c\,h(X)) = 0$. If we write $g_1(X) = b\,g(X)$ and $h_1(X) = c\,h(X)$, then we can write

$$f_1(X) = bc\,f(X) = g_1(X)h_1(X).$$

Write $\bar{k}(X) \in \mathbb{F}_p[X]$ for the reduction of a polynomial $k(X) \in \mathbb{Z}_p(X)$ modulo p. We have set things up so that $\bar{g}_1(X)$ and $\bar{h}_1(X)$ are both non-zero; it follows that $\bar{f}_1(X)$ is also non-zero, and hence that $w(f_1(X)) = w(bc\,f(X)) = 0$. Since we had already arranged things so that $w(f(X)) = 0$, it follows that $v_p(bc) = 0$, so that bc is a p-adic unit. Then we have

$$f(X) = (bc)^{-1}f_1(X) = (bc)^{-1}g_1(X) \cdot h_1(X).$$

Taking $g_0(X) = (bc)^{-1}g_1(X)$ and $h_0(X) = h_1(X)$ then gives the desired factorization. $\qquad\square$

When everything is assumed monic, the lemma is even easier:

Problem 214 Suppose that $f(X) \in \mathbb{Z}_p[X]$ is monic and factors as a product $f(X) = g(X)\,h(X)$, with $g(X)$ and $h(X) \in \mathbb{Q}_p[X]$ and monic. Show that then $g(X)$ and $h(X)$ must be in $\mathbb{Z}_p[X]$. (Hint: the main difference between this and the Lemma is that we are assuming that the factors are monic.)

In particular, we get the following:

Corollary 5.3.8 *Let $f(X) \in \mathbb{Z}_p[X]$ be a monic polynomial whose reduction modulo p is irreducible in $\mathbb{F}_p[X]$. Then $f(X)$ is irreducible over $\mathbb{Q}_p$.*

PROOF: If $f(X)$ factors over $\mathbb{Q}_p$, then it factors over $\mathbb{Z}_p$ by the Lemma; reducing the factorization modulo p gives a factorization over $\mathbb{F}_p$, which cannot exist. $\qquad\square$

Problem 215 That was pretty quick; fill in the details. For example, how do we know that the factorization modulo p is non-trivial?

Problem 216 Is the assumption that $f(X)$ is monic really necessary?

Notice that this Corollary has a kind of converse in the "second form" of Hensel's Lemma (Theorem 3.4.6), which says that, under certain conditions, factorizations over $\mathbb{F}_p$ lift to factorizations over $\mathbb{Z}_p$.

It is well known that there are many irreducible polynomials in $\mathbb{F}_p[X]$. In fact, for every n one can show[10] that there is an irreducible polynomial of degree n in $\mathbb{F}_p[X]$ whose roots generate the unique extension of degree n of $\mathbb{F}_p$. Choosing any lift of such a polynomial to a monic polynomial in $\mathbb{Z}_p[X]$ gives an irreducible polynomial of degree n in $\mathbb{Q}_p[X]$. Adjoining a root of this polynomial then gives an extension of $\mathbb{Q}_p$ of degree n, which in some sense "comes from" the extension of $\mathbb{F}_p$. So we have proved that

Corollary 5.3.9 *For each integer $n \geq 1$ there is an extension of $\mathbb{Q}_p$ which has degree exactly n and which "comes from" the unique extension of degree n of the finite field $\mathbb{F}_p$.*

In particular,

Corollary 5.3.10 *The algebraic closure $\overline{\mathbb{Q}}_p$ is an infinite extension of $\mathbb{Q}_p$.*

We should note the contrast, at this point, between $\mathbb{R}$ and $\mathbb{Q}_p$. The algebraic closure of $\mathbb{R}$ is $\mathbb{C}$, which is an extension of degree two, and is therefore complete with respect to the ∞-adic absolute value. This is a point, then, at which the p-adic and the classical theories diverge quite sharply.

Before we consider the algebraic closure in more detail, we need a better grasp of the properties of finite extensions of $\mathbb{Q}_p$. That is the point of the next section. Before we delve in, however, we prove one final result about polynomials that gives us still more finite extensions of $\mathbb{Q}_p$.

Proposition 5.3.11 (Eisenstein Irreducibility Criterion) *Let*

$$f(X) = a_n X^n + \cdots + a_1 X + a_0 \in \mathbb{Z}_p[X]$$

be a polynomial satisfying the conditions

i) $|a_n| = 1$,

ii) $|a_i| < 1$ for $0 \leq i < n$, and

iii) $|a_0| = 1/p$.

Then $f(X)$ is irreducible over $\mathbb{Q}_p$.

PROOF: Suppose $f(X)$ is reducible. By the Lemma, it is then reducible over $\mathbb{Z}_p$, i. e., there exist $g(X), h(X) \in \mathbb{Z}_p[X]$ such that

$$f(X) = g(X) \, h(X).$$

---

[10]This is one of the few facts about finite fields that we will need to make use of in this chapter. Most of them are easily proved—see any introductory book on "abstract algebra" for the details. What we are using here is the fact that, for each $n \geq 1$, the finite field $\mathbb{F}_p$ has a unique extension of degree n. This extension is a field with p^n elements. It is usually denoted by $\mathbb{F}_{p^n}$, and, since it is a separable extension, there exists a polynomial $f(X)$ such that $\mathbb{F}_{p^n}$ is obtained by adjoining a root of $f(X)$ to $\mathbb{F}_p$.

Write

$$g(X) = b_r X^r + \cdots + b_1 X + b_0$$

and

$$h(X) = c_s X^s + \cdots + c_1 X + c_0,$$

with $r + s = n$; since $|a_n| = 1$ and $a_n = b_r c_s$, we must have $|b_r| = |c_s| = 1$. As above, use bars to denote reduction modulo p; we have $\bar{f}(X) = \bar{g}(X) \bar{h}(X)$. On the other hand, the hypotheses imply that $\bar{f}(X) = \bar{a}_n X^n$. Then we must have $\bar{g}(X) = \bar{b}_r X^r$ and $\bar{h}(X) = \bar{c}_s X^s$. In particular, both b_0 and c_0 must be divisible by p. But then $a_0 = b_0 c_0$ will be divisible by p^2, so that $|a_0| \leq 1/p^2$, contradicting our third assumption. This shows that $f(X)$ must be irreducible. $\qquad\square$

The reader will note that the irreducible polynomials furnished by the Eisenstein criterion (we might call them *Eisenstein polynomials*) are certainly reducible modulo p (*very* reducible: modulo p, they look essentially like a power of X). In other words, the irreducible polynomials provided by this criterion are very different from the ones we found before. So what we have here is *another* infinite family of finite extensions of $\mathbb{Q}_p$.

Problem 217 Given that Eisenstein polynomials factor modulo p, why can't we use Hensel's Lemma to factor them in $\mathbb{Z}_p$?

To conclude this section, here are a few more problems about polynomials:

Problem 218 Is the function w defined in the proof of the Lemma above a valuation? (Hint: the difficult bit is to show that $w(f(X)g(X)) = w(f(X)) + w(g(X))$. Notice that the proof of the Lemma would be greatly simplified if we could use this identity.)

Problem 219 (This needs Galois theory.) In the situation of Corollary 5.3.9, show that the extension of $\mathbb{Q}_p$ is normal, and that its Galois group is isomorphic to the Galois group of the corresponding extension of $\mathbb{F}_p$.

Problem 220 Use Lemma 5.3.7 above to show "Gauss's Lemma," which says that if a polynomial $f(X) \in \mathbb{Z}[X]$ factors over $\mathbb{Q}$, then it factors over $\mathbb{Z}$. (This may be taken as another example of how to use "local"—i.e., p-adic—methods to prove "global" results.)

Problem 221 Can the Eisenstein criterion also be turned into a "global" result? In other words, does it give us a way to determine irreducibility over $\mathbb{Q}$?

Problem 222 Does the Lemma about factorizations over $\mathbb{Q}_p$ and $\mathbb{Z}_p$ extend to polynomials in several variables? If so, does Problem 220 also extend to that case?

5.4   Properties of Finite Extensions

The point of this section is to gather information about finite extensions of
$\mathbb{Q}_p$. On one level, what we want to say is that much of the structure we
have found in $\mathbb{Q}_p$ extends without effort. Our main interest, however, is to
see what information this gives us about finite extensions of $\mathbb{Q}_p$. To help us
understand, we will keep a few standard examples in mind as we go along;
at each step, we will consider (usually in a problem) how the result that has
just been proved looks in the particular case of our examples.

Here are the three examples:

   i) Let $p = 5$; we have checked that 2 is not a square in $\mathbb{Q}_5$, so we let
       $F_1 = \mathbb{Q}_5(\sqrt{2})$. This is an extension of degree 2, with basis $\{1, \sqrt{2}\}$.

   ii) Again, let $p = 5$. It is clear that 5 itself is not a square in $\mathbb{Q}_5$. We let
       $F_2 = \mathbb{Q}_5(\sqrt{5})$. This is also an extension of degree 2, with basis $\{1, \sqrt{5}\}$.

   iii) Our third example is more complicated. We let $p = 3$. We adjoin to
       $\mathbb{Q}_3$ a cube root of unity and a square root of 2: $F_3 = \mathbb{Q}_3(\zeta, \sqrt{2})$, where
       $\zeta^3 = 1$ but $\zeta \neq 1$. F_3 is an extension of $\mathbb{Q}_3$ of degree 4; both $\mathbb{Q}_3(\zeta)$ and
       $\mathbb{Q}_3(\sqrt{2})$ are subextensions of degree 2.

In all of this section, K will be a finite extension of degree n of $\mathbb{Q}_p$, and we
will write $|\ | = |\ |_p$ for the p-adic absolute value (extended to K as above). We
already know that the absolute value makes K a locally compact topological
field, that K is complete with respect to its absolute value, and that the
absolute value on K is given by the formula

$$|x| = \sqrt[n]{|\mathbf{N}_{K/\mathbb{Q}_p}(x)|_p}.$$

Our next step is to show that this absolute value is "discrete."

Recall that in $\mathbb{Q}_p$, the absolute value of any non-zero element was always
of the form p^v, with v an integer; in fact, this is what allowed us to define
the p-adic valuation v_p. Looking at the formula for the absolute value on K,
we immediately see that the absolute value of any non-zero $x \in K$ is of the
form p^v, where $v \in \frac{1}{n}\mathbb{Z}$, since it is the n-th root of the absolute value of some
element of $\mathbb{Q}_p$. This spurs us on to define:

Definition 5.4.1 *Let K be a finite extension of $\mathbb{Q}_p$, and let $|\ |$ be the p-adic
absolute value on K. For any $x \in K$, $x \neq 0$, we define the p-adic valuation
$v_p(x)$ to be the unique rational number satisfying*

$$|x| = p^{-v_p(x)}.$$

We extend the definition formally by setting $v_p(0) = +\infty$.

It is easy to see that v_p is a valuation, in the sense we defined:

i) $v_p(x + y) \geq \min\{v_p(x), v_p(y)\}$, and

ii) $v_p(xy) = v_p(x) + v_p(y)$.

As before, we use the standard conventions about how to interpret these equations when one of x, y, or $x + y$ is zero.

It is useful to notice that since we know exactly how to compute the p-adic absolute value of an element of K, we also know how to compute v_p. Here is the formula: for any $x \in K^\times$,

$$v_p(x) = \frac{1}{n} v_p\left(\mathbf{N}_{K/\mathbb{Q}_p}(x)\right).$$

This reduces computing v_p to computing norms.

Problem 223 Let $x = 1 + 3\sqrt{2} \in F_1$. Compute $v_5(x)$. Do the same for $x = \sqrt{2}$, $x = 1 + 5\sqrt{2}$, and $x = 5\sqrt{2}$. (Hint: the easiest way is probably to consider the images under automorphisms to compute the norm, and then use the basic formula.)

Problem 224 Let $x = 4 + \sqrt{5} \in F_2$ Compute $v_5(x)$. Do the same for $x = \sqrt{5}$, $x = 5 + \sqrt{5}$, x=$10 - 3\sqrt{5}$, $x = 1 + \sqrt{5}$.

Problem 225 Let $x = \sqrt{2} \in F_3$ Compute $v_3(x)$. Do the same for $x = \zeta$, $x = 1 - \zeta$ (be careful!), $x = 10 - 3\sqrt{2}$, $x = 2 + 3\zeta$. (This problem is a little harder than the previous two problems.)

We know that the image of v_p is contained in $\frac{1}{n}\mathbb{Z}$ (in fact, that is obvious from the formula above). But we do not yet know exactly what it is. The next result tells us what *kind* of subset of $\mathbb{Q}$ it is.

Proposition 5.4.2 *The p-adic valuation v_p is a homomorphism from the multiplicative group $K^\times$ to the additive group $\mathbb{Q}$. Its image is of the form $\frac{1}{e}\mathbb{Z}$, where e is a divisor of $n = [K : \mathbb{Q}_p]$.*

PROOF: That v_p is a homomorphism is just property (ii) above; its image is therefore an additive subgroup of $\mathbb{Q}$. We already know that the image is contained in $\frac{1}{n}\mathbb{Z}$. We also know that the image contains all of $\mathbb{Z}$, since the image of v_p on $\mathbb{Q}_p^\times$ does. Let d/e (with d and e relatively prime) be in the image, chosen so that the denominator e is the largest possible. (This makes sense because it is clear that e must be a divisor of n, so that the range of possible denominators is bounded.) Now, since d and e are relatively prime, there must be a multiple of d which is congruent to 1 modulo e, i.e., we can find r and s such that $rd = 1 + se$. But then

$$r\frac{d}{e} = \frac{1 + se}{e} = \frac{1}{e} + s$$

is in the image; since $s \in \mathbb{Z}$ is in the image, it follows that $1/e$ is in the image. Since e was chosen to be the largest possible denominator in the image, it follows that the image must be exactly $\frac{1}{e}\mathbb{Z}$, and we are done.   □

The image of the p-adic valuation on a field K is called the *value group* of K. The number e is an invariant of the field extension $K/\mathbb{Q}_p$, and therefore we give it a name:

Definition 5.4.3 *Let $K/\mathbb{Q}_p$ be a finite extension, and let $e = e(K/\mathbb{Q}_p)$ be the unique positive integer (dividing $n = [K : \mathbb{Q}_p]$) defined by*

$$v_p(K^\times) = \frac{1}{e}\mathbb{Z}.$$

We call e the ramification index *of K over $\mathbb{Q}_p$. We say the extension $K/\mathbb{Q}_p$ is* unramified *if $e = 1$. We say the extension is* ramified *if $e > 1$, and* totally ramified *if $e = n$. Finally, we write $f = f(K/\mathbb{Q}_p) = n/e$.*

Problem 226 Compute e for the fields F_1, F_2, and F_3. (Hint: we made sure to have one example each of unramified, totally ramified, and ramified-but-not-totally extensions.)

The notations e and f are traditional for these two numbers. Notice that at this point f has simply been defined as the "other factor" of n. We will soon give it a more interesting interpretation. Before we do that, however, we need to explore the structure of K a little further.

In $\mathbb{Q}_p$, the number p played a special role, due to the fact that it was an element of smallest positive valuation, $v_p(p) = 1$. This meant that any element $x \in \mathbb{Z}_p$ with $v_p(x) > 0$ was divisible by p, and in fact, we could interpret $v_p(x)$ as a multiplicity: any $x \in \mathbb{Q}_p$ can be written as $x = p^{v_p(x)}u$, where u is a p-adic unit, i.e., satisfies $v_p(u) = 0$. To do something similar in K, we need an element whose valuation is exactly $1/e$.

Definition 5.4.4 *Let $K/\mathbb{Q}_p$ be a finite extension, and let $e = e(K/\mathbb{Q}_p)$. We say an element $\pi \in K$ is a* uniformizer *if $v_p(\pi) = 1/e$.*

Notice that there are many uniformizers, just as there are many elements of $\mathbb{Z}_p$ whose valuation is exactly 1. In what follows, we will *choose* a uniformizer π, and fix it throughout the discussion. We should remark that in the *unramified* case, we have $e = 1$, and we can (and usually will) simply take $\pi = p$.

Problem 227 Find uniformizers for F_1, F_2, and F_3. (Only F_3 takes some thought.)

Having set this up, we can describe the algebraic structure of K. First of all, recall that we defined the valuation ring

$$\mathcal{O} = \mathcal{O}_K = \{x \in K : |x| \leq 1\} = \{x \in K : v_p(x) \geq 0\}$$

and its maximal ideal

$$\mathfrak{p} = \mathfrak{p}_K = \{x \in K : |x| < 1\} = \{x \in K : v_p(x) > 0\}.$$

This, as we saw in Chapter 2, is a local ring, and the residue field is the quotient

$$\mathbf{k} = \mathcal{O}_K/\mathfrak{p}_K.$$

The basic facts about these rings are easy to describe.

Proposition 5.4.5 *Let notations be as above, and fix a uniformizer π in K. Then:*

i) *The ideal $\mathfrak{p}_K \subset \mathcal{O}_K$ is principal, and π is a generator.*

ii) *Any element $x \in K$ can be written in the form $x = u\pi^{ev_p(x)}$, where $u \in \mathcal{O}_K^{\times}$ is a unit, and therefore satisfies $v_p(u) = 0$. In particular, $K = \mathcal{O}_K[\frac{1}{\pi}]$.*

iii) *The residue field $\mathbf{k}$ is a finite extension of $\mathbb{F}_p$ whose degree is less than or equal to the degree $[K : \mathbb{Q}_p]$. In particular, the number of elements in $\mathbf{k}$ is a power of p. (The exact number of elements will be determined below.)*

iv) *Any element of $\mathcal{O}_K$ is the root of a monic polynomial with coefficients in $\mathbb{Z}_p$.*

v) *Conversely, if $x \in K$ is the root of a monic polynomial with coefficients in $\mathbb{Z}_p$, then $x \in \mathcal{O}_K$.*

vi) *$\mathcal{O}_K$ is a compact topological ring. The sets $\pi^n \mathcal{O}_K$, $n \in \mathbb{Z}$, form a fundamental system of neighborhoods of zero in K, which is a totally disconnected, Hausdorff, locally compact topological space.*

vii) *Let $A = \{0, c_1, c_2, \ldots, c_f\}$ be a fixed set of representatives for the cosets of $\mathfrak{p}_K$ in $\mathcal{O}_K$. Then any $x \in K$ has a unique representation as a p-adic expansion*

$$x = \sum_{i=-m}^{\infty} a_i \pi^i = a_{-m}\pi^{-m} + \cdots + a_0 + a_1\pi + a_2\pi^2 + \ldots,$$

where each $a_i \in A$. In other words, every element of K has a unique expansion in powers of π with coefficients chosen from the "digits" $0, c_1, c_2, \ldots, c_f$.

Problem 228 Prove the proposition. (Some hints: (i) is a matter of computing v_p; (ii) is pretty much the same. For (iii), the crucial observation is that if a set of elements of $\mathcal{O}$ is linearly dependent over $\mathbb{Q}_p$, then the set of their reductions modulo π is linearly dependent over $\mathbb{F}_p$. Item (iv) was pretty much proved when we constructed the absolute value, and (v) is immediate from that construction. The rest is identical to what we did for $\mathbb{Q}_p$.)

Problem 229 Work out $\mathcal{O}_K$ and $\mathbf{k}$ for each of our running examples.

Given that $\mathbb{k}$ is a finite extension of $\mathbb{F}_p$, its degree is another natural invariant of the extension $K/\mathbb{Q}_p$. It turns out, however, to be a number we have already introduced.

Proposition 5.4.6 *Still using the notations above, let* $f = f(K/\mathbb{Q}_p)$ *be the "other factor" of the degree* n *(see Definition 5.4.3). Then* $[\mathbb{k} : \mathbb{F}_p] = f$, *so that* $\mathbb{k} = \mathbb{F}_{p^f}$ *is the finite field with* p^f *elements.*

PROOF: Let $m = [\mathbb{k} : \mathbb{F}_p]$, and let $e = e(K/\mathbb{Q}_p)$ be the ramification index. What the proposition says is that $e \cdot m = n = [K : \mathbb{Q}_p]$. First of all, choose elements $\alpha_1, \alpha_2, \ldots \alpha_m \in \mathcal{O}_K$ such that their images $\bar{\alpha}_1, \bar{\alpha}_2, \ldots \bar{\alpha}_m \in \mathbb{k}$ are a basis of $\mathbb{k}$ over $\mathbb{F}_p$. (In particular, they must be non-zero, so that the α's are actually in $\mathcal{O}_K^\times$.) As we noted above, the α's are clearly linearly independent over $\mathbb{Q}_p$ (given a dependence relation, scale so that the coefficients are integral and at least one is a unit, then reduce modulo π; this gives a dependence relation over $\mathbb{F}_p$). To prove the Proposition, we show how to complete this set to a basis of K over $\mathbb{Q}_p$.

The idea is to use the uniformizer π. Consider the elements

$$\alpha_1, \alpha_2, \ldots, \alpha_m,$$
$$\pi\alpha_1, \pi\alpha_2, \ldots, \pi\alpha_m,$$
$$\pi^2\alpha_1, \pi^2\alpha_2, \ldots, \pi^2\alpha_m,$$
$$\ldots,$$
$$\pi^{e-1}\alpha_1, \pi^{e-1}\alpha_2, \ldots, \pi^{e-1}\alpha_m.$$

We claim these form a basis of K over $\mathbb{Q}_p$. Note that, if so, the proposition follows, since we then have $n = e \cdot m$.

Proving our claim requires several steps, but is not hard. First of all, if every element of $\mathcal{O}_K$ is a $\mathbb{Q}_p$-linear combination of the $\pi^i\alpha_j$, then so is every element of K, since for any $x \in K$ we can find a power of p such that $p^r x \in \mathcal{O}_K$.

Now consider $x \in \mathcal{O}_K$. We will show that x is a $\mathbb{Z}_p$-linear combination of the elements listed above. First, reducing modulo π, we can write $\bar{x}$ as a combination of the $\bar{\alpha}_j$; in other words, we have

$$x = x_{0,1}\alpha_1 + x_{0,2}\alpha_2 + \cdots + x_{0,m}\alpha_m + \text{a multiple of } \pi,$$

with $x_{0,j} \in \mathbb{Z}_p$. Now repeat the same reasoning to the multiple of π to get

$$x = x_{0,1}\alpha_1 + x_{0,2}\alpha_2 + \cdots + x_{0,m}\alpha_m +$$
$$+ x_{1,1}\pi\alpha_1 + x_{1,2}\pi\alpha_2 + \cdots + x_{1,m}\pi\alpha_m +$$
$$+ \text{a multiple of } \pi^2.$$

Repeating this e times, and noticing that π^e and p differ by a unit (because they have the same valuation!), we see that our x can be written as

$$
\begin{aligned}
x = {}& x_{0,1}\alpha_1 + x_{0,2}\alpha_2 + \cdots + x_{0,m}\alpha_m + \\
& + x_{1,1}\pi\alpha_1 + x_{1,2}\pi\alpha_2 + \cdots + x_{1,m}\pi\alpha_m + \\
& + \ldots \\
& + x_{e-1,1}\pi^{e-1}\alpha_1 + x_{e-1,2}\pi^{e-1}\alpha_2 + \cdots + x_{e-1,m}\pi^{e-1}\alpha_m + \\
& + px',
\end{aligned}
$$

where all the coefficients $x_{i,j}$ are in $\mathbb{Z}_p$ and $x' \in \mathcal{O}_K$. Now apply the same reasoning to x'. This will give new coefficients $x_{i,j} + px'_{i,j}$, for which the equality holds modulo p^2. Continuing in this fashion produces Cauchy sequences in $\mathbb{Q}_p$ (in fact, in $\mathbb{Z}_p$) for each coefficient; taking the limit, we get the expression we want for x as a linear combination of the $\pi^i\alpha_j$, which are therefore a generating set.

To show that they are independent, suppose we have a linear dependence relation

$$\sum x_{i,j}\pi^i\alpha_j = 0$$

with $x_{i,j} \in \mathbb{Q}_p$. After scaling, we may assume that the $x_{i,j}$ are all in $\mathbb{Z}_p$ and that at least one is not divisible by p. Reducing this equation modulo π gives a dependence relation for the $\bar{\alpha}_j$ over $\mathbb{F}_p$; this must be trivial, hence the $x_{0,j}$ must reduce to zero, i.e., must be divisible by p. This makes the whole relation divisible by π; divide through. Notice that $x_{0,j}/\pi$ will still be divisible by π, since we know that $x_{0,j}$ is divisible by p, and p is "like" π^e. Now reduce modulo π again. We know that most of the equation is still divisible by π, and using the same reasoning as before, we can conclude that the $x_{1,j}$ must all be divisible by p. Continuing in this fashion, we get that *all* the $x_{i,j}$ are divisible by p, which contradicts our initial assumption. It follows that no such linear dependence relation exists, and we are finally done.   □

After that long proof, it is well to remind ourselves of what we have obtained. We have shown that the degree $n = [K : \mathbb{Q}_p]$ of a finite extension of $\mathbb{Q}_p$ breaks up as a product $n = e \cdot f$, where e measures the change of the image of the p-adic valuation v_p and $f = [\mathbf{k} : \mathbb{F}_p]$ measures the change in the residue field.

Problem 230 This was a messy proof. It's probably wise to work out the precise details.

Our next result is a partial description of the totally ramified extensions of $\mathbb{Q}_p$. It is a standard result in field theory that any extension of a field of characteristic zero (such as $\mathbb{Q}_p$) is generated by adjoining the root of an irreducible polynomial. In the case of totally ramified extensions, we can say exactly what kind of polynomial.

Proposition 5.4.7 *Let $K/\mathbb{Q}_p$ be a totally ramified finite extension of $\mathbb{Q}_p$, so that $e(K/\mathbb{Q}_p) = n = [K : \mathbb{Q}_p]$. Then $K = \mathbb{Q}_p(\pi)$, where π, as above, is a uniformizer. Furthermore, π is a root of a polynomial*

$$f(X) = X^n + a_{n-1}X^{n-1} + \cdots + a_1 X + a_0$$

which satisfies the conditions of the Eisenstein criterion, i.e., $p|a_i$ for $0 \leq i < n$ and $p^2 \nmid a_0$.

PROOF: Let π be a uniformizer, so that $v_p(\pi) = 1/n$, or, equivalently, $|\pi| = p^{-1/n}$. Take $f(X)$ to be the minimal polynomial for π over $\mathbb{Q}_p$. Recall that we can compute the absolute value of π in terms of its norm. It goes like this: if the degree of $f(X)$ is s (which must be a divisor of n) and its last coefficient is a_0, we set $r = n/s$, and then the norm of π is $(-1)^n a_0^r$. Once we know the norm, we can compute the absolute value; this gives the equation

$$p^{-1/n} = |\pi| = \sqrt[n]{|a_0^r|} = \sqrt[s]{|a_0|}.$$

Now, since a_0 is in $\mathbb{Q}_p$, its absolute value is an integral power of p. Looking at the equation, we see that we must have $s = n$ (so that $f(X)$ is of degree n) and $|a_0| = p^{-1}$.

The fact that the degree is n shows that $K = \mathbb{Q}_p(\pi)$, and $|a_0| = p^{-1}$ is exactly what we claimed about this coefficient of $f(X)$. It remains to show our claim about the other coefficients. For this, let $\pi_1 = \pi$, π_2, $\ldots$, π_n be the roots of $f(X)$. Note, first, that all of the roots have the same minimal polynomial, hence the same norm, hence the same absolute value. In particular, we have $|\pi_i| < 1$ for every i. Now, the coefficients of $f(X)$ are combinations of the roots (write $f(X)$ as the product of the $(X - \pi_i)$ and expand); it follows that we must have $|a_j| < 1$ for $1 \leq j \leq n$, and we are done. $\square$

Problem 231 In the case of F_2, which is totally ramified, what is the Eisenstein polynomial?

Problem 232 Let $p = 3$ and let $K = \mathbb{Q}_3(\zeta)$ be the field obtained by adjoining a cube root of unity. Check that this is a totally ramified extension of degree 2, and find the Eisenstein polynomial given by the Proposition.

This is quite a remarkable result, since it gives a rather precise description of the ramified extensions of $\mathbb{Q}_p$. Once we have it, it is natural to look for a similar result for unramified extensions. It turns out that those are even simpler, but to be able to prove that we will have to have one more tool in our kit. That tool is Hensel's Lemma.

Theorem 5.4.8 (Hensel's Lemma) *Let K be a finite extension of $\mathbb{Q}_p$, and let π be a uniformizer. Let $F(X) = a_0 + a_1 X + a_2 X^2 + \cdots + a_n X^n$ be a polynomial whose coefficients are in $\mathcal{O}_K$. Suppose that there exists an "integer" $\alpha_1 \in \mathcal{O}_K$ such that*

$$F(\alpha_1) \equiv 0 \pmod{\pi}$$

and

$$F'(\alpha_1) \not\equiv 0 \pmod{\pi},$$

where $F'(X)$ is the (formal) derivative of $F(X)$. Then there exists an integer $\alpha \in \mathcal{O}_K$ such that $\alpha \equiv \alpha_1 \pmod{\pi}$ and $F(\alpha) = 0$.

Recall that π is a generator of the maximal ideal $\mathfrak{p}_K$, so that we can also write the conditions as congruences modulo $\mathfrak{p}_K$, or in terms of absolute values. The proof is identical to the one we gave in Chapter 3.

Problem 233 Prove Theorem 5.4.8.

Problem 234 Formulate and prove a version of Theorem 3.4.6 (the second form of Hensel's Lemma) that works over K.

Problem 235 Formulate and prove a version of Problem 112 (the stronger form of Hensel's Lemma to which we occasionally needed to resort) that works over K.

The crucial observation, for all three problems, is that there is really nothing to do: *exactly* the same proofs work.

As before, we can use Hensel's Lemma to obtain roots of unity in K. The point is that the non-zero elements of the residue field $\Bbbk$ (which, remember, has p^f elements) form a cyclic group[11] with $p^f - 1$ elements. This means that for each m dividing $p^f - 1$, there are exactly m roots of $F_m(X) = X^m - 1$ in $\Bbbk^\times$. Choosing any lift of these to $\mathcal{O}_K^\times$ gives us m non-congruent "approximate roots." This sets us up for Hensel's Lemma, since the derivative is $F_m'(X) = mX^{m-1}$, which will be non-zero (m is not divisible by p, and our approximate roots are units). Hensel's Lemma gives us m non-congruent (and therefore m different) m-th roots of unity in $\mathcal{O}_K^\times$. Since this is true for any m dividing $p^f - 1$, it means that K contains the full cyclic group of $(p^f - 1)$-st roots of unity. In other words:

Corollary 5.4.9 *Let $K/\mathbb{Q}_p$ be a finite extension, and let $f = f(K/\mathbb{Q}_p)$. Then $\mathcal{O}_K^\times$ contains the cyclic group of $(p^f - 1)$-st roots of unity.*

Problem 236 Describe what roots of unity are given by this Corollary in each of the fields F_1, F_2, and F_3. In each case, can you decide whether there are any other roots of unity?

Of course, if K contains the $(p^f - 1)$-st roots of unity, then it also contains the m-th roots of unity for any m dividing $p^f - 1$. We can also turn this around: given an m which is not divisible by p, one can always find an f such that $p^f \equiv 1 \pmod{m}$ (since the group of invertible elements of $\mathbb{Z}/m\mathbb{Z}$

---

[11]That the non-zero elements of any field form a group is sort of obvious. That the non-zero elements of a finite field form a *cyclic* group follows from the fact that any *finite* subgroup of a field is cyclic, which the reader may have met in her abstract algebra course, and which otherwise is a very nice and challenging exercise.

is, after all, finite), which means that m divides $p^f - 1$. So by taking fields with larger and larger f, we get all the prime-to-p-th roots of unity.

Except for p-power roots of unity, this description is complete. First, if $f = f(K/\mathbb{Q}_p)$ and K contains any other roots of unity, that is, m-th roots of unity for some m which is relatively prime to $p^f - 1$, then they must be 1-units, since their reduction modulo π must be equal to 1.

Problem 237 (This problem just asks you to verify carefully what we have just asserted.) Suppose $x \in K$ satisfies $x^m = 1$.

i) Show that $x \in \mathcal{O}_K^\times$, i.e, that x is a unit in K.

ii) Show that if m is relatively prime to $p^f - 1$, then $x \equiv 1 \pmod{\pi}$, so that $x \in 1 + \mathfrak{p}_K$.

Next, a 1-unit can be an m-th root of unity only if m is a power of p. We show this by a direct argument. First we make the following useful remark:

Lemma 5.4.10 *If $x \equiv 1 \pmod{\pi}$, then $x^p \equiv 1 \pmod{\pi^2}$, and, more generally, $x^{p^r} \equiv 1 \pmod{\pi^{r-1}}$.*

PROOF: An easy exercise on the binomial theorem. Notice that unless $e = 1$ we can in fact do much better than is stated. □

Now it's easy: if ζ is a 1-unit, and $\zeta^m = 1$ for some m prime to p, then we begin with

$$\zeta \equiv 1 \pmod{\pi}.$$

Now choose any r such that $p^r \equiv 1 \pmod{m}$ (this certainly exists, as we observed above). Then, taking p^r-th powers, we get

$$\zeta = \zeta^{p^{r'}} \equiv 1 \pmod{\pi^{r-1}}.$$

Iterating (or just replacing r by a multiple), we see that in fact ζ is congruent to 1 modulo an arbitrarily large power of π. It follows that $\zeta = 1$. (Otherwise, what would be the valuation of $\zeta - 1$?)

Problem 238 Prove the lemma.

Problem 239 Push the argument above a little harder to show that if m is prime to p then two different m-th roots of 1 will never be congruent modulo π.

Problem 240 We outline an alternative way to show that no 1-units can be m-th roots of unity if m is prime to p. Suppose ζ is a 1-unit, and $\zeta^m = 1$ for m prime to p. Taking a power of ζ, we get an ℓ-th root of unity ζ_1, where ℓ is a prime not equal to p. Let $x_1 = 1 - \zeta_1 \in \mathfrak{p}_K$. Then we have

$$(1 - x_1)^\ell - 1 = 0.$$

Expand the left hand side, and rearrange to get a contradiction.

One interesting way to read the last few paragraphs is to see that they tell us something about structure of the 1-units, i.e., the elements of $U_1 = 1 + \pi \mathcal{O}_K$. This is clearly a group, since

$$(1 + \pi x)(1 + \pi y) = 1 + \pi x + \pi y + \pi^2 xy,$$

and

$$(1 + \pi x)^{-1} = 1 - \pi x + (\pi x)^2 - (\pi x)^3 + \dots$$

which clearly converges and belongs to U_1. Similarly, each of the sets $U_n = 1 + \pi^n \mathcal{O}_K$ are subgroups.

Problem 241 Show that for any n the quotient U_n/U_{n+1} is a p-group (i.e., its order is a power of p). (Hint: you need to show that it is a finite abelian group, and that the order of any element is a power of p.)

The upshot is that we have obtained an almost complete description of the roots of unity in K: if we set $f = f(K/\mathbb{Q}_p)$, then K contains $p^f - 1$ non-congruent $(p^f - 1)$-st roots of unity, and possibly some p-power roots of unity. These last will be 1-units.

We are now ready to go back to what started us on this roots-of-unity aside, that is, to describe the unramified extensions of $\mathbb{Q}_p$.

Proposition 5.4.11 *For each f there is exactly one unramified extension of degree f. It can be obtained by adjoining to $\mathbb{Q}_p$ a primitive $(p^f - 1)$-st root of unity.*

PROOF: Let $\bar{\alpha}$ be a generator of the cyclic group of non-zero elements of $\mathbb{F}_{p^f}$. Then $\mathbb{F}_{p^f} = \mathbb{F}_p(\bar{\alpha})$ is an extension of degree f (check the usual references on abstract algebra for the details); let

$$\bar{g}(X) = X^f + \bar{a}_{f-1}X^{f-1} + \cdots + \bar{a}_1 X + \bar{a}_0$$

be the minimal polynomial for $\bar{\alpha}$ over $\mathbb{F}_p$. Lifting $\bar{g}(X)$ to $g(X) \in \mathbb{Z}_p[X]$ any way we like, we get an irreducible polynomial over $\mathbb{Q}_p$. If α is a root of $g(X)$, then $K = \mathbb{Q}_p(\alpha)$ is an extension of degree f. The residue field $\mathbb{k}$ of K clearly contains a root of $\bar{g}(X)$ (the reduction of α modulo $\mathfrak{p}_K$), hence we must have $[\mathbb{k} : \mathbb{F}_p] \geq f$; since, on the other hand, the degree of the residue field is at most equal to the degree of $K/\mathbb{Q}_p$, we have $[\mathbb{k} : \mathbb{F}_p] \leq [K : \mathbb{Q}_p] = f$, it follows that $[\mathbb{k} : \mathbb{F}_p] = f = [K : \mathbb{Q}_p]$, so that $K/\mathbb{Q}_p$ is unramified. We also see that $\mathbb{k} = \mathbb{F}_{p^f}$.

This shows that there always *exists* an unramified extension of degree f (it is, in fact, the extension we considered at the end of the previous section). We still need to show the uniqueness. To do that, we will show that any extension $K/\mathbb{Q}_p$ which is unramified and of degree f will have to be equal to the extension obtained by adjoining a primitive $(p^f - 1)$-st root of unity.

By the Corollary above, we already know that K must contain all the $(p^f - 1)$th roots of unity. Hence, to show the equality we want, all we need to do is show that the smallest field extension of $\mathbb{Q}_p$ which contains the $(p^f - 1)$-st roots of unity is already of degree f, and hence must be all of K. So choose β to be a primitive $(p^f - 1)$th root of unity in K. Then we have

$$\mathbb{Q}_p \subset \mathbb{Q}_p(\beta) \subset K.$$

Now, the powers of β are exactly all the $(p^f - 1)$ roots of unity, and we know, from the Corollary, that they are all distinct modulo π. This means that $\bar{\beta}$ is a $(p^f - 1)$th root of unity, so that the residue field of the extension $\mathbb{Q}_p(\beta)/\mathbb{Q}_p$ contains $\mathbb{F}_{p^f} = \Bbbk$. Since the degree of the residue field extension is certainly less than or equal to the degree of the extension of $\mathbb{Q}_p$, it follows that the degree of $\mathbb{Q}_p(\beta)$ over $\mathbb{Q}_p$ is at least f. Since $K/\mathbb{Q}_p$ is of degree f, it follows that $K = \mathbb{Q}_p(\beta)$, and we are done. $\square$

Problem 242 For $p = 5$, consider the extensions $F_1 = \mathbb{Q}_5(\sqrt{2})$ and $K = \mathbb{Q}_5(\sqrt{3})$. Show that they are both unramified and of degree 2. Conclude that they are equal. How can this be? The theorem also says that either extension is the same as the one obtained by adjoining a primitive 24-th root of unity. Can you find a few terms of the 5-adic expansion of a primitive 24-th root of unity? (Remember that we can take $\pi = 5$ as uniformizer, because we know the extension is unramified, so that the first problem is to choose a convenient set of "digits.")

Problem 243 Find the largest subfield of F_3 which is an unramified extension of $\mathbb{Q}_3$.

Problem 244 Let $K = \mathbb{Q}_3(\sqrt[3]{2})$ be the extension of $\mathbb{Q}_3$ obtained by a adjoining a cube root of 2. Show that this extension is totally ramified.

One nice thing we can do with this result is to consider the union of all these extensions. This will be an infinite extension of $\mathbb{Q}_p$, and will contain all the unramified extensions of $\mathbb{Q}_p$. It is called *the maximal unramified extension of $\mathbb{Q}_p$*, sometimes denoted by $\mathbb{Q}_p^{\text{unr}}$.

Problem 245 Let m be an integer which is not divisible by p. Show that the maximal unramified extension of $\mathbb{Q}_p$ contains the m-th roots of unity. Conclude that we can describe $\mathbb{Q}_p^{\text{unr}}$ as being obtained by adjoining to $\mathbb{Q}_p$ all the prime-to-p-th roots of unity.

The following two problems ask you to obtain important information on $\mathbb{Q}_p^{\text{unr}}$ and $\overline{\mathbb{Q}}_p$. Make sure you either solve them or check the answers in the back of the book.

Problem 246 The p-adic absolute value and the p-adic valuation v_p make sense, of course, on $\mathbb{Q}_p^{\text{unr}}$. What is the image of v_p? What is the residue field?

Problem 247 The last problem also makes perfect sense if we replace $\mathbb{Q}_p^{\text{unr}}$ by an algebraic closure $\overline{\mathbb{Q}}_p$ of $\mathbb{Q}_p$. Do the answers change in that case?

The two main results of this section, describing the totally ramified and the unramified extensions of $\mathbb{Q}_p$, in fact yield a rather good description of arbitrary extensions. We won't go into it here in detail; basically, one shows that any extension is obtained by first taking an unramified extension, and then taking a totally ramified extension of the resulting field. This is proved in, for example, [Kob84]. The following problem is also a step in the direction of getting more precise descriptions:

Problem 248 Let $K/\mathbb{Q}_p$ be a totally ramified extension of degree e which satisfies the extra condition that p does not divide e (such extensions are called *tamely ramified*). Show that K can be obtained by adjoining to $\mathbb{Q}_p$ a root of a polynomial of the form $X^e - pu$, where $u \in \mathbb{Z}_p^\times$ is a p-adic unit.

Problem 249 Let K be a finite extension of $\mathbb{Q}_p$. Is there an analogue of the Eisenstein Criterion for polynomials with coefficients in K? If so, state it and prove it.

5.5 Analysis

Just as in the case of $\mathbb{Q}_p$, once we have a field with an absolute value we can do elementary analysis. In fact, all we need to point out is that *all of what we did in Chapter 4 extends without any difficulty*, because we were careful, when we proved our results, never to use anything that specifically requires that the field is $\mathbb{Q}_p$ rather than an extension. The only changes we have to keep in mind are those that have to do with ramification: we might need to use a uniformizer π where we used p before, and we have a larger range of possible absolute values.

In other words, we already know a lot of things, which we list:

i) A sequence (a_n) in K is Cauchy if and only if

$$\lim_{n \to \infty} |a_{n+1} - a_n| = 0.$$

ii) If a sequence (a_n) converges to a non-zero limit a, then we have $|a_n| = |a|$ for sufficiently large n.

iii) A series $\sum a_n$ in K converges if and only if its general term tends to zero.

iv) Proposition 4.1.4 holds for double series in K.

v) A power series $f(X) = \sum a_n X^n$ with coefficients $a_n \in K$ defines a continuous function on an open ball of radius $\rho = 1/\limsup \sqrt[n]{|a_n|}$; the function extends to the closed ball of radius ρ if $|a_n|\rho^n \to 0$ as $n \to \infty$.

vi) Proposition 4.3.2 and Theorem 4.3.3 are true for power series with coefficients in K.

vii) Functions defined by power series are differentiable, and their derivatives are defined by the formal derivative of the original series.

viii) If $f(X) = \sum a_n X^n$ and $g(X) = \sum b_n X^n$ are power series with coefficients in K, x_m is a convergent sequence contained in the intersection of the disks of convergence of f and g, and we have $f(x_m) = g(x_m)$ for all m, then $a_n = b_n$ for all n.

ix) Strassman's Theorem holds without any change beyond replacing $\mathbb{Q}_p$ by K and $\mathbb{Z}_p$ by $\mathcal{O}_K$.

x) The corollaries to Strassman's Theorem therefore also extend.

xi) The usual power series defines a p-adic logarithm function

$$\log_p : B \longrightarrow K,$$

where

$$B = \{x \in \mathcal{O}_K : |x - 1| < 1\} = 1 + \pi\mathcal{O}_K.$$

This function satisfies the functional equation

$$\log_p(xy) = \log_p(x) + \log_p(y) \quad \text{for any } x, y \in B.$$

xii) The usual power series defines an exponential function

$$\exp_p : D \longrightarrow K,$$

where

$$D = \{x \in \mathcal{O}_K : |x| < p^{-1/(p-1)}\}.$$

This function satisfies the functional equation

$$\exp_p(x + y) = \exp_p(x)\exp_p(y) \quad \text{for any } x, y \in D.$$

(Notice that when e is big there will certainly be elements in $\mathcal{O}_K$ whose absolute values are less than 1 but not less than $p^{-1/(p-1)}$, so that the restriction in the domain is more serious for finite extensions that it was for $\mathbb{Q}_p$ itself.)

xiii) If $x \in D$, then $\exp_p(x) \in B$ and we have

$$\log_p(\exp_p(x)) = x.$$

xiv) If $|x - 1| < p^{-1/(p-1)}$ (i.e., $x \in 1 + D$), then $\log_p(x) \in D$ and we have

$$\exp_p(\log_p(x)) = x.$$

xv) The p-adic logarithm gives a homomorphism from the multiplicative group $B = 1 + \pi\mathcal{O}_K$ to the additive group $\mathfrak{p}_K = \pi\mathcal{O}_K$.

xvi) The p-adic logarithm gives an isomorphism from the multiplicative group $1 + D$ to the additive group D, which is itself isomorphic to the additive group of $\mathcal{O}_K$.

xvii) For each $\alpha \in \mathbb{Z}_p$, the binomial series $(1+x)^\alpha = \mathbf{B}(\alpha, x)$ converges whenever $|x| < 1$ (i.e., for $x \in \pi\mathcal{O}_K = \mathfrak{p}_K$). (We need to keep the condition $\alpha \in \mathbb{Z}_p$ because we used the fact that $\mathbb{Z}$ is dense in $\mathbb{Z}_p$ to conclude that the binomial coefficients were p-adic integers. $\mathbb{Z}$ is certainly not dense in $\mathcal{O}_K$.) In other words, u^α is well defined whenever $u \in 1 + \mathfrak{p}_K$ is a 1-unit in $\mathcal{O}_K$ and $\alpha \in \mathbb{Z}_p$ is a p-adic integer.

This pretty much transports all of the elementary analysis which we developed in Chapter 4 to finite extensions of $\mathbb{Q}_p$. In fact, we will later want to extend it to infinite extensions as well (which clearly won't be a problem, except for the possibility that infinite extensions are not complete...). We conclude this section with the obvious exercise:

Problem 250 Satisfy yourself that the assertions we enumerated are all correct.

5.6   Example: Adjoining a p-th Root of Unity

The discussion in the previous sections was mostly theoretical. It may be helpful to apply it now to a concrete case. We consider, in this section, the field $K = \mathbb{Q}_p(\zeta)$, where ζ is a p-th root of unity and $p \neq 2$. (The case $p = 2$ is, clearly, a bit trivial.) In other words, ζ satisfies $\zeta^p = 1$ but $\zeta \neq 1$, and is therefore a root of the polynomial

$$\Phi_p(X) = \frac{X^p - 1}{X - 1} = X^{p-1} + X^{p-2} + \cdots + X + 1,$$

which is known as the p-th cyclotomic polynomial. The first thing we need to do, then, is to check that this polynomial is irreducible. For that, we use the Eisenstein criterion:

Lemma 5.6.1 *The polynomial*

$$\Phi_p(X) = X^{p-1} + X^{p-2} + \cdots + X + 1$$

is irreducible over $\mathbb{Q}_p$.

PROOF: The polynomial $\Phi_p(X)$ itself certainly does not satisfy the conditions for the Eisenstein criterion. So we use a little trick.

Let $F(X) = \Phi_p(X + 1)$. It is easy to see that $\Phi_p(X)$ is irreducible if and only if $F(X)$ is. We claim that $F(X)$ does satisfy the conditions in the Eisenstein criterion. To see that, we need to check two things: that all the coefficients except the first are divisible by p, and that the last coefficient is not divisible by p^2.

For the first, recall that, modulo p, taking p-th powers distributes over sums:

$$(a + b)^p \equiv a^p + b^p \pmod{p}.$$

This allows us to compute:

$$F(X) = \Phi_p(X + 1)$$

$$= \frac{(X + 1)^p - 1}{(X + 1) - 1} = \frac{(X + 1)^p - 1}{X}$$

$$\equiv \frac{X^p + 1 - 1}{X} \equiv X^{p-1} \pmod{p},$$

so that, except for the first, all the coefficients of $F(X)$ are divisible by p, as we claimed.

As for the last coefficient, it is equal to $F(0) = \Phi_p(1) = p$, which is certainly not divisible by p^2. The Eisenstein criterion then says that $F(X)$ is irreducible, which proves our assertion.   □

In particular, we can deduce the following things:

- $K = \mathbb{Q}_p(\zeta)$ is an extension of $\mathbb{Q}_p$ of degree $p - 1$ (since that is the degree of the minimal polynomial for ζ).

- Looking at the minimal polynomial, we see that $\mathbf{N}_{K/\mathbb{Q}_p}(\zeta) = 1$, and therefore that $|\zeta| = 1$. (Another way to see this is to note that ζ belongs to $\mathcal{O}_K$, and that so does $\zeta^{-1} = \zeta^{p-1}$, which shows that ζ must be a unit in $\mathcal{O}_K$.)

- The polynomial $F(X) = \Phi_p(X + 1)$ is the minimal polynomial for $\zeta - 1$. Therefore, we have $\mathbf{N}_{K/\mathbb{Q}_p}(\zeta - 1) = p$, and therefore

$$|\zeta - 1| = p^{-1/(p-1)}.$$

- K is totally ramified, and $\pi = \zeta - 1$ is a uniformizer in K.

- We have $\zeta \equiv 1 \pmod{\pi}$; in other words, ζ is a 1-unit in $\mathcal{O}_K$.

- The fact that ζ is in $\mathcal{O}_K$ shows that any polynomial $a_0 + a_1\zeta + \cdots + a_{p-2}\zeta^{p-2}$, with $a_i \in \mathbb{Z}_p$, is in $\mathcal{O}_K$ (it's clearly unnecessary to consider polynomials in ζ of higher degree, because ζ is a root of $\Phi_p(X)$). In other words, $\mathbb{Z}_p[\zeta] \subset \mathcal{O}_K$. This inclusion is actually an equality—see below.

Since K is totally ramified, we have $e = p - 1$, $f = 1$, and the residue field $\mathcal{O}_K/\pi\mathcal{O}_K$ of K is just $\mathbb{F}_p$. As usual, we can choose the integers $0, 1, \ldots p - 1$ as coset representatives, and it follows that the elements of K can all be written as π-adic expansions of the form

$$a_{-n}\pi^{-n} + a_{-n+1}\pi^{-n+1} + \cdots + a_0 + a_1\pi + \cdots + a_m\pi^m + \ldots$$

where the a_i are integers between 0 and $p-1$. This is very nice, except for a slight problem: suppose we are given the p-adic expansion of an element of $\mathbb{Q}_p$; it is not immediately clear how to obtain its π-adic expansion in a simple way. For example,

Problem 251 What is the π-adic expansion of the integer p?

We said above that it is easy to see that $\mathbb{Z}_p[\zeta] \subset \mathcal{O}_K$, and that in fact we have an equality. To see why, remember that in the proof of Proposition 5.4.6 we showed that any element of $\mathcal{O}_K$ could be written as a $\mathbb{Z}_p$-linear combination of the elements

$$\alpha_1, \alpha_2, \ldots, \alpha_f,$$
$$\pi\alpha_1, \pi\alpha_2, \ldots, \pi\alpha_f,$$
$$\pi^2\alpha_1, \pi^2\alpha_2, \ldots, \pi^2\alpha_f,$$
$$\ldots,$$
$$\pi^{e-1}\alpha_1, \pi^{e-1}\alpha_2, \ldots, \pi^{e-1}\alpha_f$$

where $\alpha_1, \ldots, \alpha_f$ were a set of elements of $\mathcal{O}_K$ reducing to a basis of the residue field $\Bbbk$ over $\mathbb{F}_p$. In our case, however, $f = 1$, and $\Bbbk$ is equal to $\mathbb{F}_p$, so we need only one element in the basis: $\alpha_1 = 1$. The result then says that any element of $\mathcal{O}_K$ is a $\mathbb{Z}_p$-linear combination of $1, \pi, \pi^2, \ldots \pi^{p-2}$ (since $e = p-1$). Remembering that we have taken $\pi = \zeta - 1$ and substituting in, this says that any element of $\mathcal{O}_K$ can be written as a polynomial in ζ, and hence that $\mathbb{Z}_p[\zeta] = \mathcal{O}_K$.

To conclude this section, we will point out some interesting things about the field K. First of all, since we have $|(\zeta-1)| < 1$, the series for the logarithm of ζ will converge. Since $\zeta^p = 1$, we must have $p\log_p(\zeta) = \log_p(1) = 0$, so that $\log_p(\zeta) = 0$. Writing out the series, this says that

$$\sum_{n=1}^{\infty} (-1)^{n+1} \frac{(\zeta-1)^n}{n} = 0,$$

which we can rearrange slightly into

$$\sum_{n=1}^{\infty} \frac{(1-\zeta)^n}{n} = 0,$$

which is a rather remarkable formula. (We've met it before in the case $p = 2$.)

Another interesting result is that there is a $(p-1)$-st root of $-p$ in K. To see why one might want to look for such a thing, remember that $v_p(\pi^e) = 1$, so that π^e differs from p by a unit in $\mathcal{O}_K$. One might argue that the "nicest" choice of π would be one where this unit were the simplest possible unit. What we are about to show is that there is a $\pi_1 \in \mathcal{O}_K$ for which $\pi_1^e = -p$, so that the unit in this case is simply -1. That's pretty nice!

It's also rather tricky: begin by recalling that the norm of $(1 - \zeta)$ is precisely p (look at the minimal polynomial for $\zeta - 1$, which we found above, and notice that since the degree of K is even, $\mathbf{N}_{K/\mathbb{Q}_p}(x) = \mathbf{N}_{K/\mathbb{Q}_p}(-x)$.). The norm, remember, can be obtained as the product of the images of our element $1 - \zeta$ under the various automorphisms of K over $\mathbb{Q}_p$. There are $p - 1$ such automorphisms, and they are given by

$$\sigma_i : \zeta \mapsto \zeta^i$$

for $i = 1, 2, \ldots, p - 1$. This means that the images of $1 - \zeta$ under the various σ_i are the $1 - \zeta^i$, and the fact that the norm is p gives the equation

$$(1 - \zeta)(1 - \zeta^2) \cdots (1 - \zeta^{p-1}) = p.$$

(We could also get this equality by setting $X = 1$ in the p-th cyclotomic polynomial.) Now, we want to make $(p - 1)$-st powers appear, so we do it by brute force, rewriting the equation as

$$(1 - \zeta)^{p-1} \cdot \frac{1 - \zeta^2}{1 - \zeta} \cdots \cdots \frac{1 - \zeta^{p-1}}{1 - \zeta} = p.$$

Notice that $(1 - \zeta)^{p-1}$ has the same valuation as p, which suggests that the other factors are units (to be precise: it shows that the *product* of all the other factors is a unit, and this *suggests* that each of the other factors is a unit). To see that this is indeed the case, suppose we can show that the factors are all in $\mathcal{O}_K$. Then their valuations would all be greater than or equal to zero. But the sum of their valuations is the valuation of the product, which is zero. Hence, each of the factors must have valuation zero. In other words, if we can show that all the factors are in $\mathcal{O}_K$, then it will follow that they are all units. But the algebraic identity

$$\frac{1 - \zeta^i}{1 - \zeta} = 1 + \zeta + \cdots + \zeta^{i-1}$$

shows that the factors are indeed in $\mathcal{O}_K$ (since they are polynomials in ζ). Hence, each of the fractions $(1 - \zeta^i)/(1 - \zeta)$ is a unit in $\mathcal{O}_K$.

There is one more thing we can get from the equation

$$\frac{1 - \zeta^i}{1 - \zeta} = 1 + \zeta + \cdots + \zeta^{i-1}.$$

Since $\zeta \equiv 1 \pmod{\pi}$ (because, after all, $\pi = \zeta - 1 \ldots$), and since there are i summands on the right hand side, we get

$$\frac{1 - \zeta^i}{1 - \zeta} \equiv i \pmod{\pi}.$$

Multiplying all of these gives something congruent modulo π to the product of the integers from 2 to $p - 1$. In other words, we get

$$\frac{1 - \zeta^2}{1 - \zeta} \cdot \cdots \cdot \frac{1 - \zeta^{p-1}}{1 - \zeta} \equiv (p - 1)! \quad (\mathrm{mod}\ \pi).$$

Now remember that

$$(p - 1)! \equiv -1 \quad (\mathrm{mod}\ p)$$

(this is "Wilson's Theorem" in elementary number theory). Changing sign, and using the previous formula, we see that

$$-\frac{1 - \zeta^2}{1 - \zeta} \cdots \frac{1 - \zeta^{p-1}}{1 - \zeta}$$

is a 1-unit, i.e., is congruent to 1 modulo π. This gives an equation of the form

$$(1 - \zeta)^{p-1} \cdot (\text{a 1-unit}) = -p.$$

What we are after, remember, is to show that $-p$ has a $(p - 1)$-st root in K; from this equation, we will be done if we can show that we can always take a $(p - 1)$-st root of a 1-unit in $\mathcal{O}_K$. But this follows easily from Hensel's Lemma:

Problem 252 Let $u \in 1 + \pi \mathcal{O}_K$, so that u is a 1-unit. Show, using Hensel's Lemma, that the polynomial $X^{p-1} - u$ has a root in $\mathcal{O}_K$.

The upshot: there exists an element $\pi_1 \in \mathcal{O}_K$ such that $\pi_1^{p-1} = -p$. This is interesting in itself, but it also gives an example of the situation described in Problem 248, since we have $e = p - 1$ prime to p, and of course $K = \mathbb{Q}_p(\pi_1)$, where π_1 is a root of $X^{p-1} + p$.

Problem 253 This was a long-drawn-out argument. Can you give a simpler proof that K contains a $(p - 1)$-st root of $-p$?

Problem 254 Even nicer than our π_1 would be a uniformizer π_2 such that $\pi_2^{p-1} = p$ (i.e., the unit is just 1). Show that K in general *does not* contain such a π_2. Can it happen, for a specific prime, that such a π_2 does exist? If so, give an example of a prime for which it does exist.

This example shows how powerful an array of tools we have already put together to study $\mathbb{Q}_p$ and its algebraic extensions. The combination of algebraic and analytic techniques is very effective!

One last bit of fun. Consider the problem of finding the roots of the equation $\log_p(x) = 0$ in K (of course, what this really means is that we want to look for roots $x \in 1 + \pi \mathcal{O}_K$, since otherwise $\log_p$ is not defined). This amounts to looking for the zeros of the logarithm series

$$\log(1 + X) = \sum \frac{(-1)^{n+1} X^n}{n}$$

in $\pi \mathcal{O}_K$, and we can do this with Strassman's theorem by changing variables. Write

$$f(X) = \log(1 + \pi X) = \sum \frac{(-1)^{n+1} \pi^n X^n}{n}.$$

Clearly $f(x)$ converges when $x \in \mathcal{O}_K$, and then Strassman's theorem says that the number of roots of $f(X)$ in $\mathcal{O}_K$ (which is the number of roots of $\log_p$ in $1 + \pi \mathcal{O}_K$) is bounded by the integer N defined by the two conditions

$$\left| \frac{\pi^N}{N} \right| = \max_{n \geq 1} \left| \frac{\pi^n}{n} \right| \quad \text{and} \quad \left| \frac{\pi^N}{N} \right| > \left| \frac{\pi^n}{n} \right| \quad \text{if } n > N.$$

So we need to estimate the absolute value

$$\left| \frac{\pi^n}{n} \right|$$

as a function of n. Let's do it with valuations this time: clearly, $v_p(\pi^n) = n/(p-1)$; that's how we chose π to begin with. So

$$v_p \left(\frac{\pi^n}{n} \right) = \frac{n}{p-1} - v_p(n).$$

We need to find the n for which the absolute value is largest; in valuation terms, we want to find the n which makes the valuation smallest. To help us get our bearings, we can tabulate the first few values: see table 5.1

The table suggests that the smallest value is $1/(p-1)$, which occurs only when $n = 1$ and when $n = p$. This means $N = p$, so that $\log_p$ has at most p roots in $1 + \pi \mathcal{O}_K$. Since we already know p roots, namely the roots of unity $1, \zeta, \zeta^2, \ldots, \zeta^{p-1}$, we already know all the roots. In particular, this tells us that the roots of unity contained in K are exactly the cyclic group of $p(p-1)$-st roots of unity (the p-th roots we just found, plus the $(p-1)$-st roots that are provided by Hensel's Lemma when $f = 1$).

Problem 255 Prove that our surmise from the table is correct, i.e., that the smallest value for $v_p(\pi^n/n)$ is $1/(p-1)$, and that it occurs last when $n = p$.

Problem 256 Investigate what would change if instead of adjoining a p-th root of unity we adjoin a p^n-th root of unity for some $n > 1$.

5.7   On to $\mathbb{C}_p$

We now want to go on to consider the algebraic closure $\overline{\mathbb{Q}}_p$ in earnest. We have already constructed its absolute value, and the next order of business is to show that it is *not* complete with respect to this absolute value. We will then go to its completion, of course, and we will be able to prove that the completion *is* algebraically closed.

n	$v_p(n)$	$v_p(\dfrac{\pi^n}{n})$
$n = 1, \ldots, p-1$	0	$\dfrac{n}{p-1}$
$n = p$	1	$\dfrac{p}{p-1} - 1 = \dfrac{1}{p-1}$
$n = p+1, \ldots, 2p-1$	0	$\dfrac{n}{p-1}$
$n = 2p$	1	$\dfrac{2p}{p-1} - 1 = 1 + \dfrac{2}{p-1}$
$\ldots$	$\ldots$	$\ldots$
$n = p^2$	2	$\dfrac{p^2}{p-1} - 2 = p - 1 + \dfrac{1}{p-1}$

Table 5.1: Computing $v_p(\pi^n/n)$

To do all that, we need to know a little more about its elements, and we begin by proving a few useful facts. The first of these is known as "Krasner's Lemma." The first thing we need in order to be able to state it is to remind ourselves of what it means for two elements of $\overline{\mathbb{Q}}_p$ to be "conjugate." This is a concept that really comes from Galois theory, but we state it here in a minimalistic fashion.

Definition 5.7.1 *Let K be a subfield of $\overline{\mathbb{Q}}_p$. Two elements a and a' of $\overline{\mathbb{Q}}_p$ are called* conjugate *over K when they are roots of the same monic irreducible polynomial with coefficients in K.*

As we have pointed out above, an equivalent way of saying this is to say that two elements are conjugate when there exists an automorphism $\sigma : \overline{\mathbb{Q}}_p \longrightarrow \overline{\mathbb{Q}}_p$ which induces the identity map on K and sends a to a'. It is clear, from either characterization, that conjugate elements have the same absolute value. (Yes? Good.)

It is worth pointing out that this definition has nothing "p-adic" about it: it works just as well for an arbitrary field K of characteristic zero, provided we replace $\overline{\mathbb{Q}}_p$ by an algebraic closure of K.

What Krasner's lemma says is that if an element b is "close enough" to a (what this means is defined by the statement of the Lemma, in terms of the conjugates of a), then a belongs to the field generated by b. Perhaps we can use the description "b is more complicated than a" to mean that adjoining b gives a field which contains the field generated by a. In that language, the Lemma says that b can only be "very close" to a if it is more complicated than a.

This is a somewhat surprising conclusion, since to say that the field generated by a is contained in the field generated by b amounts to saying that a can be written as a polynomial in b. Viewed at from this angle, Krasner's Lemma looks like the prototypical p-adic theorem: it deduces an algebraic fact (a can be written as a polynomial in b) from an analytic fact (b is very close to a).

Here is the precise statement, where we have taken care to be very general in order to be able to use the Theorem at a crucial juncture below. The reader should feel free to replace "K" by "$\mathbb{Q}_p$" everywhere in this and the following result if the added generality proves to be a hindrance.

Theorem 5.7.2 (Krasner's Lemma) *Let K be a non-archimedean complete valued field of characteristic zero, and let a and b be elements of the algebraic closure of K. Let $a_1 = a, a_2, \ldots, a_n$ be the conjugates of a over K. Suppose that b is closer to a than any of the conjugates of a, i.e.,*

$$|b - a| < |a - a_i|$$

for $i = 2, 3, \ldots, n$. Then $K(a) \subset K(b)$.

PROOF: This is short and sweet, but uses field theory a bit more seriously than other results we have proved. Let $L = K(b)$ and suppose the theorem is false, that is, that $a \notin L$. Well, then look at $L(a)$ (which, remember, is the smallest extension of L which contains a). Since we are assuming that $a \notin L$, the degree $m = [L(a) : L]$ is bigger than one. Now, there must be m homomorphisms $\sigma : L(a) \longrightarrow \overline{K}$ which send L to itself (and a to one of its conjugates, of course). There is at least one such σ for which $\sigma(a) \neq a$ (because if $\sigma(a) = a$ then σ is the identity on $L(a)$, and we're assuming that there's at least one other σ besides the identity); call it σ_0. Since we know, by the uniqueness of the extension of an absolute value, that $|\sigma(x)| = |x|$ for any σ and any $x \in \overline{K}$, we have

$$|\sigma_0(b) - \sigma_0(a)| = |b - a|.$$

But σ_0 fixes L, and b is in L, so $\sigma_0(b) = b$, and the last equality now says

$$|b - \sigma_0(a)| = |b - a|.$$

But then

$$|a - \sigma_0(a)| \leq \max\{|a - b|, |b - \sigma_0(a)|\} = \max\{|a - b|, |b - a|\} = |b - a|,$$

and that's not allowed, since our assumption was that b was closer to a than any of its conjugates, one of which is $\sigma_0(a)$. The contradiction shows that our assumption was wrong, that is, that a *does* belong to L, and that shows that $K(a) \subset K(b)$.   □

Problem 257 Use Krasner's Lemma to give another proof that the field $\mathbb{Q}_p(\zeta_p)$ obtained by adjoining a p-th root of unity contains a $(p-1)$-st root of $-p$.

A really important corollary of Krasner's Lemma is the following:

Corollary 5.7.3 *Let K be a non-archimedean complete valued field of characteristic zero. Let*

$$f(X) = X^n + a_{n-1}X^{n-1} + \cdots + a_1 X + a_0 \in K[X]$$

be a monic irreducible polynomial of degree n with coefficients in K, let λ be a root of $f(X)$, and let $L = K(\lambda)$ be the extension of K obtained by adjoining that root. Then there exists a real number $\varepsilon > 0$ such that the following holds:

- *if $g(X) = X^n + b_{n-1}X^{n-1} + \cdots + b_1 X + b_0$ is any monic polynomial of degree n for which we have*

$$|a_i - b_i| < \varepsilon \qquad \text{for all } i = 0, 1, \dots, n-1,$$

  then $g(X)$ is irreducible over K and has a root in L.

That's a bit complicated, so let's paraphrase: what the corollary says is that any polynomial which is close enough to our $f(X)$ shares the two main properties of $f(X)$: it is irreducible, and it has a root in L. How close is "close enough" may depend on the specific $f(X)$, of course.

PROOF: The proof (which is based on the one given in [Ami75]) has two parts. First, we establish that under certain conditions the conclusion holds, and then we show that we can choose ε so that the conditions must hold.

Let $\lambda_1 = \lambda, \lambda_2, \dots, \lambda_n$ be the roots of $f(X)$ in $\overline{K}$, and let

$$r = \min_{i \neq j} |\lambda_i - \lambda_j|.$$

Let $g(X)$ be as in the statement: a monic irreducible polynomial of degree n. Let $\mu_1, \mu_2, \dots, \mu_n$ be the roots (listed with multiplicities, of course) of $g(X)$ in $\overline{K}$, so that $g(X) = \prod(X - \mu_j)$. Let

$$D = \prod_i g(\lambda_i) = \prod_{i,j}(\lambda_i - \mu_j).$$

Claim 1: If $|D| < r^{n^2}$, then $g(X)$ is irreducible over K and has a root in $L = K(\lambda)$.

Proof of claim 1: If $|D| < r^{n^2}$, then at least one of the n^2 factors of D must have absolute value less than r. In other words, there must be a pair (i, j)

such that $|\lambda_i - \mu_j| < r$. Since r is the minimum distance between λ_i and its conjugates, we can apply Krasner's Lemma to conclude that $K(\lambda_i) \subset K(\mu_j)$; but this says that $K(\mu_j)$ must have degree at least n over K, and since μ_j is a root of a polynomial of degree n, the only way this can happen is when the polynomial is irreducible and the degree is exactly n. Since both fields are then of degree n, and one is contained in the other, they must in fact be equal.

Thus, we have shown that $g(X)$ is irreducible, and that $K(\lambda_i) = K(\mu_j)$. If $i = 1$, this is what we wanted to prove. If not, there is a little step more: there is certainly an automorphism of $\overline{K}$ that sends λ_i to λ, and this automorphism must send μ_j to some other root $\mu = \mu_{j_1}$ of $g(X)$. Applying the automorphism to the equality $K(\lambda_i) = K(\mu_j)$ gives $L = K(\lambda) = K(\mu)$, so that $g(X)$ has a root μ that belongs to L, which is what we wanted to prove.

Claim 2: There exists a real number $\varepsilon > 0$ such that if $|a_i - b_i| < \varepsilon$, then $|D| < r^{n^2}$.

Proof of claim 2: We leave this one to the reader. It is a matter of expressing how D depends on the coefficients of the polynomials involved. $\square$

Problem 258 Prove "Claim 2." Suggestion: try to show that the function that maps

$$(a_0, a_1, \ldots, a_{n-1}, b_0, b_1, \ldots, b_{n-1}) \mapsto D$$

is a continuous function. Why does this do the trick?

Problem 259 (For those confident of their abstract algebra) In the last two results we imposed a litany of conditions on our field: "non-archimedean complete valued field of characteristic zero." Which of these conditions are seriously needed? Are there weaker forms of these results which are valid in more generality?

One way of grasping what the corollary says is to think of it as saying that at least some aspect of the "root structure" of a polynomial varies "continuously" as a function of the coefficients of the polynomial. Specifically, it says (or, to be precise, its proof shows) that when two irreducible polynomials are "close enough" (in the sense that the coefficients are close) there will be a root of one "close" to any root of the other. The next problem gives a result in the same spirit.

Problem 260 The point of this problem is to state (and have the reader prove) a version of the statement that "the roots of a polynomial are continuous functions of the coefficients." For this, let $f(X) = \sum a_i X^i$ be a polynomial of degree n whose roots in $\overline{\mathbb{Q}}_p$ are all distinct. Show that given an $\varepsilon > 0$ there exists a $\delta > 0$ such that for any other polynomial $g(X) = \sum b_i X^i$ of degree n such that $|a_i - b_i| < \delta$ for $i = 0, 1, \ldots, n$ and every root λ of $f(X)$ there exists exactly one root μ of $g(X)$ satisfying $|\mu - \lambda| < \varepsilon$. (Uff! That's quite a mouthful!)

Problem 261 Is the hypothesis (in both the corollary and the previous problem) that the polynomials have the same degree really necessary?

Problem 262 Is the "continuity of the roots as functions of the coefficients" true for polynomials with real and/or complex coefficients?

We now proceed to the big result in this section.

Theorem 5.7.4 *The algebraic closure $\overline{\mathbb{Q}}_p$ is not complete with respect to the (extended) p-adic absolute value.*

PROOF: (This is a long one.)

Proving the theorem is going to require us to come up with a Cauchy sequence in $\overline{\mathbb{Q}}_p$ which does not converge. Since finite extensions of $\mathbb{Q}_p$ *are* complete, this sequence must involve numbers from bigger and bigger extensions as it proceeds (because otherwise we could find a finite extension containing all the terms of the sequence, and it would have to converge). So this is going to be a complicated sequence!

In fact, it is going to be an infinite series whose general term tends to zero, but which does not converge. That is good enough, of course, since we've already shown that the partial sums of such a series form a Cauchy sequence. We'll construct our example slowly and carefully, so that no one gets lost on the way. Each term in our series will belong to an *unramified* extension of $\mathbb{Q}_p$, so that our example will in fact show that the union $\mathbb{Q}_p^{\mathrm{unr}}$ of all the unramified extensions of $\mathbb{Q}_p$ is already not a complete field.

Enough preliminaries: here goes (following [Cas86]). Remember that one gets unramified extensions of $\mathbb{Q}_p$ by adjoining roots of unity of order prime to p. We begin by putting together a large list of these. We choose $\zeta_1 = 1$, and then choose a sequence of ζ_i, $i = 2, 3, \ldots$ such that:

- each ζ_i is a root of unity of order prime to p,

- for each i, $\zeta_{i-1} \in \mathbb{Q}_p(\zeta_i)$, and

- the degree of $\mathbb{Q}_p(\zeta_i)$ over $\mathbb{Q}_p(\zeta_{i-1})$ is bigger than i.

In symbols, we have

$$\zeta_i^{m_i} = 1 \qquad \text{for some } m_i \text{ not divisible by } p,$$

$$\mathbb{Q}_p(\zeta_{i-1}) \subset \mathbb{Q}_p(\zeta_i) \qquad \text{and}$$

$$[\mathbb{Q}_p(\zeta_i) : \mathbb{Q}_p(\zeta_{i-1})] > i.$$

Problem 263 But is it possible? Find a sequence of m_i such that the primitive m_i-th roots of unity form a sequence satisfying our two conditions.

Now construct the series

$$\sum_{i=0}^{\infty} \zeta_i\, p^i.$$

The partial sums of this series,

$$c_n = \sum_{i=0}^{n} \zeta_i\, p^i,$$

clearly form a Cauchy sequence in $\overline{\mathbb{Q}}_p$ (and in fact even in $\mathbb{Q}_p^{\mathrm{unr}}$). We want to prove that this sequence does not have a limit in $\overline{\mathbb{Q}}_p$.

Well, suppose it did, and call the limit $c \in \overline{\mathbb{Q}}_p$. Whatever it is, c must be a root of some irreducible polynomial over $\mathbb{Q}_p$, since it is an element of the algebraic closure. Say that this polynomial has degree d, so that $[\mathbb{Q}_p(c) : \mathbb{Q}_p] = d$. Then consider the d-th partial sum:

$$c_d = \sum_{i=0}^{d} \zeta_i\, p^i.$$

Since

$$c - c_d = \sum_{i=d+1}^{\infty} \zeta_i\, p^i$$

and the ζ's are units, we have

$$|c - c_d| \le p^{-(d+1)}.$$

Now take any automorphism $\sigma : \overline{\mathbb{Q}}_p \longrightarrow \overline{\mathbb{Q}}_p$ inducing the identity on $\mathbb{Q}_p$. Any such σ preserves absolute values, so we must have

$$|\sigma(c) - \sigma(c_d)| \le p^{-(d+1)}.$$

We are aiming for a contradiction, of course, and the first step to get there is to choose a nice bunch of σ's. Remember that we chose our ζ's so that $[\mathbb{Q}_p(\zeta_i) : \mathbb{Q}_p(\zeta_{i-1})] > i$. Applying this to $i = d$, we see that there are (at least) $d+1$ automorphisms $\sigma_1, \sigma_2, \ldots, \sigma_{d+1}$ which are the identity mapping on $\mathbb{Q}_p(\zeta_{d-1})$ (and hence fix $\zeta_1, \zeta_2, \ldots, \zeta_{d-1}$) but such that the images of ζ_d are all distinct.

Now if $i \ne j$ we get

$$\sigma_i(c_d) - \sigma_j(c_d) = \left(\sum_{i=0}^{d-1} \zeta_i p^i + \sigma_i(\zeta_d)p^d \right) - \left(\sum_{i=0}^{d-1} \zeta_i p^i + \sigma_j(\zeta_d)p^d \right)$$

$$= \Big(\sigma_i(\zeta_d) - \sigma_j(\zeta_d) \Big) p^d.$$

Notice that $\sigma_i(\zeta_d)$ and $\sigma_j(\zeta_d)$ are distinct m_d-th roots of unity; we saw above that they cannot be congruent modulo p. This means that the element

$\sigma_i(\zeta_d) - \sigma_j(\zeta_d)$ is not divisible by p, which lets us compute the absolute value:

$$|\sigma_i(c_d) - \sigma_j(c_d)| = p^{-d}.$$

We're almost there: now we apply the σ's to c, and put together all the information we have gathered. We have

$$|\sigma_i(c_d) - \sigma_i(c)| \le p^{-(d+1)},$$

$$|\sigma_j(c_d) - \sigma_j(c)| \le p^{-(d+1)},$$

and

$$|\sigma_i(c_d) - \sigma_j(c_d)| = p^{-d}.$$

Together, these show that

$$|\sigma_i(c) - \sigma_j(c)| = p^{-d},$$

(use "all triangles are isosceles" twice). In particular, it follows that $\sigma_i(c) \ne \sigma_j(c)$.

In other words, we have found $d+1$ automorphisms $\sigma_1, \sigma_2, \ldots, \sigma_{d+1}$ of $\overline{\mathbb{Q}}_p$ which are the identity on $\mathbb{Q}_p$ and such that the images of c under the σ_j are all distinct. But then the minimal polynomial for c must have at least $d+1$ roots (the various images), which is a contradiction, since, after all, it had degree d.

The contradiction shows that in fact c cannot be the root of a polynomial with coefficients in $\mathbb{Q}_p$, hence does not belong to $\overline{\mathbb{Q}}_p$. This proves what we wanted: the algebraic closure of $\mathbb{Q}_p$ is not complete with respect to the (extended) p-adic absolute value. $\qquad\square$

In fact, since all the ζ_i were roots of unity of order prime to p, we have also proved:

Theorem 5.7.5 *The maximal unramified extension $\mathbb{Q}_p^{unr}$ of $\mathbb{Q}_p$, obtained by adjoining all the roots of unity of order prime to p, is not complete with respect to the (extended) p-adic absolute value.*

Notice that our proof has in fact constructed an explicit element that is transcendental over $\mathbb{Q}_p$. One can, in fact, use the basic idea of the proof to obtain general method for showing that the sums of certain series are transcendent.

Problem 264 Explain the statement: "the expression of c_n as a series is actually its p-adic expansion." Does this observation make the proof/construction easier to understand?

Well, "c'est la vie." Since $\overline{\mathbb{Q}}_p$ is not complete, we need to construct a completion. This is done exactly as in the case of $\mathbb{Q}_p$, by playing with the ring of all Cauchy sequences in $\overline{\mathbb{Q}}_p$. (In fact, our construction in Chapter 3 clearly works in utter generality.) The upshot is the following:

Proposition 5.7.6 *There exists a field* $\mathbb{C}_p$ *and an absolute value* $|\ |$ *on* $\mathbb{C}_p$ *such that:*

- $\mathbb{C}_p$ *contains* $\overline{\mathbb{Q}}_p$, *and the restriction of* $|\ |$ *to* $\overline{\mathbb{Q}}_p$ *coincides with the p-adic absolute value;*

- $\mathbb{C}_p$ *is complete with respect to* $|\ |$; *and*

- $\overline{\mathbb{Q}}_p$ *is dense in* $\mathbb{C}_p$.

Recall that whenever we have a convergent sequence $x_n \to x \neq 0$ in a non-archimedean field, there exists an N such that $|x_n| = |x|$ for $n \geq N$ (this is Lemma 3.2.10, except that we are saying that it works for any non-archimedean field... which it clearly does). This means that the set of possible absolute values in $\mathbb{C}_p$ is exactly the same as in $\overline{\mathbb{Q}}_p$. In other words,

Proposition 5.7.7 *If* $x \in \mathbb{C}_p$, $x \neq 0$, *then there exists a rational number* $v \in \mathbb{Q}$ *such that* $|x| = p^{-v}$. *In other words, the p-adic valuation* v_p *extends to* $\mathbb{C}_p$, *and the image of* $\mathbb{C}_p^\times$ *under* v_p *is* $\mathbb{Q}$.

Problem 265 Convince yourself that the last two Propositions are true. (In both cases, it is just a matter of seeing that arguments we presented before in the case of $\mathbb{Q}_p$ extend without any difficulty.)

Problem 266 Since we have a valuation, we have a valuation ring, its valuation ideal, etc. Give explicit definitions. Can you describe the residue field? Is the valuation ideal principal? Does the concept of a uniformizer still make sense?

We write $\mathfrak{O}$ for the valuation ring of $\mathbb{C}_p$, i.e.,

$$\mathfrak{O} = \{x \in \mathbb{C}_p : |x| \leq 1\}.$$

This contains the valuation ideal

$$\mathfrak{P} = \{x \in \mathbb{C}_p : |x| < 1\}.$$

As always, $\mathfrak{O}$ is a local ring.

Problem 267 Show that any element of $\mathbb{C}_p$ can be written as a product of (i) a root of unity, (ii) a 1-unit, and (iii) a fractional power of p.

$\mathbb{C}_p$ is an enormous field, gotten by a series of complicated operations: start with $\mathbb{Q}$ and the p-adic absolute value, take a completion, take the algebraic closure of the result, and then complete once again! One might wonder whether the process will ever stop, i.e., whether one might not need to take another algebraic closure, and so proceed without ever ending. On the contrary:

Proposition 5.7.8 $\mathbb{C}_p$ *is algebraically closed.*

PROOF: We give a jazzy proof, and ask the reader to come up with a more direct proof in a problem.

Take an irreducible polynomial $f(X)$ with coefficients in $\mathbb{C}_p$. Since $\overline{\mathbb{Q}}_p$ is dense in $\mathbb{C}_p$, we can find polynomials of the same degree and with coefficients in $\overline{\mathbb{Q}}_p$ whose coefficients are as close as we like to the coefficients of $f(X)$. By Proposition 5.7.3, if we choose such an $f_0(X)$ with coefficients close enough to those of $f(X)$, it will be irreducible over $\mathbb{C}_p$, and *a fortiori* also irreducible over $\overline{\mathbb{Q}}_p$. Since $\overline{\mathbb{Q}}_p$ is algebraically closed, this means that $f_0(X)$ will have degree one. Since $f(X)$ and $f_0(X)$ have the same degree, it follows that $f(X)$ has degree one.

This shows that any irreducible polynomial in $\mathbb{C}_p$ must be of degree one, which means that $\mathbb{C}_p$ is algebraically closed.   □

Problem 268 Here's an idea for a more direct proof (which might be technically more difficult). Take any polynomial $f(X) \in \mathbb{C}_p[X]$; we can assume it has no repeated roots (do you see why?). Build a sequence of polynomials $f_i(X) \in \overline{\mathbb{Q}}_p[X]$, all of the same degree, whose coefficients approach those of $f(X)$. Show, using Problem 260, that one can choose a root of each of the $f_i(X)$ so as to form a Cauchy sequence in $\overline{\mathbb{Q}}_p$ which converges to a root of $f(X)$ in $\mathbb{C}_p$.

Problem 269 (Hard) Show that $\mathbb{C}_p$ is not locally compact.

In fact, one can show that any locally compact (and therefore complete) field of characteristic zero must be isomorphic to either $\mathbb{R}$, $\mathbb{C}$, or a finite extension of $\mathbb{Q}_p$. One can even[12] start the whole thing from here, i.e., start with a locally compact field of characteristic zero, and reconstruct the absolute value from the Haar measure on that field.

The real home of p-adic analysis is $\mathbb{C}_p$, which except for not being locally compact, is closely analogous to the field of complex numbers. The next chapter will take a look at some aspects of analysis in $\mathbb{C}_p$, without trying to be in any way exhaustive.

---

[12]Mumbo-jumbo alert: this sentence talks about high-powered stuff which we really don't think our readers know about.

6 Analysis in $\mathbb{C}_p$

This chapter tries to give the reader a taste of what analysis in $\mathbb{C}_p$ is like. Rather than attempt to be exhaustive, which would violate the goals of this book, we try to touch on a few remarkable points: the theory of Newton polygons, the p-adic Weierstrass Preparation Theorem, the description of entire functions. As usual, the first step is to re-appropriate all the results we obtained earlier. We then go on to consider how to extend the p-adic valuation to polynomials and power series. This will yield a norm on the spaces of polynomials and of power series, which will prove to be an important tool. We then go on to proving the main theorems themselves.

Before we start, recall the notation we introduced above: we write

$$\mathfrak{O} = \{x \in \mathbb{C}_p : |x| \le 1\}$$

for the valuation ring in $\mathbb{C}_p$ (we might want to call it the ring of integers in $\mathbb{C}_p$) and

$$\mathfrak{P} = \{x \in \mathbb{C}_p : |x| < 1\}$$

for the valuation ideal. The ideal $\mathfrak{P}$ is not principal, and the residue field $\mathbb{F} = \mathfrak{O}/\mathfrak{P}$ is an algebraic closure of $\mathbb{F}_p$.

6.1 Almost Everything Extends

As we have already pointed out, most of the results in Chapter 4 did not really depend on the fact that we were working over $\mathbb{Q}_p$; in fact, they hold just as well for more general fields. In particular, we can repeat (and enlarge) our list of "things that extend:"

i) A sequence (a_n) in $\mathbb{C}_p$ is Cauchy if and only if

$$\lim_{n \to \infty} |a_{n+1} - a_n| = 0.$$

ii) If a sequence (a_n) converges to a non-zero limit a, then we have $|a_n| = |a|$ for sufficiently large n.

iii) A series $\sum a_n$ in $\mathbb{C}_p$ converges if and only if its general term tends to zero.

iv) Proposition 4.1.4 holds for double series in $\mathbb{C}_p$.

v) A power series $f(X) = \sum a_n X^n$ with coefficients $a_n \in \mathbb{C}_p$ defines a continuous function on an open ball of radius $\rho = 1/\limsup \sqrt[n]{|a_n|}$; the function extends to the closed ball of radius ρ if $|a_n|\rho^n \to 0$ as $n \to \infty$. Note that in contrast to what happens in $\mathbb{Q}_p$ or even in its finite extensions, we can characterize ρ by saying that $\sum a_n x^n$ converges for $|x| < \rho$ and diverges for $|x| > \rho$.

vi) Therefore, given a power series $f(X) = \sum a_n X^n$ with radius of convergence ρ, we can define a function on the open (and perhaps also the closed) ball of radius ρ around $\alpha \in \mathbb{C}_p$ by putting

$$f(x) = \sum a_n (x - \alpha)^n$$

for any $x \in B(\alpha, \rho)$ (or $\overline{B}(\alpha, \rho)$).

vii) Proposition 4.3.2 and Theorem 4.3.3 are true for power series with coefficients in $\mathbb{C}_p$.

viii) Functions defined by power series are differentiable, and their derivatives are defined by the formal derivative of the original series.

ix) If $f(X) = \sum a_n X^n$ and $g(X) = \sum b_n X^n$ are power series with coefficients in $\mathbb{C}_p$, x_m is a convergent sequence contained in the intersection of the disks of convergence of f and g, and $f(x_m) = g(x_m)$ for all m, then $a_n = b_n$ for all n.

x) Strassman's Theorem holds without any change beyond replacing $\mathbb{Q}_p$ with $\mathbb{C}_p$ and $\mathbb{Z}_p$ with $\mathfrak{O}$.

xi) The corollaries to Strassman's Theorem therefore also extend.

xii) The usual power series defines a p-adic logarithm function

$$\log_p : B \longrightarrow \mathbb{C}_p,$$

where

$$B = \{x \in \mathfrak{O} : |x - 1| < 1\} = B(1, 1) = 1 + \mathfrak{P}.$$

This function satisfies the functional equation

$$\log_p(xy) = \log_p(x) + \log_p(y)$$

for any $x, y \in B$.

xiii) The usual power series defines an exponential function $\exp_p : D \longrightarrow \mathbb{C}_p$, where

$$D = \{x \in \mathfrak{O} : |x| < p^{-1/(p-1)}\} = B(0, p^{-1/(p-1)}).$$

This function satisfies the functional equation

$$\exp_p(x + y) = \exp_p(x) \exp_p(y)$$

for any $x, y \in D$.

xiv) If $x \in D$, then $|\exp_p(x) - 1| < p^{-1/(p-1)}$; in particular, $\exp_p(x) \in B$, and we have

$$\log_p(\exp_p(x)) = x.$$

xv) If $|x - 1| < p^{-1/(p-1)}$ (i.e., $x \in 1 + D$), then $\log_p(x) \in D$, and we have

$$\exp_p(\log_p(x)) = x.$$

xvi) The p-adic logarithm gives a homomorphism from the multiplicative group $B = B(1,1) = 1 + \mathfrak{P}$ into the additive group $\mathfrak{P} = B(0,1)$.

xvii) The p-adic logarithm gives an isomorphism from the multiplicative group

$$1 + D = B(1, p^{-1/(p-1)})$$

to the additive group

$$D = B(0, p^{-1/(p-1)}).$$

xviii) For each $\alpha \in \mathbb{Z}_p$, the binomial series $(1 + x)^\alpha = \mathbf{B}(\alpha, x)$ converges whenever $|x| < 1$ (i.e., for $x \in \mathfrak{P} = B(0,1)$). In other words, u^α is well defined whenever $u \in B(1,1) = 1 + \mathfrak{P}$ is a 1-unit in $\mathfrak{O}$ and $\alpha \in \mathbb{Z}_p$ is a p-adic integer.

Problem 270 Make sure you understand how to prove the claims above. Pay particular attention to where the situation in $\mathbb{C}_p$ differs from the ones we considered before. For example, why is the statement characterizing ρ in item (v) true?

Problem 271 In assertion *xiv*, is it true that the additive group of D is isomorphic to the additive group of $\mathfrak{O}$? (The analogous statement *was* true in $\mathbb{Q}_p$...)

Another important class of results that can be extended to $\mathbb{C}_p$ are the several variants of Hensel's Lemma. The trick here is not to use versions which refer to uniformizers, since there are no uniformizers in $\mathbb{C}_p$. (A uniformizer would be an element with the largest possible absolute value which was still less than one, but in $\mathbb{C}_p$ we have elements with absolute value p^r for any $r \in \mathbb{Q}$, so there is no such thing.) Still, one can either replace "mod p" with "mod $\mathfrak{P}$" throughout, or simply state things in terms of absolute values. So here are two versions of Hensel's Lemma:

Theorem 6.1.1 (Hensel's Lemma in $\mathbb{C}_p$) *Let*

$$F(X) = a_0 + a_1 X + a_2 X^2 + \cdots + a_n X^n$$

be a polynomial whose coefficients are in $\mathfrak{O}$. Suppose that there exists an $\alpha_1 \in \mathfrak{O}$ such that

$$|F(\alpha_1)| < 1 \qquad and \qquad |F'(\alpha_1)| = 1$$

where $F'(X)$ is the (formal) derivative of $F(X)$. Then there exists an $\alpha \in \mathfrak{O}$ such that $|\alpha - \alpha_1| < 1$ and $F(\alpha) = 0$.

PROOF: We need to be just a little bit careful in order to avoid trouble. (The problem is that there is no uniformizer in $\mathbb{C}_p$, and all our proofs up to here depended explicitly on their being one. What we do is to use a convenient element of small valuation to replace the uniformizer.) Let $\delta = |F(\alpha_1)| < 1$, and choose $\pi \in \mathbb{C}_p$ such that $|\pi| = \delta$. Then the argument in the original proof of Hensel's Lemma (Theorem 3.4.1), with p replaced everywhere by π, will allow us to find α_2 such that $|\alpha_1 - \alpha_2| \leq \delta$ and $|F(\alpha_2)| \leq \delta^2$. Proceeding inductively, we get a sequence α_n which converges to the root α. $\square$

Problem 272 Check the details!

And now the other version, for polynomials: as before, one can state this with absolute values or by talking of reduction modulo the ideal $\mathfrak{P}$. We choose the second path here, simply so that we don't have to talk about "the maximum of the absolute values of the differences of the coefficients" of two polynomials.

Theorem 6.1.2 (Hensel's Lemma, Second Form, for $\mathbb{C}_p$) *Let $f(X) \in \mathfrak{O}[X]$ be a polynomial with coefficients in $\mathfrak{O}$, and assume that there exist polynomials $g_1(X)$ and $h_1(X)$ in $\mathfrak{O}[X]$ such that*

i) $g_1(X)$ is monic

ii) $g_1(X)$ and $h_1(X)$ are relatively prime modulo $\mathfrak{P}$, and

iii) $f(X) \equiv g_1(X)h_1(X)$ (mod $\mathfrak{P}$) (understood coefficient-by-coefficient).

Then there exist polynomials $g(X)$, $h(X) \in \mathfrak{O}[X]$ such that

i) $g(X)$ is monic,

ii) $g(X) \equiv g_1(X)$ (mod $\mathfrak{P}$) and $h(X) \equiv h_1(X)$ (mod $\mathfrak{P}$), and

iii) $f(X) = g(X)h(X)$.

Problem 273 Give a proof of the second form of Hensel's Lemma. The same caution we used for the first version will be necessary, and the relevant δ will be the maximum of the absolute values of the differences between coefficients of $f(X)$ and coefficients of $g_1(X) h_1(X)$.

Problem 274 In the statement of the second form of Hensel's Lemma, we make the assumption that "$g_1(X)$ and $h_1(X)$ are relatively prime modulo $\mathfrak{P}$." Show, using the fact that the residue field $\mathbb{F} = \mathfrak{O}/\mathfrak{P}$ is algebraically closed, that this can be replaced by "$g_1(X)$ and $h_1(X)$ have no common roots in $\mathbb{F}$."

Problem 275 In this version of Hensel's Lemma, can we conclude that $\deg g(X) = \deg g_1(X)$?

6.2   Deeper Results on Polynomials and Power Series

The main goal of this section is to prove a theorem that has become known as the "p-adic Weierstrass Preparation Theorem." This is, of course, a p-adic version of a classical theorem due to Weierstrass which dealt with power series in several variables and is an important tool in the theory of functions of several complex variables. One can find the statement in many texts; for example, see [Vit89, p. 22]. There are also versions of the theorem which apply to formal power series in several variables; these versions are useful in algebraic geometry. For this version, see, for example, [ZS75, Vol. 2, VII.1, Thm. 5]. The p-adic version gives fundamental information on p-adic functions defined by power series. Our account of this theorem follows the one in [Cas86].

We will approach power series by way of polynomials. In other words, we will want to think of a power series as a limit of polynomials, and our results will be proved first for polynomials, then for power series.

While we will constantly keep $\mathbb{C}_p$ in mind, the results we will obtain are true, interesting, and useful when we work over other fields, too. In fact, since some of them describe how polynomials factor, they are especially interesting when the field is not algebraically closed. On the other hand, we will often want to interpret what the theorems say in terms of the roots of the polynomials, in which case it's most convenient to place ourselves in $\mathbb{C}_p$. So we'll switch back and forth between these two situations. Just to fix notation, let's let K be some extension of $\mathbb{Q}_p$ which is complete; in practice, we will always be working either with a finite extension of $\mathbb{Q}_p$ or with $\mathbb{C}_p$. Let's write $\mathcal{O}$ for the valuation ring $\mathcal{O} = \{x \in K : |x| \le 1\}$, $\mathfrak{p}$ for its maximal ideal, and $\mathbb{F}$ for the residue field. It might be good to remind ourselves that $\mathbb{F}$ is a finite field if K is a finite extension of $\mathbb{Q}_p$ and that it is the algebraic closure of $\mathbb{F}_p$ when $K = \mathbb{C}_p$.

The first step is to define absolute values (or norms) in the spaces we are interested in studying. Let's look first at polynomials. The most obvious way to define a norm on the space of polynomials is to simply look at the coefficients. This works, but we actually want to do something a little more subtle.

Suppose we are interested in understanding how the values of a polynomial

$$f(X) = a_0 + a_1 X + a_2 X^2 + \cdots + a_n X^n$$

vary when we plug in numbers belonging to the closed ball of radius c around the origin. Then, if $x \in \overline{B}(0, c)$, we have $|x| \le c$, and

$$
\begin{aligned}
|f(x)| &= |a_0 + a_1 x + a_2 x^2 + \cdots + a_n x^n| \\
&\le \max\{|a_0|, |a_1 x|, |a_2 x^2|, \ldots, |a_n x^n|\} \\
&\le \max\{|a_0|, |a_1|c, |a_2|c^2, \ldots, |a_n|c^n\} \\
&\le \max_i |a_i| c^i
\end{aligned}
$$

If $c = 1$, this last number is just the "obvious" measure of the size of $f(X)$: the absolute value of the largest coefficient. For other values of c, it turns out that we may still use this number as a good measure of the size of $f(X)$.

Theorem 6.2.1 *Let $c > 0$ be an arbitrary positive real number. Define a function $\| \ \|_c : K[X] \longrightarrow \mathbb{R}_+$ as follows: for each polynomial*

$$f(X) = a_0 + a_1 X + a_2 X^2 + \cdots + a_n X^n,$$

set

$$\|f(X)\|_c = \max_i |a_i| c^i.$$

Then we have

i) $\|f(X)\|_c = 0$ *if and only if $f(X)$ is identically zero.*

ii) $\|f(X) + g(X)\|_c \le \max\{\|f(X)\|_c, \|g(X)\|_c\}$.

iii) $\|f(X)g(X)\|_c = \|f(X)\|_c \|g(X)\|_c$.

iv) *If $f(X) = a_0$ is a polynomial of degree zero, then $\|f(X)\|_c = |a_0|$. In other words, $\| \ \|_c$ induces the p-adic absolute value on the constants.*

v) *If $|x| \le c$, then $|f(x)| \le \|f(X)\|_c$.*

In particular, $\| \ \|_c$ defines a non-archimedean absolute value on the field of rational functions $K(X)$.

PROOF: A lot of this is easy, and can be safely left to the reader. In fact, the first and second statements follow at once from the properties of the p-adic absolute value, and the last two follow at once from the definition of $\| \ \|_c$. The statement about multiplicativity is the hard one.

Let $f(X)$ and $g(X)$ be two polynomials. Write

$$f(X) = a_0 + a_1 X + a_2 X^2 + \cdots + a_n X^n$$
$$g(X) = b_0 + b_1 X + b_2 X^2 + \cdots + b_n X^n$$

where of course we may very well have some zero coefficients (nobody said $f(X)$ and $g(X)$ had the same degree). Then each coefficient of the product $f(X)g(X)$ looks like a sum

$$\sum_{i+j=k} a_i b_j,$$

and we can estimate the absolute value by

$$\left| \sum_{i+j=k} a_i b_j \right| c^k \le \max_{i+j=k} |a_i||b_j| c^k = \max_{i+j=k} \left(|a_i| c^i \right) \left(|b_j| c^j \right),$$

which is certainly less than or equal to

$$\left(\max_i |a_i|c^i\right)\left(\max_j |b_j|c^j\right) = \|f(X)\|_c\|g(X)\|_c.$$

This shows that

$$\|f(X)g(X)\|_c \leq \|f(X)\|_c\|g(X)\|_c.$$

Proving the reverse inequality is much trickier. (Part of the reason is that the estimate above could afford to be really "sloppy," since we were looking only for an upper bound. To get the converse, we must be careful about exactly what gets multiplied with what.) Here is a proof.

Let's begin by giving names to things. Choose I so that

$$|a_I|c^I = \|f(X)\|_c \quad \text{and} \quad |a_i|c^i < \|f(X)\|_c \quad \text{for } i < I.$$

In other words, I is chosen to be the smallest exponent for which $|a_i|c^i$ achieves its maximum. Similarly, choose J so that $|b_J|c^J$ achieves the maximum:

$$|b_J|c^J = \|g(X)\|_c \quad \text{and} \quad |b_j|c^j < \|g(X)\|_c \quad \text{for } j < J.$$

(These are clearly the coefficients to keep track of!)

Now look at the coefficient of X^{I+J} in the product $f(X)g(X)$. It is given by the horrible formula

$$\sum_{i+j=I+J} a_i b_j.$$

We want to estimate each term of this sum. There are three cases to consider:

Suppose i < I. In this case, we know that

$$|a_i|c^i < \|f(X)\|_c \quad \text{and} \quad |b_j|c^j \leq \|g(X)\|_c.$$

Putting these two together gives

$$|a_i b_j| < c^{-i-j}\|f(X)\|_c\|g(X)\|_c = c^{-I-J}\|f(X)\|_c\|g(X)\|_c.$$

(Remember that $i + j = I + J$ in our sum!)

Suppose j < J. This is similar to the previous case; just switch the roles of i and j to get

$$|a_i b_j| < c^{-i-j}\|f(X)\|_c\|g(X)\|_c = c^{-I-J}\|f(X)\|_c\|g(X)\|_c.$$

Finally, if i = I and j = J, we get $|a_I|c^I = \|f(X)\|_c$ and $|b_J|c^J = \|g(X)\|_c$, so we get an equality:

$$|a_I b_J| = c^{-I-J}\|f(X)\|_c\|g(X)\|_c.$$

This means that in the sum

$$\sum_{i+j=I+J} a_i b_j$$

there is one largest term: the one with $i = I$ and $j = J$. Since we are in a non-archimedean field, the absolute value of the sum will be equal to the absolute value of the largest term. In other words,

$$\left| \sum_{i+j=I+J} a_i b_j \right| = c^{-I-J} \|f(X)\|_c \|g(X)\|_c$$

which we can rewrite as

$$\left| \sum_{i+j=I+J} a_i b_j \right| c^{I+J} = \|f(X)\|_c \|g(X)\|_c.$$

Now, to compute $\|f(X)g(X)\|_c$, one has to take the maximum over all coefficients of the product; this last inequality says that the $I+J$-th coefficient already gives something equal to $\|f(X)\|_c \|g(X)\|_c$. The maximum can only be bigger. In other words, we have proved

$$\|f(X)g(X)\|_c \geq \|f(X)\|_c \|g(X)\|_c.$$

Putting this together with the opposite inequality (proved just above), we get what we claimed:

$$\|f(X)g(X)\|_c = \|f(X)\|_c \|g(X)\|_c.$$

As promised, the other statements are left to the reader. $\square$

Problem 276 Prove the remaining statements in the theorem.

Problem 277 Suppose we have two different complete fields $K_1 \subset K_2$. A polynomial $f(X) \in K_1[X]$ also belongs to $K_2[X]$. Show that the value of $\|f(X)\|_c$ does not depend of which ring we put it in. In other words, we have really defined a norm on $\mathbb{C}_p[X]$, and its restriction to $K[X]$ gives the norm for polynomials in $K[X]$.

Problem 278 We can interpret polynomials as $\mathbb{C}_p$-valued functions on K (and also on any extension of K, and even on $\mathbb{C}_p$), and in particular as functions on the closed ball $\overline{B}(0,c) \subset K$. This means we can define a norm on the space of polynomials using the "sup norm" from classical analysis:

$$\|f(X)\| = \sup_{\substack{|x| \leq c \\ x \in K}} |f(x)|.$$

Show that we have $\|f(X)\| \leq \|f(X)\|_c$. Does the equality hold? (Hint: the answer is easier to get if $K = \mathbb{C}_p$.)

Problem 279 Now that we have norms on the space of polynomials, it is not difficult to restate the second form of Hensel's Lemma (Theorem 6.1.2) in terms of the $\| \ \|_1$ norm. Do so. Does a version using the $\| \ \|_c$ norm with $c \neq 1$ work?

The existence of the absolute values $\| \ \|_c$ on the ring of polynomials is a useful tool; for example, it can be used to give simpler proofs of some of the results in the previous chapter (such as Lemma 5.3.7 and the Eisenstein Irreducibility Criterion, Theorem 5.3.11). Some examples will also appear below.

Problem 280 Suppose c is a real number of the form p^r with $r \in \mathbb{Q}$, and let α be an element of $\mathbb{C}_p$ such that $|\alpha| = c$. (Why does one exist?) Show that the map $\phi : \mathbb{C}_p[X] \longrightarrow \mathbb{C}_p[X]$ defined by $X \mapsto \alpha X$ satisfies the condition

$$\|f(X)\|_c = \|\phi(f(X))\|_1.$$

Notice that ϕ is clearly a ring isomorphism. What does this tell us about the relation between $\| \ \|_1$ and the various $\| \ \|_c$?

Problem 281 How would one have to restate the previous problem in order to get something that is true over some finite extension of $\mathbb{Q}_p$?

Problem 282 Can one do anything like the previous problems in the case where c is *not* of the form p^r with $r \in \mathbb{Q}$?

Lemma 6.2.2 *Let* $\| \ \| = \| \ \|_c$ *for some* $c > 0$, *and let* $f(X) \in K[X]$ *be any polynomial. Let*

$$g(X) = b_0 + b_1 X + b_2 X^2 + \cdots + b_N X^N$$

be a polynomial of degree N *with coefficients in* K *satisfying the condition*

$$\|g(X)\| = |b_N|c^N.$$

(In other words, the maximum of the $|b_n|c^n$ *is realized at the very last co-efficient.) Let* $q(X)$ *and* $r(X)$ *be the quotient and the remainder which we obtain when we divide* $f(X)$ *by* $g(X)$, *so that*

$$f(X) = g(X)\,q(X) + r(X) \qquad and \qquad \deg r(X) < N.$$

Then we have both

$$\|f(X)\| \geq \|q(X)\|\,\|g(X)\| \qquad and \qquad \|f(X)\| \geq \|r(X)\|.$$

PROOF: Rather than plod through a million inequalities (which is elementary but difficult), here is an attempt at a conceptual proof. See [Cas86] for a more direct method. What we will do is handle the case $c = 1$ first, then move on to the general case by means of the ideas in the last three problems.

1. **If c=1,** then we have $|b_N| = \max|b_i|$. Multiplying $g(X)$ by some element in K if necessary, we may assume that $|b_N| = 1$. Again, we can multiply the whole equation $f(X) = q(X)\,g(X) + r(X)$ by some element in K in order to get $\max\{\|q(X)\|, \|r(X)\|\} = 1$, which implies $\|f(X)\| \leq 1$. What the theorem says after both reductions is that in fact $\|f(X)\| = 1$.

Well, suppose not. Then every coefficient of $f(X)$ has absolute value less than 1, which means that they belong to the valuation ideal $\mathfrak{p}$. If we use bars to denote reduction modulo $\mathfrak{p}$, we get an equation

$$0 = \bar{f}(X) = \bar{g}(X)\,\bar{q}(X) + \bar{r}(X).$$

But now, since $|b_N| = 1$ (and here is where we seriously use the assumption about $g(X)$),

$$\deg\bar{g}(X) = N > \deg r(X) \geq \deg\bar{r}(X);$$

this forces $\bar{q}(X) = 0$, which then implies $\bar{r}(X) = 0$, which contradicts the assumption that $\max\{\|q(X)\|, \|g(X)\|\} = 1$. This proves the lemma when $c = 1$.

Before we go on, notice that the polynomials $q(X)$ and $r(X)$ necessarily have coefficients in K. Their norms, however, do not depend on the field, as we pointed out above (Problem 277). Since our theorem is about the norms, we might as well assume $K = \mathbb{C}_p$, and we will.

2. **If $c \neq 1$, but c is of the form p^r for $r \in \mathbb{Q}$,** choose an element $\alpha \in \mathbb{C}_p$ with $|\alpha| = c$. Then consider the polynomials $f_1(X) = f(\alpha X)$ and $g_1(X) = g(\alpha X)$. It's easy to see that $\|f_1(X)\|_1 = \|f(X)\|_c$, and similarly for the g's. Applying part 1 to $f_1(X)$ and $g_1(X)$ then gives the inequality we want for the c-norms.

3. **If c is not of the form p^r for $r \in \mathbb{Q}$,** then we have to resort to black magic. Since the numbers of the form p^r are dense in $\mathbb{R}$ (can you see why?), there is a sequence c_i of real numbers such that $c_i = p^{r_i}$ with $r_i \in \mathbb{Q}$ and $c_i \to c$ as $i \to \infty$. Then clearly we have $\|f(X)\|_{c_i} \to \|f(X)\|_c$ as $i \to \infty$, so we can get the estimate we want by using part 2 for each of the c_i and taking the limit. $\square$

Now here's an interesting example of the kind of result we are led to. It is a version for polynomials of the Weierstrass preparation theorem.

Proposition 6.2.3 *Let $c > 0$ be some real number, and $\|\ \| = \|\ \|_c$. Let*

$$f(X) = a_0 + a_1 X + a_2 X^2 + \cdots + a_n X^n$$

be a polynomial in $K[X]$, and suppose that there exists an integer N such that $0 < N < n$ for which we have

$$\|f\| = |a_N|c^N \qquad \text{and} \qquad \|f\| > |a_j|c^j \quad \text{for any } j > N.$$

Then there exist polynomials $g(X)$, of degree N, and $h(X)$, of degree $n - N$, with coefficients in K, such that $f(X) = g(X)\,h(X)$. Furthermore, we have

$$\|g(X)\| = \|f(X)\| \qquad \text{and} \qquad \|h(X) - 1\| < 1.$$

PROOF: For the case $c = 1$, it would not be hard to give a direct proof using Theorem 6.1.2. Instead, we give a general argument (in roughly the same spirit) that works for all choices of c. The point is to start with an approximate factorization and then improve it. If we prove, by induction, that this can always be done, it will produce a convergent sequence of polynomials which, in the limit, give the factorization we want. In order to make the structure of the proof as clear as possible, we separate it into two pieces. The first will describe the general induction step. The next will show how the induction starts. Putting the two pieces together gives the proof.

Step: Let δ be a fixed real number, $\delta < 1$. Suppose that at some stage we have found polynomials $g_i(X)$ and $h_i(X)$ satisfying the following conditions:

  i) $\deg g_i(X) = N$ and $\deg h_i(X) \leq n - N$,

  ii) $\|f(X) - g_i(X)\|_c \leq \delta \|f(X)\|_c$ and $\|h_i(X) - 1\|_c \leq \delta$, and

  iii) $\|f(X) - g_i(X)h_i(X)\|_c \leq \delta^i \|f(X)\|_c$.

(Never mind, for now, where such things might come from.) Let's describe how to get two polynomials that give a still better approximation.

First of all, because $\| \ \|$ satisfies the non-archimedean property ("all triangles are isosceles"), the condition $\|f(X) - g_i(X)\|_c \leq \delta \|f(X)\|_c$ implies (since δ is less than 1) that $\|f(X)\|_c = \|g_i(X)\|_c$. Now we find a way to bring in the estimates in the previous lemma.

If we divide $f(X) - g_i(X)h_i(X)$ by $g_i(X)$, we get

$$f(X) - g_i(X)h_i(X) = q(X)\,g_i(X) + r(X),$$

where $\deg r(X) < N$, and hence $\deg q(X) \leq n - N$. The previous lemma gives inequalities for the absolute values of $q(X)$ and of $r(X)$:

$$\|q(X)\|_c \leq \frac{\|f(X) - g_i(X)h_i(X)\|_c}{\|g_i(X)\|_c} = \frac{\|f(X) - g_i(X)h_i(X)\|_c}{\|f(X)\|_c} = \delta^i < \delta$$

and

$$\|r(X)\|_c \leq \|f(X) - g_i(X)h_i(X)\|_c = \delta^i \|f(X)\|_c < \delta \|f(X)\|_c.$$

Now let

$$g_{i+1}(X) = g_i(X) + r(X) \qquad \text{and} \qquad h_{i+1}(X) = h_i(X) + q(X).$$

We claim that these will do the job.

To begin with, since $\deg r(X) < N$, we will have $\deg g_{i+1}(X) = N$. Similarly, since $\deg q(X) \leq n - N$, we get $\deg h_{i+1} \leq n - N$. This shows our new polynomials still satisfy our first condition.

Next, we have

$$\begin{aligned}
\|f(X) - g_{i+1}(X)\|_c &= \|f(X) - g_i(X) - r(X)\| \\
&\leq \max\{\|f(X) - g_i(X)\|_c, \|r(X)\|\} \\
&\leq \delta\|f(X)\|_c,
\end{aligned}$$

since we have shown that $\|r(X)\|_c < \delta\|f(X)\|_c$. Similarly,

$$\begin{aligned}
\|h_{i+1}(X) - 1\|_c &= \|h_i(X) - 1 + q(X)\|_c \\
&\leq \max\{\|h_i(X) - 1\|_c, \|q(X)\|_c\} \\
&\leq \delta,
\end{aligned}$$

since we have shown that $\|q(X)\|_c < \delta$. This shows that the second condition above still holds.

Finally, we check that this gives a better approximate factorization:

$$\begin{aligned}
f(X) - g_{i+1}(X)h_{i+1}(X) &= f(X) - (g_i(X) + r(X))(h_i(X) + q(X)) \\
&= f(X) - g_i(X)h_i(X) - q(X)g_i(X) - r(X)h_i(X) - r(X)q(X) \\
&= r(X) - r(X)h_i(X) - r(X)q(X) \\
&= r(X)(1 - h_i(X)) - r(X)q(X),
\end{aligned}$$

which gives

$$\begin{aligned}
\|f(X) - g_{i+1}(X)h_{i+1}(X)\|_c &= \|r(X)\|_c\|(1 - h_i(X)) - q(X)\|_c \\
&\leq \delta^i \max\{\|1 - h_i(X)\|_c, \|q(X)\|_c\} \\
&\leq \delta^{i+1}\|f(X)\|_c.
\end{aligned}$$

This means that $g_{i+1}(X)$ and $h_{i+1}(X)$ satisfy all our conditions, with δ^i replaced by δ^{i+1}.

To check that these functions actually form the sort of sequence we want, notice that the inequality $\|q(X)\|_c \leq \delta^i$ translates into

$$\|g_i(X) - g_{i+1}(X)\|_c \leq \delta^i.$$

Similarly, the inequality $\|r(X)\|_c \leq \delta^i\|f(X)\|_c$ translates to

$$\|h_i(X) - h_{i+1}(X)\|_c \leq \delta^i\|f(X)\|.$$

Since $\delta < 1$, these inequalities show that both the sequence of the $g_i(X)$ and the sequence of the $h_i(X)$ are Cauchy sequences with respect to the $\|\ \|_c$ norm.

Start: To start the process, we need to find a $\delta < 1$ and an initial pair $g_1(X)$ and a $h_1(X)$. But those are relatively easy to find. The assumption is that $\|f(X)\|_c = |a_N|c^N$ and that the terms of higher degree are smaller. This means that if we subtract off the part of $f(X)$ up to degree N the remaining polynomial will have smaller norm. In other words, we will have

$$\|f(X) - \sum_{i=0}^{N} a_i X^i\|_c < \|f\|_c.$$

Let δ be a measure of how much smaller:

$$\|f(X) - \sum_{i=0}^{N} a_i X^i\|_c = \delta\|f(X)\|_c.$$

Then, of course, $0 < \delta < 1$.

Then let

$$g_1(X) = \sum_{i=0}^{N} a_i X^i = a_0 + a_1 X + a_2 X^2 + \cdots + a_N X^N$$

and let $h_1(X) = 1$. It is very easy to check that all of our conditions are satisfied.

Convergence: We're almost there. We've got a Cauchy sequence of polynomials *of bounded degree*. And that's enough, by the next problem, to guarantee convergence. Taking the limit, we get $g(X)$ and $h(X)$ as specified by the Proposition. $\qquad\square$

Problem 283 Show that a Cauchy sequence of polynomials of bounded degree is convergent with respect to the $\| \ \|_c$. Show that the hypothesis that the degree is bounded is essential.

Problem 284 The limit of a sequence of polynomials of degree N is a polynomial of degree at most N. Why can we assert that $g(X)$ is actually of degree N?

Problem 285 Show that the polynomial $g(X)$ obtained in our proof has the property that $\|f(X) - g(X)\|_c < \|f(X)\|_c$. This can sometimes be a useful extra bit of information.

Problem 286 Show that the polynomial $g(X) = \sum b_i X^i$ of degree N obtained in the theorem satisfies the condition that $\|g(X)\| = |b_N|c^N$, i.e., the maximum of the $|b_i|c^i$ is attained at the very last term. In particular, this means that we cannot re-apply the Proposition to factor $g(X)$ itself.

Problem 287 What can be said about the speed of the convergence of the sequences of polynomials $g_i(X)$ and $h_i(X)$?

Problem 288 Suppose $c = 1$. Give a proof of the Proposition for this case that is just a direct application of Theorem 6.1.2.

From polynomials, we go on to power series. Of course, convergence questions now become important. It will be useful to remember that the power series

$$\sum a_n X^n$$

converges for $|x| \leq c$ if and only if $\lim |a_n| c^n = 0$. This suggests that the same idea used for polynomials will make sense here. We first define appropriate subrings of the ring of power series with coefficients in K.

Definition 6.2.4 *Let $c > 0$ be an arbitrary positive real number. We define A_c to be the ring of power series $\sum a_n X^n \in K[[X]]$ which satisfy the condition $\lim |a_n| c^n = 0$.*

Notice that, if $f(X) \in A_c$, then $f(X)$ converges for x in the closed unit ball of radius c. The next couple of problems check that this works as advertised.

Problem 289 Show that A_c is indeed a ring, and that it is also a vector space over K.

Problem 290 We have avoided indicating the base field in the notation for A_c so that the notation does not become too heavy. For this problem, however, write $A_c(K)$ for the ring we get when the field of coefficients is K. It's clear that if $K_1 \subset K_2$, then $A_c(K_1) \subset A_c(K_2)$. Show that in fact we have

$$A_c(K_1) = A_c(K_2) \cap K_1[[X]].$$

(In other words: the fact that the series is in A_c is independent of the field to which we think its coefficients belong.) We will use this fact, as before, to move up and down between smaller and bigger fields.

Problem 291 Suppose c can be written as a rational power of p, i.e., that there exists $r \in \mathbb{Q}$ such that $c = p^r$. Show that a power series $f(X)$ belongs to A_c if and only if it converges in the closed ball in $\mathbb{C}_p$ with center 0 and radius c. Is the same true without the assumption on c?

Problem 292 If $c_1 > c_2$, show that $A_{c_1} \subset A_{c_2}$.

Now we put norms on our spaces:

Theorem 6.2.5 *Let $c > 0$ be an arbitrary positive real number. Define a function $\| \; \|_c : A_c \longrightarrow \mathbb{R}_+$ as follows: for each power series*

$$f(X) = a_0 + a_1 X + a_2 X^2 + \cdots + a_n X^n + \ldots$$

belonging to A_c, set

$$\|f(X)\|_c = \max_n |a_n| c^n.$$

Then we have

i) $\|f(X)\|_c = 0$ *if and only if* $f(X)$ *is identically zero.*

ii) $\|f(X) + g(X)\|_c \le \max\{\|f(X)\|_c, \|g(X)\|_c\}.$

iii) $\|f(X)g(X)\|_c \le \|f(X)\|_c \|g(X)\|_c.$

iv) $\| \ \|_c$ *induces the p-adic absolute value on the constant power series.*

v) *If* $|x| \le c$, *then* $|f(x)| \le \|f(X)\|_c.$

PROOF: This is all easy. Notice that the definition makes sense because we know that $|a_n|c^n$ tends to zero as n grows, and hence there must be a maximum. □

Problem 293 Prove the theorem. (It is all very straightforward.)

Problem 294 The game we played above relating the sup-norm for functions on $\overline{B}(0, c)$ with the c-norm on the polynomials makes sense also for power series in A_c. Does anything change?

Problem 295 Once again, the norm does not depend on the base field: show that if $K_1 \subset K_2$ then $\| \ \|_c$ on $A_c(K_2)$ restricts to $\| \ \|_c$ on $A_c(K_1)$. In particular, when we are interested only in computing the norm of some power series, we might as well think of it as having coefficients in $\mathbb{C}_p$.

Problem 296 Suppose $K = \mathbb{C}_p$. Use the idea we used above for polynomials to show that if c is of the form p^r for some rational number r then there is an isomorphism $\phi : A_c \longrightarrow A_1$ which is an isometry, i.e., which satisfies $\|\phi(f(X))\|_1 = \|f(X)\|_c$. What happens when c is not of this form?

Problem 297 Suppose $c_1 > c_2$. Consider the map

$$\iota : A_{c_1} \longrightarrow A_{c_2}$$

that maps each $f(X)$ to itself (this makes sense because, as we showed in a previous problem, $A_{c_1} \subset A_{c_2}$ in this case.). Give A_{c_1} the topology defined by $\| \ \|_{c_1}$ and give A_{c_2} the topology defined by $\| \ \|_{c_2}$. Is the map ι continuous?

We are now ready to begin to work toward the proof of the main result in this section, the Weierstrass Preparation Theorem. This can be viewed as a direct extension of Strassman's Theorem from Chapter 4. The goal is to get a very close relation between functions defined by power series and functions defined by polynomials.

We will work with the norms described above, but at first will stick to $c = 1$. Hence, we will be working with power series which converge in the closed unit ball around the origin, i.e., for any x such that $|x| \le 1$. A series $\sum a_n X^n$ will have this property if and only if $a_n \to 0$ as $n \to \infty$, and this is what we assume. (Notice that this was also what we needed for Strassman's Theorem.)

Theorem 6.2.6 (p-adic Weierstrass Preparation Theorem) *Let*

$$f(X) = \sum a_n X^n$$

be a power series with coefficients in K such that $a_n \to 0$ as $n \to \infty$, so that $f(x)$ converges for $x \in \mathcal{O}$. Let N be the number defined by the conditions

$$|a_N| = \max |a_n| \quad and \quad |a_n| < |a_N| \quad for \; all \; n > N.$$

Then there exists a polynomial

$$g(X) = b_0 + b_1 X + \cdots + b_N x^N$$

of degree N and with coefficients in K, and a power series

$$h(X) = 1 + c_1 X + c_2 X^2 + \ldots$$

with coefficients in K, satisfying:

i) $f(X) = g(X)\,h(X)$,

ii) $|b_N| = \max |b_n|$, i.e., $\|g(X)\|_1 = |b_N|$,

iii) $\lim\limits_{n \to \infty} c_n = 0$, so that $h(x)$ converges for $x \in \mathcal{O}$ (i.e., $h(X) \in A_1$),

iv) $|c_n| < 1$ for all $n \geq 1$, i.e., $\|h(X) - 1\|_1 < 1$, and

v) $\|f(X) - g(X)\|_1 < 1$

In particular, $h(X)$ has no zeros in $\mathcal{O}$.

This clearly is closely related to Strassman's Theorem. In fact, since $h(X)$ has no zeros in $\mathcal{O}$, it is clear that the zeros of $f(X)$ in $\mathcal{O}$ are exactly the same as the zeros of $g(X)$. Since $g(X)$ is a polynomial of degree N, there are at most N of these, and we get Strassman's Theorem. If we move to $\mathbb{C}_p$ we can say more: since $\mathbb{C}_p$ is algebraically closed, we get that, counting multiplicities, $g(X)$ has exactly N zeros in $\mathbb{C}_p$, and the condition on its coefficients means that all of them are in $\mathcal{O}$ (see the problem below). So we know that, counting multiplicities, $f(X)$ has exactly N zeros in $\mathcal{O}$, which gives a stronger version of Strassman's Theorem.

Problem 298 Suppose the polynomial $g(X) = b_0 + b_1 X + \cdots + b_N X^N$ satisfies the condition in the theorem: $|b_N| = \max |b_n|$. Show that if $g(\alpha) = 0$, then $|\alpha| \leq 1$.

Proving the Weierstrass Preparation Theorem will take a while, and will require some effort. We will do it by means of a series of lemmas of various kinds. To begin to set everything up, recall that A_1 is the ring of power series that converge in $\mathcal{O}$; in other words, a power series

$$f(X) = \sum a_n X^n = a_0 + a_1 X + a_2 X^2 + \ldots$$

belongs to A_1 if and only if $a_n \to 0$ as $n \to \infty$. We already know that defining

$$\|f(X)\| = \|f(X)\|_1 = \max_n |a_n|$$

gives a norm on A_1 (we will drop the subscript 1, since this is the only norm we'll be working with in this proof). The first step of the proof is to prove that A_1, with this norm, has very nice properties.

Lemma 6.2.7 A_1 *is complete with respect to the norm* $\| \ \|$.

PROOF: We have to show that a Cauchy sequence in A_1 (with respect to the norm $\| \ \|$) converges. So consider a sequence of power series

$$f_i(X) = a_{i0} + a_{i1}X + a_{i2}X^2 + a_{i3}X^3 + \dots$$

Saying this sequence is Cauchy amounts to saying that for each $\varepsilon > 0$ there exists an M such that we have $\|f_i(X) - f_j(X)\| < \varepsilon$ whenever $i, j > M$. Translating that inequality, we get that

$$\max_n |a_{in} - a_{jn}| < \varepsilon \qquad \text{whenever } i, \, j > M$$

which certainly implies that

$$|a_{in} - a_{jn}| < \varepsilon \qquad \text{for each } n, \text{ whenever } i, \, j > M$$

In other words, each of the sequences $(a_{in})_i$ is Cauchy. Since K is complete, that means they are all convergent.

So, for each n, let

$$a_n = \lim_{i \to \infty} a_{in},$$

and consider the series

$$g(X) = a_0 + a_1 X + a_2 X^2 + a_3 X^3 + \dots$$

We obviously want to say that it is the limit of the sequence of series. To see that, we need two things: first, we need to estimate

$$\|f_i(X) - g(X)\| = \max_n |a_{in} - a_n|$$

and show that it goes to zero; next, we need to show that $g(X)$ is actually in A_1.

The first part is easy: we know that if $i, \, j > M$ we have $|a_{in} - a_{jn}| < \varepsilon$ for every n. Letting $j \to \infty$, it follows that if $i > M$ we have $|a_{in} - a_n| \leq \varepsilon$ for all n, which means that

$$\|f_i(X) - g(X)\| = \max_n |a_{in} - a_n| \leq \varepsilon,$$

so that $f_i(X) \to g(X)$ with respect to $\| \ \|$.

For the second part, we use what we have just proved. For $i > M$, we know that $|a_{in} - a_n| < \varepsilon$ for every n. Now, since $f_i(X) \in A_1$, we know that $a_{in} \to 0$ as $n \to \infty$, i.e., that for each i there exists an M_i such that $|a_{in}| < \varepsilon$ for all $n > M_i$. Choose any $i > M$. Then if n is greater than the corresponding M_i we have

$$|a_n| \leq |a_{in} - a_n| + |a_{in}| < \varepsilon + \varepsilon.$$

It follows that $a_n \to 0$ as $n \to \infty$.

Thus, $f_i(X) \to g(X)$ and $g(X) \in A_1$; since this works for any Cauchy sequence of power series, it shows that A_1 is complete.   □

Problem 299 Go through that proof and make sure it works as advertised. Make sure you understand the different roles of i and n.

Problem 300 Consider the sequence $f_n(X) = 1 + X + X^2 + \cdots + X^n$. Clearly all the $f_n(X)$ belong to A_1. Does this sequence converge?

Problem 301 Is A_c complete with respect to the norm $\| \ \|_c$?

The complete space A_1 contains the polynomials as subspace. It is natural to guess that they are in fact a dense subspace. This does turn out to be the case:

Lemma 6.2.8 *The space of polynomials $K[X]$ is dense in A_1.*

PROOF: We need to show that any power series is the limit of a sequence of polynomials. Let

$$f(X) = a_0 + a_1 X + a_2 X^2 + a_3 X^3 + \ldots$$

be a power series in A_1, so that $a_n \to 0$ as $n \to \infty$. We need to get a sequence of polynomials which approximate $f(X)$ (with respect to the $\| \ \|_1$ norm). The obvious choice is to take the truncations of $f(X)$. So let

$$f_0(X) = a_0$$
$$f_1(X) = a_0 + a_1 X$$
$$f_2(X) = a_0 + a_1 X + a^2 X^2$$
$$\ldots$$
$$f_k(X) = a_0 + a_1 X + a_2 X^2 + \cdots + a_k X^k$$

Then we have

$$\|f(X) - f_k(X)\|_1 = \max_{n>k} |a_n|,$$

which tends to zero because $a_n \to 0$. Then $f_k(X) \to f(X)$, so that $f(X)$ is a limit of polynomials.   □

Problem 302 Check that if we have $a_n \to 0$, then also

$$\lim_{k \to \infty} \max_{n > k} |a_n| = 0.$$

Problem 303 Will this proof work if we replace $\| \ \|_1$ by $\| \ \|_c$?

We will often use the fact that the polynomials are dense in the space of convergent power series to prove things about power series by using facts about polynomials. The main issue in such a proof will be to check that the properties we are interested in are preserved when taking limits. The next lemma is an example of this: it is a version for series of a lemma we proved for polynomials:

Lemma 6.2.9 *Let $f(X) \in A_1$ be a power series converging in the closed unit disk, and let*

$$g(X) = b_0 + b_1 X + \cdots + b_N X^N$$

be a polynomial with coefficients in K satisfying

$$|b_N| = \max_i |b_i|.$$

Then there exist a power series $q(X) \in A_1$ and a polynomial $r(X) \in K[X]$, of degree less than N, such that

$$f(X) = g(X) \, q(X) + r(X)$$

where $q(X)$ and $r(X)$ satisfy

$$\|f(X)\| \geq \|g(X)\| \, \|q(X)\| \qquad and \qquad \|f(X)\| \geq \|r(X)\|.$$

PROOF: The idea of the proof is to use the statement for polynomials to obtain the statement for series, using the fact that any power series is the $\| \ \|_1$-limit of polynomials. So let $f_k(X)$ be a sequence of polynomials converging to $f(X)$. By Lemma 6.2.2, one can find polynomials $q_k(X)$ and $r_k(X)$ such that

$$f_k(X) = g(X) \, q_k(X) + r_k(X) \qquad and \qquad \deg r(X) < \deg g(X)$$

and which satisfy the conditions

$$\|f_k(X)\| \geq \|q_k(X)\| \, \|g(X)\| \qquad and \qquad \|f_k(X)\| \geq \|r_k(X)\|.$$

We need to show that as $k \to \infty$ the sequences $q_k(X)$ and $r_k(X)$ converge. Since we have already shown that the space A_1 is complete, what we need to do is show that these sequences are Cauchy.

To see that, consider the equations

$$f_k(X) = g(X) \, q_k(X) + r_k(X) \qquad and \qquad f_{k+1}(X) = g(X) \, q_{k+1}(X) + r_{k+1}(X)$$

Subtracting one from the other gives

$$f_{k+1}(X) - f_k(X) = g(X)\left(q_{k+1}(X) - q_k(X)\right) + \left(r_{k+1}(X) - r_k(X)\right).$$

Now, since both $r_k(X)$ and $r_{k+1}(X)$ have degree less than N, so does their difference. What that means is that $q_{k+1}(X) - q_k(X)$ is the quotient and $r_{k+1}(X) - r_k(X)$ is the remainder when we divide $f_{k+1}(X) - f_k(X)$ by $g(X)$. Using Lemma 6.2.2 yields the estimates

$$\|q_{k+1}(X) - q_k(X)\| \le \|g(X)\|^{-1}\|f_{k+1}(X) - f_k(X)\|$$

and

$$\|r_{k+1}(X) - r_k(X)\| \le \|f_{k+1}(X) - f_k(X)\|.$$

Finally, remember that the sequence $f_k(X)$ is convergent, hence Cauchy, so that

$$\lim_{k \to \infty} \|f_{k+1}(X) - f_k(X)\| = 0.$$

It follows that

$$\lim_{k \to \infty} \|r_{k+1}(X) - r_k(X)\| = \lim_{k \to \infty} \|q_{k+1}(X) - q_k(X)\| = 0,$$

which, since the norm is non-archimedean, means that both sequences are Cauchy, hence convergent. Letting $r(X) = \lim r_k(X)$ and $q(X) = \lim q_k(X)$ gives the equation we want; furthermore, since each $r_k(X)$ is a polynomial of degree less than N, so is $r(X)$. Finally, the estimates on the norms are clearly preserved by passing to the limit, so that we are done.   □

We now see how to prove the Weierstrass Preparation Theorem. The point is to notice that it is just the power series version of Proposition 6.2.3. The proof of that proposition was a direct application of the Lemma preceding it, whose power series version is the Lemma we have just proved. Hence...

PROOF OF THE WEIERSTRASS PREPARATION THEOREM: Mimic the proof of Proposition 6.2.3 replacing calls to Lemma 6.2.2 with calls to Lemma 6.2.9.□

Problem 304 Make sure you understand how to prove the Theorem. How does one prove the various statements about $g(X)$ and $h(X)$?

Problem 305 Why is it, in the statement of the Weierstrass Preparation Theorem, that the conditions on the power series $h(X)$ imply that it has no zeros in $\mathcal{O}$?

The power of the Weierstrass Preparation Theorem will only become clear from its applications. To get some idea of how it is used, let

$$f(X) = a_0 + a_1 X + a_2 X^2 + a_3 X^3 + \dots$$

be a power series converging in the closed unit disk, so that $a_n \to 0$ (in other words, $f(X) \in A_1$). Let N be chosen as in the theorem:

$$|a_N| = \max_n |a_n| \qquad \text{and} \qquad |a_N| > |a_j| \quad \text{if } j > N.$$

Then, according to the Theorem, $f(X)$ can be factored as $g(X)\,h(X)$, where $g(X)$ is of degree N and $h(X)$ is a power series with no zeros of absolute value ≤ 1. We want to consider the roots of $g(X)$, so we move, for a while, to $\mathbb{C}_p$. Since $\mathbb{C}_p$ is algebraically closed, we can factor $g(X)$ as a product

$$g(X) = b_0 + b_1 X + b_2 X^2 + \cdots + b_N X^N$$
$$= b_N(X - \alpha_1)(X - \alpha_2)\ldots(X - \alpha_N),$$

where $\alpha_1, \alpha_2, \ldots \alpha_N$ are the roots of $g(X)$ (counted with multiplicities). This shows that $f(X)$ will have exactly N zeros in $\mathfrak{O}$, counted with multiplicities, and also gives a precise sense to the "multiplicity" of a zero of a p-adic power series converging on $\mathfrak{O}$: it is just the multiplicity of that zero in the polynomial appearing in the Weierstrass factorization.

Problem 306 This problem (taken from [Cas86]) gives an alternative definition for the multiplicity of a zero. Let $f(X) \in A_1$ be a power series converging on $\mathfrak{O}$. Consider the successive derivatives $f(X)$, $f'(X)$, $f''(X)$, ..., $f^{(n)}(X)$. Show that for any $x \in \mathfrak{O}$ there must exist an n such that

$$f(x) = f'(x) = \cdots = f^{(n-1)}(x) = 0 \qquad \text{but} \qquad f^{(n)}(x) \neq 0.$$

Show that n is equal to the multiplicity, as defined above, of x as a zero of $f(X)$. Conclude that the sum of the multiplicities of all the zeros is exactly N. Why is Cassels' definition nicer than the one given above?

Just as we did for Strassman's Theorem, we can easily apply the Weierstrass Preparation Theorem to functions defined on bigger or smaller balls around zero by scaling the variable appropriately.

Problem 307 Let $c = p^r$ for some $r \in \mathbb{Q}$. Let $f(X)$ be a power series converging in the closed ball of radius c around zero. Explain how to use the Weierstrass Preparation Theorem to count the number of zeros of $f(X)$. Does anything change if we take more general values for c?

While the previous problem shows that the Weierstrass Preparation Theorem, as given above, can be applied in a large number of situations, it is tidier to find a version that applies not only to the $\|\ \|_1$ norm but also to the other norms $\|\ \|_c$. The statement is not hard to find:

Theorem 6.2.10 (p-adic Weierstrass Preparation Theorem) *Let c be a positive real number, and let*

$$f(X) = \sum a_n X^n$$

be a power series with coefficients in K such that $|a_n|c^n \to 0$ as $n \to \infty$, so that $f(x)$ converges for $x \in \overline{B}(0,c)$. Let N be the number defined by the conditions

$$|a_N|c^N = \max_n |a_n|c^n = \|f(X)\|_c \quad \text{and} \quad |a_n|c^n < |a_N|c^N \text{ for all } n > N.$$

Then there exists a polynomial

$$g(X) = b_0 + b_1 X + \cdots + b_N x^N$$

of degree N and with coefficients in K, and a power series

$$h(X) = 1 + d_1 X + d_2 X^2 + \ldots$$

with coefficients in K, satisfying:

i) $f(X) = g(X) h(X)$,

ii) $|b_N| c^N = \max |b_n| c^n$, *so that* $\|g(X)\|_c = |b_N| c^N$,

iii) $h(X) \in A_c$,

iv) $|d_n| c^n < 1$ *for all $n \geq 1$, so that* $\|h(X) - 1\|_c < 1$, *and*

v) $\|f(X) - g(X)\|_c < 1$.

In particular, $h(X)$ has no zeros in $\overline{B}(0, c)$.

Problem 308 Prove the generalized p-adic Weierstrass Preparation Theorem. (Hint: you will need to generalize Lemma 6.2.9 to arbitrary c; for that, you might imitate the trick used in the proof of Lemma 6.2.2.)

Problem 309 Let

$$f(X) = \log(1 + X) = X - \frac{X^2}{2} + \frac{X^3}{3} + \ldots$$

Count the number of zeros in various balls. (We can clearly divide this by X in order to get a series whose zeroth term is 1. The resulting series converges in the open ball of radius 1, so we need to consider various closed balls of smaller radius.) How many zeros does the series have (in $\mathbb{C}_p$, of course) in the open unit ball around zero?

6.3 Entire Functions

One of the applications of the Weierstrass Preparation Theorem is the description of "entire" p-adic power series, i.e., power series which converge in all of $\mathbb{C}_p$. This is actually quite easy, but offers a nice enough example that we decided it deserved its own section.

For this section, then, let $f(X)$ be a power series which converges in all of $\mathbb{C}_p$. If we write it out,

$$f(X) = a_0 + a_1 X + a_2 X^2 + a_3 X^3 + \ldots$$

we must have $|a_n|c^n \to 0$ for *any* $c \in \mathbb{R}$. We can also write this in terms of v_p as $v_p(a_n) - kn \to +\infty$ for any $k \in \mathbb{Q}$. Dividing by n gives

$$\frac{v_p(a_n)}{n} - k \to +\infty$$

for any k, which just amounts to

$$\frac{v_p(a_n)}{n} \to +\infty.$$

In other words, a power series will be entire if $v_p(a_n)$ tends to infinity better than linearly.

Very well, suppose we have such a power series. If $a_0 = 0$, we can factor out a power of X so that the remaining series has zeroth coefficient not equal to zero. So, for our purposes, we might as well assume that $a_0 \neq 0$; in that case, we can divide by a_0 and assume that the zeroth coefficient is equal to one. So let

$$f(X) = 1 + a_1 X + a_2 X^2 + a_3 X^3 + \ldots$$

and assume that $v_p(a_n)/n \to \infty$, so that $f(X)$ is entire. Our plan to understand the zeros of $f(X)$ is to apply the Weierstrass Preparation Theorem in larger and larger balls around zero.

So begin with the closed unit ball: a straight application of the theorem says that we can factor $f(X)$ as $g_0(X) h_0(X)$ where $g_0(X)$ is a polynomial (whose degree is given precisely in the theorem, but that won't matter all that much here) and where $h_0(X)$ is a power series of the form $1 + b_1 X + b_2 X^2 + \ldots$ with $|b_i| < 1$. Since $\mathbb{C}_p$ is algebraically closed, $g_0(X)$ factors into a bunch of linear terms, which we can write as follows:

$$g_0(X) = \prod_{i=1}^{N}(1 - \lambda_i X),$$

since we are assuming that $a_0 = 1$. Notice that the λ_i are not the roots of $g_0(X)$, but rather the *reciprocals* of the roots of $g_0(X)$; the reason for this particular bit of perverseness will become clear soon. In any case, the upshot is that

$$f(X) = h_0(X) \cdot \prod_{i=1}^{N}(1 - \lambda_1 X),$$

where $\|h_0(X) - 1\|_1 < 1$.

Problem 310 Show that any polynomial $g(X)$ satisfying $g(0) = 1$ can be written in the form

$$g(X) = \prod(1 - \lambda X),$$

where λ runs through the reciprocals of the roots of $g(X)$.

Now what happens when we look at a bigger ball? Consider, say, the closed ball of radius p around the origin. To apply the ($\|\ \|_1$-form[1] of the) Weierstrass Preparation Theorem, we need to change variables: let $f_1(Y) = f(Y/p)$. Then plugging in $x \in \overline{B}(0,p)$ into $f(X)$ amounts to plugging in $y = px$ into $f_1(Y)$, so the roots of $f(X)$ in the ball we are looking at correspond to roots of $f_1(Y)$ in the unit ball. Finally, it's clear that $f_1(Y)$ is still entire (if $f(\alpha)$ converges for every α then so does $f(\alpha/p)$), and that the first coefficient is still equal to 1. Applying the theorem gives

$$f_1(Y) = g_1(Y)(1 + c_1 Y + c_2 Y^2 + \cdots),$$

with, as before, $g_1(Y)$ a polynomial and $|c_i| < 1$. To get back to $f(X)$, we just replace Y by pX to get

$$f(X) = f_1(pX) = g_1(pX)(1 + d_1 Y + d_2 Y^2 + \cdots)$$

with $|d_i| = |p^i c_i| < 1/p^i$. Now $g_1(pX)$, whatever it is, is just another polynomial, whose roots give the roots of $f(X)$ in the closed ball of radius p. (It is therefore divisible by $g_0(X)$; do you see why?) So we can repeat the trick:

$$g_1(pX) = \prod_{i=1}^{N_1} (1 - \lambda_i X),$$

where now the λ_i are the reciprocals of the roots of $f(X)$ in the closed ball of radius p. In other words, we've got $f(X)$ written as

$$f(X) = h_1(X) \cdot \prod_{i=1}^{N_1} (1 - \lambda_i X).$$

The inequalities on the d_i show that we have $\|h_1(X) - 1\|_p < 1$, and also show that $\|h_1(X) - 1\|_1 < 1/p$. Notice that this inequality implies that if $x \in \overline{B}(0,p)$ then we must have $|h_1(x) - 1| < 1$, which implies that $h_1(x) \neq 0$; in other words, the inequality shows that $h_1(X)$ has no zeros in the closed ball of radius p.

Problem 311 Explain why $g_0(X)$ is a factor of $g_1(pX)$.

We can now understand why it's nice to use the reciprocals of the roots: they do two things for us. First, they give a clean way to write a product expression for a polynomial whose independent coefficient is 1 (and whose top coefficient might be anything). Second, and more interesting, notice that as our ball grows, the λ_i get smaller: if the root has absolute value p, say, then $|\lambda_i| = 1/p$, which is cheering if we're looking for convergence.

And we are. The reader can probably see what's coming by now: we work in bigger and bigger disks. The polynomial part gives a longer and longer

---

[1] Using Theorem 6.2.10 would also work well.

product expression (since the disks are nested in each other, the roots that appear for a disk reappear in any bigger disk, so that the product is indeed growing longer rather than just changing). The λ_i that appear in the product expression are reciprocals of roots with larger and larger absolute value, so they get smaller and smaller. The other factor gets closer and closer to 1. In the limit, we get just the product! So we've proved:

Proposition 6.3.1 *Let $f(X) = 1 + a_1X + a_2X^2 + \ldots$ be a power series defining an entire function on $\mathbb{C}_p$. Then $f(X)$ has a finite number of zeros in any closed ball around the origin, and a countable number of zeros in $\mathbb{C}_p$. The reciprocals of these zeros form a sequence λ_i tending to zero, and $f(X)$ can be written as an infinite product*

$$f(X) = \prod_{i=1}^{\infty}(1 - \lambda_i X)$$

(with convergence in the $\| \ \|_c$ metric for any c).

PROOF: We've done it all except for the remark on convergence. What we showed was that the infinite product converged to $f(X)$ in the $\| \ \|_1$ metric. But $f(X)$ is entire, and as usual we can change variables to handle the $\| \ \|_c$ metrics.   □

Problem 312 Have we really "done it all"? Make sure you see that the proof is indeed to be found in the text above.

Problem 313 Explain the cryptic remark at the end of the proof. How would one prove that the product converges to $f(X)$ in the $\| \ \|_c$ topology?

Problem 314 One might also want to understand in what sense the functions given by the partial products converge to the function defined by $f(X)$. Show that the convergence is uniform in any closed ball around zero (this is easy if you know about uniform convergence). One almost wants to say that the convergence is "uniform on compact sets"... if it weren't for the slight detail that closed balls in $\mathbb{C}_p$ are not compact!

Passing from power series whose initial term is 1 to general power series is easy:

Corollary 6.3.2 *Let $f(X)$ be a power series defining an entire function on $\mathbb{C}_p$. Then $f(X)$ can be written as an infinite product*

$$f(X) = aX^r \prod_{i=1}^{\infty}(1 - \lambda_i X),$$

where $a \in \mathbb{C}_p$, r is an integer, $r \geq 0$, and λ_i ranges through the reciprocals of the nonzero roots of $f(X)$, which form a sequence tending to zero.

This is very similar to, but also simpler than, a classical result about complex entire functions; see, for example, [Ahl79, Chapter 5, Section 2.3]. It is the starting point for any serious study of p-adic entire functions.

Problem 315 We haven't really met any (non-polynomial) entire functions. Can you give an example?

Problem 316 Show that the product expansion can be used to construct entire functions: take a sequence λ_i tending to zero; does it make sense to define a function by

$$f(x) = \prod_{i=1}^{\infty} (1 - \lambda_i x)?$$

For which x does this converge? Can the resulting function be expressed as a power series? What we are aiming for, of course, is a converse of Corollary 6.3.2.)

6.4   Newton Polygons

One of the best ways to understand the theory of polynomials and power series with coefficients in a complete p-adic field K is to introduce the concept of the Newton polygon of a polynomial (and later of a power series also). This gives us a clear geometric picture that encodes much of the information we have collected about the zeros of polynomials and power series.

We begin, once again, by considering polynomials. We will define the Newton polygon and then explore its meaning in a leisurely way. As before, we will work in a field K, which will either be a finite extension of $\mathbb{Q}_p$ or equal to $\mathbb{C}_p$ (in particular, K is complete with respect to the p-adic valuation). So let $f(X) \in K[X]$ be a polynomial. Since we are mostly interested in understanding the zeros of $f(X)$ we may as well factor out any powers of X which divide $f(X)$. In other words, we may assume that $f(0) \neq 0$. Then, dividing through by $f(0)$, we may also assume that $f(0) = 1$.

Thus, we take a polynomial

$$f(X) = 1 + a_1 X + a_2 X^2 + \cdots + a_n X^n$$

with $a_i \in K$. On a set of axes, we plot the points $(0,0)$ and, for each i between 1 and n, $(i, v_p(a_i))$. (There is one caveat: if $a_i = 0$ for some i, it is not clear what $v_p(a_i)$ is to be; we just take it to be $+\infty$, and think of the point as "infinitely high." In practice this just means that we ignore that value of i.) The polygon we want to consider is, in fancy terms, the lower boundary of the convex hull of this set of points. In less fancy terms, we can think of it this way:

i) Start with the vertical half-line which is the negative part of the y-axis (i.e., the points $(0, y)$ with $y \leq 0$).

ii) Rotate that line counter-clockwise until it hits one of the points we have plotted.

iii) "Break" the line at that point, and continue rotating the remaining part until another point is hit.

iv) Continue until all the points have either been hit or lie strictly above a portion of the polygon.

(One may or may not want to think of the polygon as ending with an infinitely long vertical line going upwards; we will prefer to simply cut off the polygon at its last vertex.)

The resulting polygon is called the *Newton polygon* of the polynomial $f(X)$. Notice that, in the same spirit as before, the polygon depends only on the $v_p(a_i)$, which do not depend on which field we think the a_i belong to. In other words, the polygon belongs to the polynomial, rather than to the polynomial as an element of $K[X]$.

It may be that an example helps more at this point than any number of words. Let's take $p = 5$ and consider the polynomial

$$f(X) = 1 + 5X + \frac{1}{5}X^2 + 35X^3 + 25X^5 + 625X^6.$$

The points we want to work with are

$$(0,0) \quad (1,1) \quad (2,-1) \quad (3,1) \quad (5,2) \quad (6,4)$$

(as agreed, we simply ignore the missing term of degree 4, or think of its point as "very, very high up"). Plotting these points gives figure 6.1. The process with the rotating line gives the polygon in figure 6.2.

The first portion of this section will focus on how to extract information about the roots of the polynomial from this polygon. The crucial things in which we will be interested will be:

i) the slopes of the line segments appearing in the polygon—we will call these the "Newton slopes" of $f(X)$;

ii) the "length" of each slope (by which we mean the length of the projection of the corresponding segment on the x-axis);

iii) the "breaks," i.e., the values of i such that the point $(i, v_p(a_i))$ is a vertex of the polygon.

In our example, the slopes are $-1/2$, 1, and 2, of lengths 2, 3, and 1, respectively, and the breaks happen when $i = 0, 2, 5, 6$. Notice that the sum of all the lengths will always be equal to the degree, and that $(0,0)$ and $(n, v_p(a_n))$ will always be vertices. It is also clear from the "rotating line" construction that the slopes will form an increasing sequence.

Figure 6.1: Points for $f(X) = 1 + 5X + \frac{1}{5}X^2 + 35X^3 + 25X^5 + 625X^6$

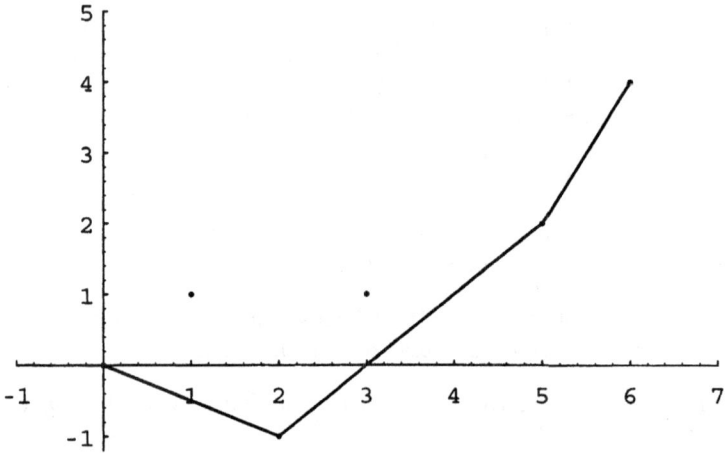

Figure 6.2: Newton polygon for $f(X) = 1 + 5X + \frac{1}{5}X^2 + 35X^3 + 25X^5 + 625X^6$

Problem 317 Let p=5. Work out the Newton polygon of the following polynomials:

i) $1 + X + X^2 + X^3 + X^4 + 2X^5 + 100X^6$

ii) $1 + X + X^2 + X^3 + X^4 + 2X^5 + \frac{1}{100}X^6$

iii) $3 + 5X + 4X^2 + 35X^3 + 40X^4 + 1250X^5 + 100X^6$ (Remember that in our discussion above we normalized things so that $f(0) = 1$. That means you must divide through by 3 before making the polygon... but must you really?)

iv) $3 + 5X + 4X^2 + 35X^3 + 40X^4 + 1250X^5 + 100X^6 + 5X^{10}$ (How does this relate to the previous one?)

Problem 318 Suppose a polynomial $F(X)$ satisfies the condition in the Eisenstein irreducibility criterion over $\mathbb{Q}_p$ (i.e., it is an "Eisenstein polynomial"). Let $f(X)$ be the polynomial obtained by dividing $F(X)$ by whatever number is necessary so that $f(0) = 1$. Describe the Newton polygon of $f(X)$.

Problem 319 It might be useful to generalize the definition in order to remove the condition $f(0) = 1$, and just assume $f(0) \neq 0$. How would the definition change? What would be the relation between the polygons of $f(X)$ and of $af(X)$ (for $a \in K^{\times}$)?

In order to begin to see what information is hidden in the Newton polygon of a polynomial, let's begin by seeing the significance of the breaks. What we want to do is to consider a polynomial $f(X) = 1 + a_1X + a_2X^2 + \cdots + a_nX^n$ and look at its norms $\|f(X)\|_c$ for many different c. Since the norm corresponding to c is essentially the sup-norm on the closed ball of radius c centered at 0 (in $\mathbb{C}_p$), it is easy to see that they satisfy

$$\text{If } c_1 > c_2, \text{ then } \|f(X)\|_{c_1} \geq \|f(X)\|_{c_2}.$$

(It's also very easy to give a direct proof of this.) So, if we start with a very small c and gradually increase it, the norms $\|f(X)\|_c$ will also increase. This observation will help interpret the breaks in the Newton polygon.

Let's look at the first segment of the Newton polygon. If this segment has slope m, it connects the point $(0,0)$ to some other point (i, mi). (So that the first Newton slope is m, and it has length i.) Let's think about what that means. First, it means that there are no points below the line $y = mx$; in other words, $v_p(a_j) \geq mj$ for every j. Second, the point (i, mi) itself tells us that $v_p(a_i) = mi$. Third, the fact that there is a break tells us that the subsequent points are really *above* the line; in other words, $v_p(a_j) > mj$ if $j > i$.

Translating from valuations to absolute values, we get

- $|a_j| \leq p^{-mj} = (p^{-m})^j$ for all j, which we can rewrite as $|a_j|(p^m)^j \leq 1$ for all j,

- $|a_i| = p^{-mi}$, which we can rewrite as $|a_i|(p^i)^m = 1$, and

- $|a_j| < p^{-mj}$ if $j > i$, which we can rewrite as $|a_j|(p^m)^j < 1$ if $j > i$.

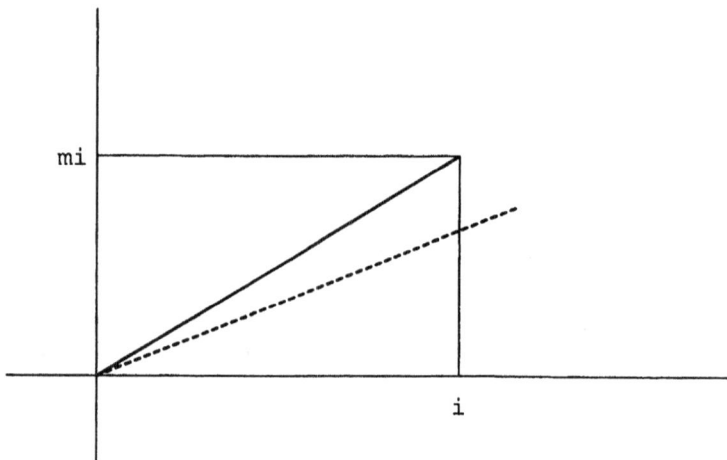

Figure 6.3: The first segment

If we now let $c = p^m$, we can read these conditions in terms of the c-norm. They say:

- $\|f(X)\|_c = 1$, and

- i is the largest integer such that $\|f(X)\|_c = |a_i| c^i$.

In other words, the fact that the first break is at (i, mi) means that if we take $c = p^m$ then $\|f(X)\|_c = 1$ and i is the distinguished number that appears in Proposition 6.2.3. In particular, we see that if i is less than the degree of $f(X)$, then $f(X)$ is divisible by a polynomial of degree i.

Just this is already quite nice. Let's follow Cassels in making the following definition:

Definition 6.4.1 *A polynomial $f(X) \in K[X]$ is called* pure *if its Newton polygon has only one slope. If this slope is m, we will say $f(X)$ is* pure of slope m.

Then we can state what we have just observed as:

Proposition 6.4.2 *Irreducible polynomials are pure.*

PROOF: A break at (i, mi) yields, by the discussion above, a factor of degree i; hence, if there is a break at $i \neq 0$, n, the polynomial is reducible. □

In fact, we can go further, by noticing that a polynomial $h(X) = 1 + b_1 X + b_2 X^2 + \cdots + b_n X^n$ will be pure of slope m, according to the discussion above, exactly when it has the the property that, for $c = p^m$, $\|f(X)\|_c = |b_n| c^n = 1$, i.e., the maximum occurs at the end and is equal to 1.

Problem 320 Prove that a polynomial $h(X) = 1 + b_1 X + b_2 X^2 + \cdots + b_n X^n$ is pure of slope m if and only if we have $\|f(X)\|_{p^m} = |b_n|p^{mn} = 1$.

Using this, we can push the analysis further:

Proposition 6.4.3 *Let*

$$f(X) = 1 + a_1 X + \cdots + a_n X^n \in K[X],$$

and assume the Newton polygon of $f(X)$ has its first break at (i, mi). Then there exist polynomials $g(X)$, $h(X) \in K[X]$ satisfying:

i) $f(X) = g(X)h(X)$,

ii) $g(X)$ has degree i and is pure of slope m,

iii) $h(X)$ has no zeros in the closed ball of radius p^m around 0.

PROOF: This has all been proved already; it's just a matter of putting all the pieces together. ☐

Problem 321 Put all the pieces together.

The connection between "pureness" and polynomial factorization is important. Here is another data-point:

Problem 322 Let $f(X)$ and $g(X)$ both be pure polynomials of slope m. Show that their product is also pure of slope m.

We are still not done thinking of the meaning of the first break... What we still need to do is understand what is the significance of the *slope* of the first segment. That isn't hard to do.

Lemma 6.4.4 *Let $f(X) = 1 + a_1 X + a_2 X^2 + \cdots + a_n X^n \in K[X]$, and assume that the first break of the Newton polygon of $f(X)$ occurs at the point (i, mi). Let c be any positive real number less than p^m. Then we have $\|f(X)\|_c = 1$ and $\|f(X) - 1\|_c < 1$.*

PROOF: If $\|f(X) - 1\|_c < 1$, then we must have $\|f(X)\|_c = 1$ by the usual "all triangles are isosceles" yoga, so we only need to prove the inequality.

A line through the origin with slope $m_1 < m$ (e.g., the dotted line in figure 6.3) passes below all the points on the polygon, touching it only at $(0, 0)$. This means that $v_p(a_j) > m_1 j$ for every $j > 0$. Translating to absolute values, this means that $|a_j| < p^{-m_1 j}$, or $|a_j|/(p^{m_1})^j < 1$ for any $j > 0$. Since $a_0 = 1$, the zeroth coefficient of $f(X) - 1$ is just 0, and therefore also $|a_0 - 1| < 1$. It follows that $\|f(X) - 1\|_c < 1$ for $c = p^{m_1}$. ☐

This already gives one way to characterize the first slope, as the next problem shows:

Problem 323 Show that if the first break happens at (i, mi), and $c_1 > p^m$, then $\|f(X)\|_{c_1} > 1$.

We can read this as saying that $c = p^m$ is the largest value of c such that $\|f(X)\|_c = 1$, which gives an interpretation of the slope of the first segment. The more interesting interpretation, however, has to do with obtaining information about zeros:

Lemma 6.4.5 Let $f(X) = 1 + a_1 X + a_2 X^2 + \cdots + a_n X^n \in K[X]$, and assume that the first break of the Newton polygon of $f(X)$ occurs at the point (i, mi). Let c be any positive real number less than p^m. Then $f(X)$ has no zeros in the closed ball in $\mathbb{C}_p$ of radius c around 0.

PROOF: The previous lemma says that $\|f(X) - 1\|_c < 1$. From that, it follows that for any x such that $|x| \le c$ we have $|f(x) - 1| < 1$, which certainly implies that $f(x) \ne 0$. Thus, $f(X)$ has no zeros in $\overline{B}(0, c)$. $\qquad\square$

So let's put it together: if there's a break at (i, mi), then

- If $c < p^m$, $f(X)$ has no roots (even in $\mathbb{C}_p$) in the closed ball of radius c.

- $f(X)$ factors as the product of a pure polynomial $g(X)$ of slope m and a polynomial which has no roots in the closed ball of radius p^m.

What about the roots of $g(X)$? Well, any root of $g(X)$ is a root of $f(X)$, so we know that $g(X)$ has no roots of absolute value less than p^m. On the other hand, if $\alpha_1, \alpha_2, \ldots, \alpha_i$ are the roots of $g(X)$ (in $\mathbb{C}_p$, with multiple roots listed repeatedly), then we must have

$$g(X) = (1 - \alpha_1^{-1}X)(1 - \alpha_2^{-1}X)\ldots(1 - \alpha_i^{-1}X),$$

and hence the top coefficient of $g(X)$ is equal to $(\alpha_1 \alpha_2 \ldots \alpha_m)^{-1}$. Since $g(X)$ is pure, this must have valuation mi; since we already know $v_p(\alpha_j) < -m$, it follows that all of the α_j have valuation exactly equal to $-m$. Translating back, all the roots of $g(X)$ have absolute value p^m.

Problem 324 Generalize the argument above to show that all the roots of any pure polynomial of slope m have absolute value p^m (or, equivalently, valuation $-m$).

Putting all the pieces together, we get:

Proposition 6.4.6 Let $f(X) = 1 + a_1 X + a_2 X^2 + \cdots + a_n X^n \in K[X]$, and assume that the first break of the Newton polygon of $f(X)$ occurs at the point (i, mi). Then $f(X)$ has no roots with absolute value less than p^m and has exactly i roots (counting multiplicities, in $\mathbb{C}_p$) with absolute value p^m.

Very well, let's move on to the second segment. In other words, let's assume that there are breaks at (i, mi) and at $(k, mi + m'(k - i))$, so that the first slope is m and has length i and the second slope is m' and has length

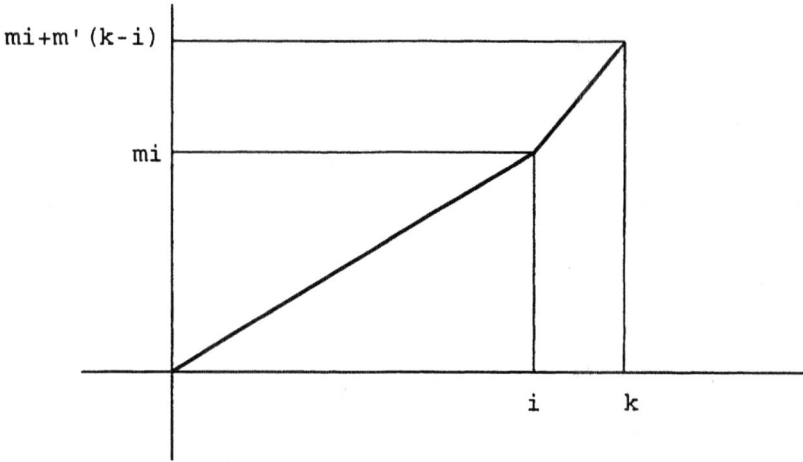

Figure 6.4: The second segment

$k - i$. All that we have obtained about the first segment still works, of course, so what we want to understand is the second segment.

For that, we first translate the fact that the second statement has slope m'. The line through (i, mi) with slope m' has equation

$$y = mi + m'(x - i),$$

and we know the following things.

- (i, mi) and $(k, mi + m'(k - i))$ are on the line; in other words, $v_p(a_i) = mi$ and $v_p(a_k) = mi + m'(k - i)$.

- All the points between i and k are on or above the line. In other words, $v_p(a_j) \geq m'(x - j) + mi$ if $i < j < k$.

- All the points beyond k are strictly above the line. In other words, $v_p(a_j) > m'(x - j) + mi$ if $j > k$. The same inequality holds for $j < i$, as is very easy to see (draw a picture!).

Now we translate all this to absolute values. Our inequalities say that

- $|a_k| = p^{-m'(k-i)-mi} = p^{-m'k}p^{(m'-m)i}$,

- For $i \leq j \leq k$, we have $|a_j| \leq p^{-m'(j-i)-mi} = p^{-m'j}p^{(m'-m)i}$,

- For $j < i$ and for $j > k$, we have $|a_j| < p^{-m'(j-i)-mi} = p^{-m'j}p^{(m'-m)i}$.

Which we rewrite once again by taking $c = p^{m'}$:

- $|a_j|c^j \leq p^{(m'-m)i}$ for all j,

- the equality holds for $j = k$, and

- the inequality is strict for $j > k$.

In other words, we get, for $c = p^{m'}$, that $\|f(X)\|_c = p^{(m'-m)i}$ (and notice that since $m' > m$ this is bigger than 1) and that k is the distinguished number in Proposition 6.2.3 (i.e., the maximum is realized at the degree k term). Using the proposition, we again find a factor $g(X)$, which now need not be pure (why?). In any case, we can go through a process completely analogous to what we did before to conclude that $f(X)$ has exactly k roots in the closed ball of radius $p^{m'}$, i of which have absolute value p^m (we knew that already), and $k - i$ of which have absolute value $p^{m'}$. Of course, we can go through a similar argument at the other breaks, and get roots with bigger absolute values. In the end, we'll get all the roots, and we'll know exactly what their absolute values should be:

Theorem 6.4.7 Let $f(X) = 1 + a_1 X + a_2 X^2 + \cdots + a_n X^n \in K[X]$ be a polynomial, and let $m_1, m_2, \ldots, m_r$ be the slopes of its Newton polygon (in increasing order). Let $i_1, i_2, \ldots, i_r$ be the corresponding lengths. Then, for each k, $1 \leq k \leq r$, $f(X)$ has exactly i_k roots (in $\mathbb{C}_p$, counting multiplicities) of absolute value p^{m_k}.

PROOF: Just repeat the arguments we went through above at each break in the polygon.   □

Notice that since the sum of all the lengths is equal to the degree, the theorem accounts for all the roots of the polynomial.

Problem 325 Fill in the complete details of the analysis of the second break, and convince yourself that the argument will indeed work at the other breaks.

Notice that one of the things that follows from the theorem is the fact that the factor $g(X)$ of $f(X)$ whose existence follows from the existence of a break (together with Proposition 6.2.3) has the same Newton polygon as $f(X)$ up to that break. This shows that the polygons and the factors they tell us about are really tightly connected.

Problem 326 Suppose the Newton polygon of $f(X)$ has breaks, as above, at i and k, with slopes m and m' of length i and $k - i$, respectively. Our discussion shows that there exists a polynomial $g_1(X)$ of degree i which is pure of slope m and divides $f(X)$, and a polynomial $g_2(X)$ of degree k whose Newton polygon coincides with that of $f(X)$ up to k. Show that $g_1(X)$ is a divisor of $g_2(X)$. Let $h(X)$ denote the quotient, so that $g_2(X) = g_1(X)h(X)$. Is $h(X)$ pure?

Problem 327 Suppose the Newton polygon of $f(X)$ starts with a segment of slope m. Let λ be a root of $f(X)$ with absolute value p^m (one exists, by the discussion above). Let $h(X)$ be the polynomial such that $f(X) = (1 - \lambda^{-1}X)h(X)$ (it exists, since λ is a root). Can you relate the Newton polygons of $f(X)$ and of $h(X)$?

Problem 328 Go back to the polygons you drew above, and explain what they tell you about the roots of their polynomials.

Problem 329 Consider the polynomials $f(X) = 1 + X + p^{300}X^{100}$ and $g(X) = 1 + X + p^{100}X^{100}$. These polynomials are "very close," since we have $\|f(X) - g(X)\|_1 = p^{-100}$. Are their Newton polygons close? What is similar in the two polygons? What is different?

Problem 330 The previous problem showed that two polynomials can be very close with respect to the $\| \ \|_1$-norm, and still have different numbers of roots in balls of radius larger than one. What condition would you need in order to be able to conclude that $f(X)$ and $g(X)$ have the same number of roots in the closed ball of radius c?

The moral of the story so far is that Newton polygons codify quite a lot of information about the zeros of polynomials. That should encourage us in the next step, which is to consider the Newton polygon of a power series.

The definition is formally identical: given a power series of the form

$$f(X) = 1 + a_1 X + a_2 X^2 + \cdots + a_n X^n + \cdots$$

we plot the points

$$(i, v_p(a_i)) \qquad \text{for } i = 0, 1, 2, \ldots,$$

ignoring, as before, any points where $a_i = 0$. The Newton polygon of $f(X)$ is again obtained by the "rotating line" procedure. In this case, however, things are more complicated than in the case of polynomials. An example will do more to explain what can happen than any number of generic descriptions.

Consider the power series

$$f(X) = 1 + pX + pX^2 + pX^3 + \cdots + pX^n + \cdots$$

The points we get are

$$(0,0), \quad (1,1), \quad (2,1), \quad (3,1), \ldots$$

Now, clearly the line can sweep unbroken until it is horizontal, but then we have the following curious situation:

- *none* of the points $(i, 1)$ are on our line (so there is nowhere to "break" it), but

- if we rotate the line ever so slightly, some points will be left behind. More precisely, for any positive slope ϵ there exists an i such that $\epsilon i > 1$, so that the point $(i, 1)$ is below the line $y = \epsilon x$.

This means that we must amend our rules for obtaining the Newton polygon to account for this possibility. So here are revised rules:

Start with the vertical half-line which is the negative part of the y-axis (i.e., the points $(0, y)$ with $y \leq 0$). Rotate that line counter-clockwise until one of the following happens:

i) The line simultaneously "hits" infinitely many of the points we have plotted. In this case, stop, and the polygon is complete.

ii) The line reaches a position where it contains only one of our points (the one currently serving as the center of rotation) but can be rotated no further without leaving behind some points. In this case, stop, and the polygon is complete.

iii) The line hits a finite number of the points. In this case, "break" the line at the last point that was hit, and begin the whole procedure again. Notice that the segment beginning at the last point hit may find itself immediately in the situation of case (ii), so that there may be no further change.

This procedure pretty much assumes that the power series is really a series, rather than a polynomial. To handle the case of a polynomial in a unified way, we would have to add one further stopping procedure: if the line reaches the vertical position (after rotating 180 degrees), we stop. The Newton polygon of a polynomial will then end with an infinite vertical segment.

Notice that there are only three ways for the procedure to end:

i) the last segment contains an infinite number of points,

ii) the last segment contains a finite number of points, but can be rotated no further,

iii) there is an infinite sequence of segments of finite length.

Problem 331 Can it happen that "the line can be rotated no further" from the very beginning, so that the Newton polygon gets reduced simply to the negative half of the y-axis?

Let's look at a simple example. Take the power series for

$$f(x) = \frac{1}{1 - pX} = 1 + pX + p^2 X^2 + p^3 X^3 + \cdots + p^n X^n + \cdots$$

The points $(i, v_p(a_i))$ are just

$$(0, 0), \quad (1, 1), \quad (2, 2), \quad (3, 3), \quad \ldots, \quad (i, i), \quad \ldots$$

We are in the first case above, and the polygon comes out to be a line of slope one which contains infinitely many points (figure 6.5).

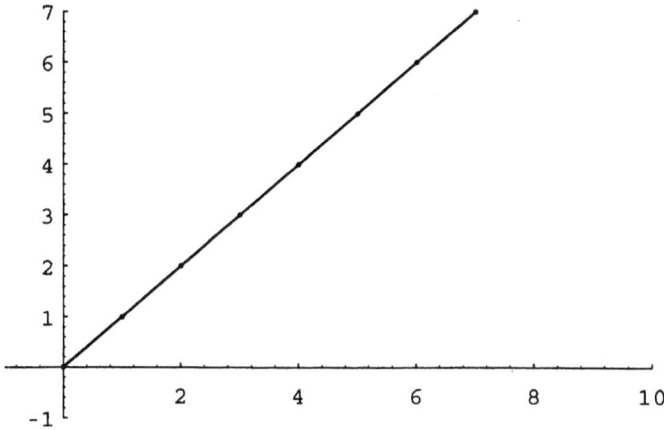

Figure 6.5: Newton polygon for $1 + pX + p^2X^2 + p^3X^3 + \cdots + p^nX^n + \cdots$

To work out the radius of convergence of this series, we need to compute

$$\limsup_{n \to \infty} \sqrt[n]{|a_n|} = \limsup_{n \to \infty} \sqrt[n]{p^{-n}} = 1/p.$$

It follows that the series converges for $|x| < p$ and diverges for $|x| > p$. To handle the remaining case, notice that if $|x| = p$, then clearly $|p^n x^n| = p^{-n}p^n = 1$, and the series does not converge. As we will soon prove, the fact that the Newton polygon ends in (in fact, is) a line of slope 1 is connected with the fact that the region of convergence is the open ball of radius p^1.

We have already seen an example of a series whose Newton polygon falls into case (*ii*) above:

$$f(X) = 1 + pX + pX^2 + pX^3 + \cdots + pX^n + \cdots$$

In this case, the Newton polygon is a horizontal line (see figure 6.6). Notice that in this case

$$\limsup_{n \to \infty} \sqrt[n]{|p|} = \lim_{n \to \infty} p^{-1/n} = 1,$$

so that the radius of convergence is $1 = p^0$. Checking the special case shows that if $|x| = 1$, then the series does not converge, so that once again the region of convergence is an open ball, this time of radius 1.

We clearly need an example where the region of convergence is a closed ball. To get one, lets define a function $\ell(n) = \lfloor \log n \rfloor$, where $\lfloor \cdot \rfloor$ is the "greatest integer" or "floor" function (in words, $\ell(n)$ is the greatest integer which is less than or equal to $\log n$). Then consider the power series

$$1 + \sum_{n=1}^{\infty} p^{\ell(n)} X^n = 1 + X + X^2 + pX^3 + \cdots$$

Figure 6.6: Newton polygon for $1 + pX + pX^2 + pX^3 + \cdots + pX^n + \cdots$

The points we want to plot are

$$(0,0), \quad (1,0), \quad (2,0), \quad (3,1), \quad \ldots, \quad (n, \ell(n)), \quad \ldots$$

If we use the rotating line procedure, we can certainly rotate the line unbroken until it becomes horizontal (at which point it hits the first three points in our list).

We claim the line can rotate no further. To see this, consider a line through the point $(2,0)$ of some small positive slope ε; this will have equation $y = \varepsilon(x - 2)$. We want to see that there are some points $(n, \ell(n))$ below this line; this translates to the assertion that we have $\ell(n) < \varepsilon(n - 2)$ for some n. To see that this is indeed the case, notice that for $n > 2$ we have

$$0 \le \frac{\ell(n)}{n - 2} \le \frac{\log n}{n - 2},$$

and remember that

$$\lim_{n \to \infty} \frac{\log n}{n - 2} = 0.$$

It follows that

$$\lim_{n \to \infty} \frac{\ell(n)}{n - 2} = 0,$$

which means that given any ε we can find an n_0 such that $\ell(n_0)/(n_0 - 2) < \varepsilon$. Rearranging, this says that $\ell(n_0) < \varepsilon(n_0 - 2)$, which is what we wanted to prove: any line of positive slope has some (most, in fact) of our points below it. The conclusion is that the Newton polygon of our series is once again a horizontal line. It is easy to see that the radius of convergence of this series is 1, and that it *does* converge when $|x| = 1$, so that in this case the region of convergence is the closed unit ball.

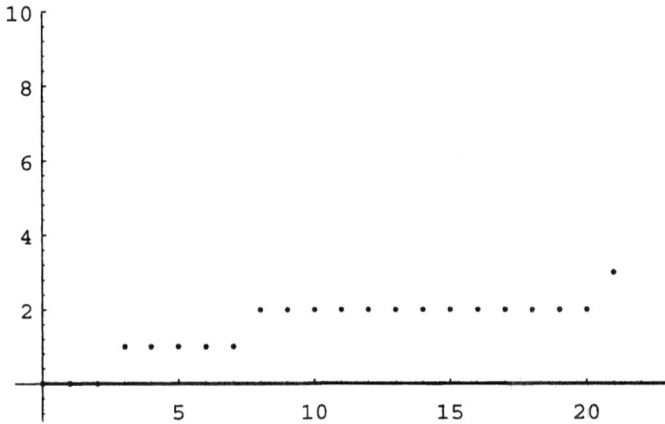

Figure 6.7: Newton polygon for $1 + \sum p^{\ell(n)} X^n$

Problem 332 Modify this last series slightly by changing the first few coefficients:

$$g(X) = 1 + \frac{1}{p} X + \frac{1}{p} X^2 + \sum_{n=3}^{\infty} p^{\ell(n)} X^n.$$

What does the Newton polygon now look like?

The next lemma gives the connection (which we have been hinting at) between the slope of the final segment and the radius of convergence.

Lemma 6.4.8 *Let m be the sup of the slopes appearing in the Newton polygon of a series $f(X) = 1 + a_1 X + a_2 X^2 + \cdots$ (so that m is either a number or is $+\infty$). Then the radius of convergence of the series is p^m (which we understand as $+\infty$ if $m = +\infty$).*

PROOF: Let $|x| = p^b$ with $b < m$. Let's show directly that

$$f(x) = 1 + a_1 x + a_2 x^2 + \cdots + a_n x^n + \cdots$$

converges. For that, we need to prove that $|a_i x^i|$ goes to zero as $i \to \infty$. Since $|x| = p^b$, we have $|a_i x^i| = |a_i| p^{bi}$; translating to valuations, this says $v_p(a_i x^i) = v_p(a_i) - bi$. To show $|a_i x^i|$ goes to zero is the same as to show its valuation goes to infinity, so we want to show that $v_p(a_i)$ gets arbitrarily larger than bi as i grows.

Now superpose the line $y = bx$ on the Newton polygon of the series (see figure 6.8). Since the slope of the polygon eventually becomes larger than b, the polygon eventually passes and then gets farther and farther above the

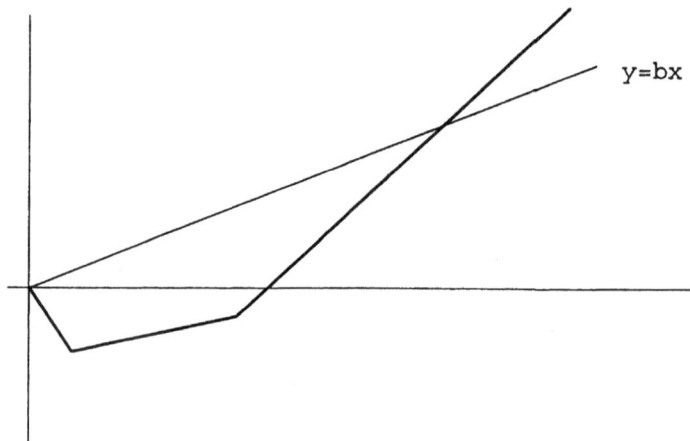

Figure 6.8: A Newton polygon and a line of slope b

line $y = bx$. The points $(i, v_p(a_i))$ are on or above the polygon, so it follows that

$$v_p(a_i) - bi \to \infty \qquad \text{as } i \to \infty,$$

and the series converges.

If $m = +\infty$, we are done. If not, to show that p^m is actually the radius of convergence, we need also to check that if $|x| = p^b$ with $b > m$ the series does not converge. We leave that to the reader (just use the same idea). □

Problem 333 Complete the proof of the lemma.

Of course, we would like to know the exact region of convergence: is it the open or the closed disk of radius p^m? That turns out to be a bit harder to decide, so we'll content ourselves with a partial answer:

Lemma 6.4.9 *Let m be the sup of the slopes appearing in the Newton polygon of a series $f(X) = 1 + a_1 X + a_2 X^2 + \cdots$. Then:*

   i) if the polygon ends in an infinite segment of slope m which contains infinitely many of the points $(i, v_p(a_i))$, then the region of convergence is the open ball of radius p^m.

   ii) if the polygon contains an infinite number of segments of finite length, then the region of convergence is the open ball of radius p^m.

PROOF: Suppose, first, that the polygon ends in an infinite segment of slope m which contains an infinite number of the points $(i, v_p(a_i))$. This means that there is a subsequence $i_1, i_2, \ldots, i_j, \ldots$ such that $v_p(a_{i_j}) = k + mi_j$, where k is some fixed constant. In absolute value notation, this says

$$|a_{i_j}| = p^{-k}p^{-mi_j}.$$

To see that the region of convergence is the open disk of radius p^m, what we need to show is that the series fails to converge if $|x| = p^m$. So suppose $|x| = p^m$. Then, along our subsequence, we would have

$$|a_{i_j}x^{i_j}| = p^{-k}p^{-mi_j}p^{mi_j} = p^{-k}.$$

Since this does not converge to zero, the series $f(x)$ does not converge, and we are done.

Now suppose the polygon has infinitely many line segments. Since the sup of all the slopes is m, all of the segments will have slopes less than m, and the slopes will form an increasing sequence converging to m. To handle this case, we can use an argument similar to the one in the previous lemma: the series will converge at a point x with $|x| = p^m$ if we have

$$\lim_{i \to \infty} |a_i|p^{mi} = 0,$$

or, in valuation notation, if $v_p(a_i) - mi$ goes to infinity as i goes to infinity. This would mean that the points in our polygon get arbitrarily far above the line $y = mx$. But that clearly cannot happen.   □

Problem 334 Convince yourself that it "clearly cannot happen."

Problem 335 Do the proofs we gave apply to the case where the polygon is just the negative y-axis (i.e., where the "rotating line" can't even leave its starting point? What conclusion should we get in that case?

Problem 336 The reader will have noticed that we avoided saying what the exact region of convergence would be if the final segment does not contain infinitely many of the points $(i, v_p(a_i))$. This case is complicated, as the two examples above show. Try to come up with a criterion to decide what happens in this case.

Problem 337 Work out the Newton polygon and the region of convergence for each of the series

i) $1 + pX + p^4X^2 + p^9X^3 + \cdots + p^{n^2}X^n + \cdots$

ii) $1 + X^p + pX^{p^2} + p^2X^{p^3} + \cdots + p^{n-1}X^{p^n} + \cdots$

iii) $1 + X + 2X^2 + 3X^3 + \cdots + nx^n + \cdots$

iv) $1 + X + \frac{1}{4}X^2 + \frac{1}{9}X^3 \cdots + \frac{1}{n^2}X^n + \cdots$

Problem 338 (More examples.) Find the Newton polygons for the power series defining the p-adic logarithm (you'll need to divide by X first in order to get the zeroth coefficient to be 1), for the power series defining the exponential, and for the series for $(1 + x)^{1/2}$ (you'll want to assume $p \neq 2$ for this one).

We now want to go on to obtain power series versions of the results describing how the Newton polygon carries information about the zeros of a power series. The crucial insight, here, will be to notice that the arguments we obtained for polynomials all work without change for power series: all we need to do is replace references to Proposition 6.2.3 to references to the Weierstrass Preparation Theorem (more precisely, to Theorem 6.2.10).

Rather than simply send the reader back to check that our arguments do work, let's re-examine the discussion of the first segment of the Newton polygon. So let

$$f(X) = 1 + a_1 X + a_2 X^2 + a_3 X^3 + \cdots$$

be a power series, and suppose that its Newton polygon has a first segment of length i and slope m. Since we are dealing with series, we need to be careful about what we want to assume about what goes on *after* this initial segment, so let's make the necessary assumptions specific. We assume that:

i) The points $(0,0)$ and (i, mi) are on the polygon, and the segment connecting them is part of the polygon, and

ii) *either* the polygon has a "break" at (i, mi) (i.e., it continues with a different slope) *or* it continues with an infinite segment of slope m which does not contain any more of the points $(j, v_p(a_j))$. In the former case, we know that the series will converge on the closed ball of radius p^m; in the latter case, we will *assume* that it does.

The reason for these assumptions is really clear: we want to relate the segment of slope m to the zeros on the closed ball of radius p^m. The assumptions simply describe the two situations in which the series converges on that closed ball.

One way to think about our special assumptions for the case when there is an infinite line of slope m is that they give a definition for the length of that segment. In other words, if the Newton polygon of a series ends in an infinite portion of slope m we will say the length of that portion is ℓ if ℓ is the distance between the x-coordinates of the first and last of the points $(n, v_p(a_n))$ which are on the line, provided that the series converges on the closed ball of radius p^m. (Recall that the convergence assumption implies that there *is* a last such point.) Otherwise, we may want to say that the length corresponding to slope m is zero.

Once we have made these assumptions we have the following. First, $f(X)$ converges on the closed ball of radius p^m. Next all the points $(j, v_p(a_j))$ are on or above the line $y = mx$, and the ones where $j > i$ are strictly above it. This translates to

- $|a_j|(p^m)^j \le 1$ for all j,

- $|a_i|(p^i)^m = 1$, and

- $|a_j|(p^m)^j < 1$ if $j > i$.

This says that $\|f(X)\|_{p^m} = 1$ and that the maximum is last realized at degree
i. In other words, it puts us exactly in the same position as in the case of polynomials: we can use the Weierstrass preparation theorem to conclude that
there is a polynomial $g(X)$ of degree i and a power series $h(X)$, satisfying
the inequality $\|h(X) - 1\|_{p^m} < 1$, such that $f(X) = g(X)h(X)$. Further-
more, we know that $\|f(X) - g(X)\|_{p^m} < 1$, which implies that $\|g(X)\|_{p^m} =$
$\|f(X)\|_{p^m} = 1$, so that $g(X)$ is pure of slope m. Then, using what we know
about Newton polygons of polynomials, it follows that the zeros of $f(X)$ in
the closed ball of radius p^m coincide with those of $g(X)$, which we already
know are all of absolute value p^m. So we've got the same result as for polynomials: if the first segment is of length i and slope m, then $f(X)$ has exactly
i zeros of absolute value p^m, and no zeros of smaller absolute value.

Thinking about what we just did suggests the following:

Proposition 6.4.10 *Let*

$$f(X) = 1 + a_1 X + a_2 X^2 + a_3 X^3 + \cdots$$

be a power series. Let m_1, m_2, $\ldots$, m_k *be the first k slopes of the Newton
polygon of $f(X)$, and assume that $f(X)$ converges on the closed ball of radius
$c = p^{m_k}$. Let N be the x-coordinate of the right endpoint of the k-th segment
of the Newton polygon. Then there exist a polynomial $g(X)$ of degree N and
a power series $h(X)$ such that*

  i) $f(X) = g(X)h(X)$,

  ii) $\|f(X) - g(X)\|_c < 1$,

  iii) $h(X)$ *converges on the closed ball of radius c,*

  iv) $\|h(X) - 1\|_c < 1$, *and*

  v) *the Newton polygon of $g(X)$ is equal to the portion of the Newton polygon of $f(X)$ contained in the region $0 \leq x \leq N$.*

PROOF: By induction on k:

If $k = 1$, then we have the situation above, and we have already proved
the existence of $g(X)$ and $h(X)$.

Now assume the proposition is true for $k - 1$. Then we know there is
a polynomial $g_1(X)$ which is a factor of $f(X)$ and whose Newton polygon
coincides with the first $k - 1$ segments of the polygon for $f(X)$. We have
$f(X) = g_1(X)h_1(X)$, and we know $h_1(X)$ has no zeros on the closed ball of
radius $p^{m_{k-1}}$. Let's go on, then, to consider the k-th segment.

First of all, the fact that the k-th segment ends at $x = N$ says that for
any $i > N$ the point $(i, v_p(a_i))$ lies above the line of slope m_k through the
point $(N, v_p(a_N))$. As in our analysis of "the second segment" of the Newton
polygon of a polynomial, it is easy to see that this means that $\|f(X)\|_c =$

$|a_N|c^N$ and that $|a_N|c^N > |a_i|c^i$ for any $i > N$. Therefore, we can apply the Weierstrass preparation theorem to get a polynomial $g(X)$. It is then easy to see that $g(X)$ is divisible by $g_1(X)$, and that its Newton polygon coincides with the relevant portion of the Newton polygon of $f(X)$. $\square$

Problem 339 Flesh out the details of the proof. The crucial point is that any zero of $g(X)$ must be either a zero of $g_1(X)$, and we know about those, or a zero of $h_1(X)$ (and hence outside the ball of radius p^{m_k-1}). One needs to show that there are no zeros of absolute value *less* than p^{m_k}, and the rest falls into place.

Corollary 6.4.11 *Let*

$$f(X) = 1 + a_1 X + a_2 X^2 + a_3 X^3 + \cdots$$

be a power series which converges on the closed ball of radius $c = p^m$. Let $m_1, m_2, \ldots, m_k$ be the slopes of the Newton polygon of $f(X)$ which are less than or equal to m, and let $i_1, i_2, \ldots, i_k$ be their lengths. Then, for each j, $f(X)$ has i_j zeros with absolute value p^{m_j}, and there are no other zeros in the closed ball of radius p^m.

PROOF: Clear, because we know this about polynomials, and the proposition says that the relevant part of the Newton polygon of $f(X)$ *is* the Newton polygon of the polynomial $g(X)$. Since $g(X)$ is a factor of $f(X)$ and the quotient $h(X)$ clearly has no zeros in the closed ball of radius p^m, the conclusion follows. $\square$

Problem 340 Is the following version of the last proposition true?
Possible Proposition Let

$$f(X) = 1 + a_1 X + a_2 X^2 + a_3 X^3 + \cdots$$

be a power series which converges on the closed ball of radius $c = p^m$. Let N be the integer defined by the conditions

$$\|f(X)\|_c = |a_N|c^N \qquad \text{and} \qquad |a_n|c^n < |a_N|c^N \quad \text{if } n > N.$$

Then there exist a polynomial $g(X)$ of degree N and a power series $h(X)$ such that

i) $f(X) = g(X)h(X)$,

ii) $\|f(X) - g(X)\|_c < 1$,

iii) $h(X)$ converges on the closed ball of radius c,

iv) $\|h(X) - 1\|_c < 1$, and

v) the Newton polygon of $g(X)$ is equal to the portion of the Newton polygon of $f(X)$ contained in the region $0 \le x \le N$.

Furthermore, all the slopes in this portion of the Newton polygon of $f(X)$ will be less than or equal to m.

6.5 Problems

We've gone about as far as we want to go, but the reader may enjoy exploring further. Here are a few random problems involving p-adic analysis. No hints will be supplied for these (it would spoil the fun!) beyond remarking that some of them are very much harder than others...

Problem 341 In $\mathbb{C}$, it is trivial to see that any analytic function (even any continuous function) is bounded on any closed ball, because closed balls in $\mathbb{C}$ are compact. In $\mathbb{C}_p$, closed balls are no longer compact. Nevertheless, the boundedness result is still true: show that if $f(X)$ is a power series converging on a closed ball of radius r, then $f(X)$ is bounded on $\overline{B}(0,r)$. In fact, show that $f(X)$ has a maximum (rather than just a sup) on $\overline{B}(0,r)$.

Problem 342 Let $f(X)$ be a power series converging on the closed ball of radius r. By the previous problem, $f(X)$ is bounded. Show that

$$\max_{x \in \overline{B}(0,r)} |f(x)| = \max_{|x|=r} |f(x)|.$$

We might want to read this as "the maximum occurs at the boundary," even though we know that the sphere is not the boundary of the closed ball. (This is the p-adic analogue of the "maximum modulus principle.")

Problem 343 Suppose $f_n(X)$ is a family of power series satisfying:

 i) All of the $f_n(X)$ converge in the closed ball of radius $\rho > 1$ around the origin.

 ii) There exists a bound B such that $\|f_n(X)\|_\rho \le B$ for all n.

 iii) There exists a power series $f(X)$ such that the series $f_n(X)$ converge to $f(X)$ with respect to the norm $\| \ \|_1$ (or, what is the same, coefficient-by-coefficient).

Show that $f(X)$ converges in the open ball of radius ρ, and that the $f_n(X)$ converge to $f(X)$ with respect to the norm $\| \ \|_c$ for any $c < \rho$.

Problem 344 How close do two power series need to be in order to allow us to conclude that they have the same number of zeros in the closed ball of radius r around 0? (This question is deliberately open-ended.)

Problem 345 (From [Par84].)

 i) Let $f(X) = 1 - X^{p-1}$, and define

$$m(f,k) = \sup\{|f(x)| : x \in \mathbb{Q}_p, |x| = p^k\}.$$

 Compute $m(f,k)$ for each $k \in \mathbb{Z}$. Does the answer change if we let $x \in \mathbb{C}_p$ instead?

 ii) Find a sequence of integers $h_1, h_2, \ldots, h_k, \ldots$ such that if we set

$$f_k(X) = f(X) \cdot (f(pX))^{h_1} \cdot (f(p^2 X))^{h_2} \cdots (f(p^k X))^{h_k},$$

 then we have

$$\sup\{|f_k(X)| : x \in \mathbb{Q}_p, |x| \le p^k\} = 1.$$

iii) Use this to construct an example of an entire function which is bounded on $\mathbb{Q}_p$. What happens if we go to $\mathbb{C}_p$?

The point, of course, is that in $\mathbb{C}$ there are no non-constant bounded entire functions.

Problem 346 Prove that $2^{p-1} \equiv 1 \pmod{p^2}$ if and only if p divides the numerator of

$$\sum_{j=1}^{p-1} \frac{(-1)^j}{j}.$$

Problem 347 Prove that for every positive integer k, we have

$$\sum_{n=0}^{\infty} n^k p^n \in \mathbb{Q}.$$

(The assertion is that the series converges, and that the sum is a rational number.)

Problem 348 One approach to defining functions on the p-adic numbers which we really have not explored is making direct appeal to the p-adic expansion. Consider for example the function (stolen from [Mah73]) $f : \mathbb{Z}_p \longrightarrow \mathbb{Z}_p$ which maps

$$x = a_0 + a_1 p + a_2 p^2 + a_3 p^3 + \cdots + a_n p^n + \cdots$$

to

$$f(x) = a_0 + a_1 p + a_2 p^4 + a_3 p^9 + \cdots + a_n p^{n^2} + \cdots$$

Is this function continuous? Is it differentiable? Can you extend it to $\mathbb{Q}_p$? To $\mathbb{C}_p$?

Problem 349 (Also from [Mah73], but originally due to Dieudonné; see [Die44].) In the same spirit as the previous problem, consider $g : \mathbb{Z}_p \longrightarrow \mathbb{Z}_p$ which maps

$$x = a_0 + a_1 p + a_2 p^2 + a_3 p^3 + \cdots + a_n p^n + \cdots$$

to

$$g(x) = a_0^2 + a_1^2 p + a_2^2 p^2 + a_3^2 p^3 + \cdots + a_n^2 p^n + \cdots$$

(Notice that this will not be a "p-adic expansion," because the coefficients are not necessarily between 0 and $p - 1$, but it clearly does converge, so that the definition makes sense.) Show that if $p \neq 2$ the function g is continuous but not differentiable on $\mathbb{Z}_p$.

Problem 350 (This problem was proposed in the *American Mathematical Monthly* by Nicholas Strauss and Jeffrey Shallit. A solution by Don Zagier, using 3-adic methods, appeared in the January, 1992 issue.)

If k is a positive integer, let $v(k) = v_3(k)$ be the 3-adic valuation. For each positive integer n, let

$$r(n) = \sum_{i=0}^{n-1} \binom{2i}{i}$$

Prove that

• $v(r(n)) \geq 2v(n)$, and

- $v(r(n)) = v\left(\binom{2n}{n}\right) + 2v(n)$.

Zagier's solution generalizes and extends this statement, and even formulates a conjecture at the end, so make sure to check it out after you've solved the problem.

Problem 351 Suppose f is a continuous function on $\mathbb{Z}_p$. Consider the values

$$a_n = \sum_{k=0}^{n} (-1)^k \binom{n}{k} f(n-k).$$

- Explain the significance of the a_n. (Notice that they depend only on $f(n)$ for n a positive integer.)
- Define a formal series

$$f^*(X) = \sum_{n=0}^{\infty} a_n \binom{X}{n}.$$

  Show that if m is a positive integer, then $f^*(m) = f(m)$.
- Suppose that $a_n \to 0$ as $n \to \infty$. Show that $f^*(x)$ converges uniformly for $x \in \mathbb{Z}_p$, and is a continuous function of x. Conclude that in this case $f^* = f$.
- Show that if f is continuous on $\mathbb{Z}_p$, then $a_n \to 0$ as $n \to \infty$. (This is quite hard.)

This problem gives an approach to the interpolation problem developed by Mahler in [Mah73]. If we know $f(n)$ for n a positive integer and we can show that the a_n tend to zero, then it gives a way of constructing an interpolating function. The last item above is quite difficult to prove; see Mahler's book for a detailed proof.

A   Hints and Comments on the Problems

This appendix contains hints and comments of several kinds for the various problems set in the main text. Expect no complete solutions here; rather, the intention is to provide a jump-off point for a solution, and perhaps to discuss the implications of some of the problems. Some of the comments even suggest further problems! The hints and partial solutions become sketchier as we move toward the latter part of the book, in the expectation that the experience and ability of the reader will increase. As advertised, we do not give hints for the problems in the last section of Chapter 6.

1 The formula for the sum of the geometric series says that

$$1 + a + a^2 + a^3 + a^4 + \cdots = \frac{1}{1-a}$$

provided that $|a| < 1$. This can be used directly for the first expansion. For the third, write $X - 1 = 1 + (X - 2)$ and use $a = -(X - 2)$ in the geometric series.

2 If the expansion is finite, it will certainly become a polynomial after we multiply by $(X - \alpha)^m$, where m is the biggest exponent appearing in a denominator.

3 The sum is easy:

$$\sum_{i \geq n_0} a_i (X - \alpha)^i + \sum_{i \geq n_0} b_i (X - \alpha)^i = \sum_{i \geq n_0} (a_i + b_i)(X - \alpha)^i$$

(why can we assume the two series start at n_0?). The product takes only a little more work:

$$\left(\sum_{i \geq n_0} a_i (X - \alpha)^i \right) \left(\sum_{i \geq n_0} b_i (X - \alpha)^i \right) = \sum_{i \geq 2n_0} c_i (X - \alpha)^i,$$

where the new coefficients are given by

$$c_i = \sum_{i_1 + i_2 = i} a_{i_1} b_{i_2},$$

which is a finite sum because the negative exponents only go back so far. Most of the field properties are easy to check; for the existence of inverses, one has to show that the equations for the coefficients of the product can be solved to find the coefficients of the inverse.

4 Well, x and $-x$ have to add up to zero, so the first digit has to be $-x = (p - a_0) + \ldots$. That way the sum of the two first digits is p, which gives a first digit of zero and a "carry." Now continue.

5 Imitate the definitions for Laurent series, but watch out for carrying. That's tricky, so here are two suggestions of ways to make it easier. One idea, which is in fact the way Hensel did it originally, is to define an "irregular p-adic number" to be *any* expansion of the form

$$a_0 + a_1 p + a_2 p^2 + \cdots + a_n p^n + \ldots,$$

with no restriction on the a_i except that they be non-negative integers. It's then very easy to define the sum and product of irregular p-adic numbers by using the same ideas as with Laurent series. Then we need a theorem that says that any irregular p-adic number can be reduced to a regular p-adic number. The process would be something like this: beginning with a_0, write out each coefficient in base p, and rearrange the series accordingly. What needs to be proved is that this is all well-defined. (A harder question is whether the process can actually be done in a finite amount of time!)

Another option is to work like this: start with a p-adic number x, and factor out a power of p so that we have $x = a_0 + a_1 p + \ldots$ (a_0 may be equal to 0, of course). It's clear that it is enough to define the sum and product of such numbers (because we then factor the powers of p back in). Now, given such an x, let x_n be its truncation at p^n, so that

$$x_n = a_0 + \cdots + a_n p^n.$$

Now note that x_n is an integer, and we know how to add and multiply integers! Then define the sum and the product of two p-adic numbers by the rule

$$(x + y)_n = (x_n + y_n)_n \qquad \text{and} \qquad (x \cdot y)_n = (x_n \cdot y_n)_n$$

(i.e., multiply the truncations, and truncate the result). Once you've checked that all these truncations of $x+y$ and $x\cdot y$ "match," you can put them together to get a p-adic number. You need to check that this number is uniquely defined, i.e., that two p-adic numbers which have the same n-truncations for every n must be equal. This gives the operations. (You've still got to check the field properties!)

Note: what is really going on here is that we want to deduce the operations in $\mathbb{Q}_p$ from operations in $\mathbb{Q}$ (and even in $\mathbb{Z}$); this is the lazy way to do it, since it frees us from having to work out the carrying business explicitly, but it forces a bit of mumbo-jumbo. The difficulty of proving the field properties when we use a formal definition of $\mathbb{Q}_p$ is one of the reasons for the more conceptual theory we'll develop in the next chapter.

6 As the hint says, follow the usual proof: if an expansion is periodic, then multiplying by a power of p and subtracting gives a *finite* expansion, and we are done. For example, if

$$x = a + bp + ap^2 + bp^3 + \dots$$

then

$$p^2 x = ap^2 + bp^3 + ap^4 + bp^5 + \dots$$

so that

$$x - p^2 x = a + bp$$

and hence

$$x = \frac{a + bp}{1 - p^2}$$

which is rational. For "eventually periodic," subtract off the non-periodic part first, then do the same. The converse (rationals have finite or periodic expansions) is also not hard: find an algorithm for computing the p-adic expansion of a rational number a/b, and show that it is periodic.

7 We really have two choices. If, say,

$$x = a_{-2}p^{-2} + a_{-1}p^{-1} + a_0 + a_1 p + a_2 p^2 + \dots$$

we can represent it either as

$$x = a_{-2}a_{-1}.a_0 a_1 a_2 \dots$$

or in reverse order as

$$x = \dots a_2 a_1 a_0 . a_{-1} a_{-2}$$

In the first case, carrying works backward: add the leftmost terms, and carry to the right (of course, in base p). In the second, carrying works as usual (and, in fact, all the usual rules for addition and multiplication work), but numbers are infinitely long to the left.

An additional problem: what happens if we consider exactly this setup, but in base 10? Show that the resulting operations do work, but that one can find two non-zero (infinitely long) numbers whose product is zero, so that the resulting object is not a field.

8 First of all, any ideal J in $\mathbb{C}[X]$ is principal, and will be maximal when its generator is irreducible. Since $\mathbb{C}$ is algebraically closed, the only irreducible polynomials are those of degree one, and we can always divide by the degree one coefficient (an invertible element of $\mathbb{C}[X]$, so the ideal doesn't change) to get a polynomial of the form $X - \alpha$. So, to a maximal ideal J we can attach the number α, and conversely. Part (ii) is even easier: just remember that $f(\alpha) = 0$ if and only if $f(X)$ is divisible by $X - \alpha$. Part (iii) is also standard.

For the rational numbers, just follow the hints as given. The order of the "pole at p" will be the largest power of p dividing the denominator of x. As to whether this is a reasonable thing to do, it turns out to be a very useful point of view in modern algebraic geometry, so I guess it must be "reasonable"... It certainly does make the analogy a little bit tighter.

9 This is standard business; see any basic text on number theory. To do it yourself, note that saying that $x^2 \equiv 25$ (mod p^n) is the same as saying that p^n divides $x^2 - 25 = (x-5)(x+5)$. Then it's a matter of showing that it's not possible for both factors to be divisible by p. For $p = 2, 5$ there can be more than two roots. For example, $X^2 \equiv 25$ (mod 25) has roots 0, 5, 10, 15, 20 (mod 25). Describing the general behavior in the "bad" cases is harder than in the case $p \neq 2, 5$.

10 $X^2 \equiv 49$ (mod 5^n) goes just like the example in the text. Similarly, $X^3 \equiv 27$ (mod 2^n) goes smoothly (there is only one root).

11 Just work it out. For every large enough n you should find *four* roots, only two of which "continue" on to the next n.

12 Standard number theory business. For $n = 1$, you are looking at an equation of degree 2 in a field. Then show that solutions modulo p^n always lift uniquely to solutions modulo p^{n+1}. Alternatively, imitate problem 9.

13 See the previous problem. This can be found in most books, too, but it's easy anyway. If we have a solution a (mod 7^n) then a "lift" $\tilde{a}$ (mod 7^{n+1}) has to be of the form $\tilde{a} = a + x7^n$, with $x = 0$ or 1 or... or 6. Now plug that into the equation, and show that one can always solve (uniquely) for x.

14 One idea is to use truncations, as in 5 above: show that $(x_1^2)_n = (2)_n$ for every n, just by tracing through where we got x_1.

15 If $x = a_0 + a_1 5 + \ldots$ and $x^2 = 2$, then $a_0^2 \equiv 2$ (mod 5).

16 For the negative statement, it's enough to check that $X^2 + 1 = 0$ has no solutions modulo 7. For the positive statement, start from the fact that it *does* have solutions modulo 5, and then use the methods we've been playing with.

17 This is just a generic version of Problem 14; the "truncation" method will work.

18 Read carefully over the last several problems, and write up your methods as a general result.

19 As long as $p \neq 2$, what we have already done solves the problem: one can always find an m which is not a square in $\mathbb{Q}$ but is a quadratic residue modulo p, hence is a square in $\mathbb{Q}_p$. For $p = 2$, consider either cubes, or roots of polynomials of the form $X^2 + X + m = 0$.

20 Repeat the last problem in reverse. For any $p \neq 2$, there is an m which is *not* a quadratic residue modulo p, hence not a square in $\mathbb{Q}_p$, which is therefore not algebraically closed. The same workaround as before handles $p = 2$.

21 The usual proof works: multiply by p, and subtract.

22 I certainly can't, but see Problem 10.10 in [Par84] and its solution.

23 Let $| \ |$ be an absolute value on $\Bbbk$. We have $|0| = 0$ by the definition. The equation $1 = 1 \cdot 1$ forces $|1| = |1| \cdot |1|$. Since $|1|$ is a strictly positive real number, it follows that $|1| = 1$. Now take any element $x \in \Bbbk$, $x \neq 0$. Since $\Bbbk$ is a finite field, there exists an integer q such that $x^q = x$ (we can take q to be equal to the number of elements in $\Bbbk$). Taking absolute values, we get $|x|^q = |x|$; since $|x|$ is real and positive, this forces $|x| = 1$. Thus, $| \ |$ must be trivial.

24 If $a/b = c/d$, then $ad = bc$. By unique prime factorization, the highest power of p that divides ad is just the sum of the highest powers dividing a and d; thus, $v_p(ad) = v_p(a) + v_p(d)$. Similarly, $v_p(bc) = v_p(b) + v_p(c)$. Then, if $ad = bc$, we have

$$v_p(a) + v_p(d) = v_p(ad) = v_p(bc) = v_p(b) + v_p(c).$$

Now rearrange.

25 It's just a matter of factoring: $v_5(400) = 2$, $v_7(902) = 0$, $v_3(123/48) = 0$, $v_5(180/3) = 1$. Try a large number: what is $v_{11}(452, 298)$?

26 First consider the case when x and y are both integers. Write $x = p^a x'$ and $y = p^b y'$ where both x' and y' are not divisible by p. Since we may interchange x and y if necessary, we can assume that $a \leq b$. Then $xy = p^{a+b} x' y'$, which shows (*i*), and

$$x + y = p^a x' + p^b y' = p^a(x' + p^{b-a} y')$$

which shows that $v_p(x + y) \geq a$, and so proves (ii). This proves both statements when x and y are integers. To get it for fractions, let $x = t/q$, $y = r/s$. Then

$$v_p(xy) \quad = v_p\left(\frac{tr}{qs}\right) = v_p(tr) - v_p(qs)$$

$$= v_p(t) + v_p(r) - v_p(q) - v_p(s)$$

$$= v_p\left(\frac{t}{q}\right) + v_p\left(\frac{r}{s}\right)$$

This proves part (i) for the general case. Part (ii) is similar (in other words, do it!).

27 $1/7$, $1/7$, 1, 343, respectively. (Notice that with respect to this absolute value, $3/686$ is big, while 35 is small...)

28 There's nothing much to do here: just straight translation. Remember that the elements of K look like a/b where a and b are in A (and $b \neq 0$), and that $a/b = c/d$ if and only if $ad = bc$. Then follow your nose.

29 First, $v(1) = 0$, so 0 is in the image. If α and β are in the image, then we must have $\alpha = v(x)$ and $\beta = v(y)$ for some non-zero x, $y \in \Bbbk$. But then $\alpha + \beta = v(xy)$ and $-\alpha = v(1/x)$, so that we have a subgroup. In the case of the p-adic valuation, the value is always an integer by definition, so the value group is $\mathbb{Z}$.

30 This is easy to see, since we defined $|p^n| = p^{-n}$, so that $|p^n| \to 0$ as $n \to \infty$. The more a number is divisible by p, the smaller it is in the p-adic world.

31 This is straightforward, since passing from the valuation properties of v_p to the properties of the absolute values is just a matter of taking powers. The obvious conjecture is that it does not matter what value of c is used, in the sense that the resulting absolute values for varying c are "similar" enough that we might as well treat them as being the same. That's exactly what happens. Why $c = p$ is a good choice is more subtle—see ahead for the Product Formula.

32 Yes, it is enough to check for polynomials (extending from polynomials to rational functions is easy—just plug into the formulas). For polynomials, both equations are well-known, if we restate them in terms of the degree. (Notice that the sum of two polynomials of the same degree can have smaller degree, so that after changing signs the $\geq$ is indeed necessary.)

33 A rational function f/g will be small with respect to $|\ |_\infty$ when $v_\infty(f/g)$ is *big*, hence when the degree of g is much bigger than the degree of f. Hence, polynomials are never small. In fact, if f is a polynomial, then $\deg(f) \geq 0$ gives $|f|_\infty \geq 1$.

34 Boring but easy. Just run through the definition of the p-adic absolute value and check that everything works. For concreteness, play with the case where $F = \mathbb{R}$ and $p(t) = 1 + t^2$, or the case $F = \mathbb{C}$, $p(t) = t - 4$.

35 In every case, it turns out to be the trivial absolute value. For $|\ |_\infty$, just notice that any non-zero constant has degree zero, and hence absolute value 1. For the $p(t)$-adic absolute values, notice that the constants in $F[t]$ are not divisible by any irreducible polynomial.

36 Every polynomial of degree n with coefficients in $\mathbb{C}$ has n roots, so that we can always write it as a product of linear terms. Hence, the only irreducible polynomials are the ones of degree one, $p(t) = t - \lambda$. The $p(t)$-adic valuation of a polynomial $f(t)$ just measures the multiplicity of λ as a root of $f(t)$. We are very close indeed to Hensel's original idea.

37 This is very hard, and, as advertised, depends on the choice of the field F. If an archimedean absolute value on $F(t)$ can be found, its restriction to the subfield of constants F will have to be an absolute value on F, and it will have to be archimedean (if you can't see why, wait till the next section). So it can't be done if (a) there are no archimedean absolute values on F, nor if (b) we require that the restriction to F be the (non-archimedean) trivial absolute value.

Here's the sneaky bit: take $F = \mathbb{Q}$, and choose a transcendental number, say π. Since π is not a root of any polynomial over $\mathbb{Q}$, the rings $\mathbb{Q}[\pi]$ and $\mathbb{Q}[t]$ are isomorphic. (Map $\mathbb{Q}[t] \to \mathbb{Q}[\pi]$ by $t \mapsto \pi$; this is obviously onto, and what can possibly be in the kernel?) It follows that the fields $\mathbb{Q}(\pi)$ and $\mathbb{Q}(t)$ are isomorphic. Now, $\mathbb{Q}(\pi)$ is contained in $\mathbb{R}$, so we can restrict the archimedean absolute value on $\mathbb{R}$ to $\mathbb{Q}(\pi)$, and then pull it back to $\mathbb{Q}(t)$ via the isomorphism to get an archimedean absolute value! (We know it'll be archimedean by computing $|2|$.)

I said it was sneaky.

38 Consider a polynomial $f(t) = a_n t^n + \cdots + a_1 t + a_0$, with $a_n \neq 0$, so that its degree is really n. We have $v_\infty(f) = -n$, by the definition above. Now, as a polynomial in $1/t$, we have

$$f(t) = t^n \left(a_n + \cdots + a_1 \left(\frac{1}{t}\right)^{n-1} + a_0 \left(\frac{1}{t}\right)^n \right)$$

$$= \frac{a_n + \cdots + a_1(\frac{1}{t})^{n-1} + a_0(\frac{1}{t})^n}{(\frac{1}{t})^n}$$

so that $v_1(f) = -n$ (the numerator above is clearly not divisible by $1/t$).

39 The construction of the π-adic valuation v_π should be routine by now. Some hints: the "good" value for c will depend on π; if π divides a rational prime p, then we want to choose either p or p^2 (can you come up with a reason?). For the last question, reading the three cases above, we'll get that $v_\pi(p)$ will equal zero for all primes except the (unique) one that is divisible by π, in which case it will always be equal to one, except if $p = 2$. (Now that is a convoluted sentence!) What is $v_{1+i}(2)$? If $\pi = x + iy$ and $\bar\pi = x - iy$ are two primes as in case (iii), what is the relation between v_π and $v_{\bar\pi}$?

40 This is all rather easy: the point is that $|x|$ is always a positive real number. For (ii), just note that if $\lambda^n = 1$ and λ is a positive real number, then $\lambda = 1$. Statement (iii) is just (ii) with $n = 2$; statement (iv) then follows from $|-x| = |-1| \cdot |x|$. Finally, in a finite field with q elements, we have $x^{q-1} = 1$ whenever $x \neq 0$, and applying (ii) shows that any absolute value must then be trivial.

41 Suppose $\sup\{|n| : n \in \mathbb{Z}\} = C$, and $C > 1$. Then there must exist an integer m whose absolute value is bigger than 1. But them $|m^k| = |m|^k$ gets arbitrarily large as k grows, so that C cannot be finite. It follows that $C \leq 1$, and since $|1| = 1$, this means $C = 1$, so that $|\ |$ is non-archimedean.

42 This is straight translation from the properties of absolute values.

43 For (i), notice that

$$d(x + y, x_0 + y_0) = |(x + y) - (x_0 + y_0)| = |(x - x_0) + (y - y_0)|$$

and use the triangle inequality. For (ii), notice that

$$xy - x_0 y_0 = x(y - y_0) + y_0(x - x_0).$$

For (iii), use

$$\frac{1}{x} - \frac{1}{x_0} = \frac{x_0 - x}{x x_0}.$$

You should work these out carefully if you find them troublesome.

44 The hints in the text should be enough to suggest the proof. For the second statement, notice that for any two positive real numbers α and β we have $\alpha + \beta \geq \max\{\alpha, \beta\}$.

45 We have

$$x - y = -\frac{1}{15} \qquad y - z = -\frac{4}{15} \qquad x - z = -\frac{5}{15} = -\frac{1}{3};$$

the first two sides have length 5, the third has length 1.

46 Open balls first. If $x \in B(a,r)$, then let $\delta = |x - a| < r$. We need to show that a small enough ball around x is completely contained in $B(a,r)$. Consider the ball around x with radius $\varepsilon = r - \delta$. If a point y belongs to this ball, then $|y - x| < \varepsilon$. But if that is the case, then

$$|y - a| \leq |y - x| + |x - a| < \varepsilon + \delta = r - \delta + \delta = r,$$

so that $y \in B(a,r)$. Make a picture if this is not clear!

Do something similar for closed balls (if you made a picture, this should be easy).

47 The missing parts are easily done, by imitating the parts that were done. We need the $r \neq 0$ condition because a closed ball of radius 0 is a point, which is *not* an open set unless the absolute value is trivial. (Why not?) By contrast, open balls of radius zero are just empty sets, which are always both closed and open anyway. (Why?)

48 $\overline{B}(0,1)$ is the set of fractions a/b such that $|a/b|_p \leq 1$, which means that $v_p(a/b) \geq 0$. This will happen when, after putting the fractions into lowest terms, the denominator is not divisible by p. So the closed unit ball around 0 consists of the fractions a/b where p does not divide b.

$B(3,1)$ is the set of fractions a/b such that $a/b - 3$ has absolute value less than one. Reasoning as before, this means that the denominator is not divisible by p, but the numerator *is*. If we assume that a/b is in lowest terms (so that a and b have no common factors, then

$$\frac{a}{b} - 3 = \frac{a - 3b}{b}$$

will also be in lowest terms (check!), so the conditions are that p does not divide b but does divide $a - 3b$. To find out what integers satisfy this condition we set $b = 1$; the first condition is automatically true, and the second condition says that p divides $a - 3$, i.e., that $a \equiv 3 \pmod{p}$.

49 Well, as we found out in Problem 48, the closed unit ball is the set of all fractions a/b where p does not divide b. Look at the numbers

$$a, \qquad a - b, \qquad a - 2b, \qquad a - 3b, \qquad \ldots \qquad a - (p-1)b.$$

It is easy to see that exactly one of these numbers will be divisible by p (the easiest way to see it is to note that these are p integers, and no two of them can be congruent modulo p, because if p divides the difference of any two, then it divides b, which it doesn't). But if $a - ib$ is divisible by p, then

$$\left| \frac{a}{b} - i \right| = \left| \frac{a - ib}{b} \right| < 1,$$

so that a/b is in the open ball $B(i, 1)$ of center i and radius 1. This proves the equality. The disjointness amounts to the statement that *only one* of the numbers listed can be divisible by p.

50 The point is that the 5-adic absolute value can only take values of the form 5^n, where n is an integer. So saying that a 5-adic absolute value is less than one, less than $1/2$, or less than or equal to $1/5$ all amount to the same thing.

51 To see that the sphere is closed, notice that it is the intersection of the closed unit ball with the complement of the open unit ball. These sets are both closed, and so their intersection is closed. (Notice that this part doesn't depend on the absolute value being non-archimedean.)

To see that the sphere is also open, take x in the sphere, so that $|x-a| = r$, and choose $\varepsilon < r$. Then if $|x - y| < \varepsilon$ we must have $|y - a| = r$ because all triangles are isosceles. Hence, the open ball around x of radius ε is completely contained in the sphere. (Notice that this part does.)

For the sphere to be the boundary of the open ball, any open ball centered on a point on the sphere should intersect the open ball. But that can't happen if the sphere is an open set.

52 To go one way, we need to take $A = S \cap U_1$ and $B = S \cap U_2$. To go the other, we need to find two open sets; we might try to let U_1 be the complement of the closed set $\overline{B}$, and U_2 be the complement of the closed set $\overline{A}$, but these won't be disjoint. Instead, note that the distance from a point in A to the points in B has a lower bound, use this to cover A with a huge number of little open balls, and take U_1 to be their union; do the same for U_2. By the way, requiring U_1 and U_2 to be disjoint is not really necessary, and is a mistake if we're working in a general topological space. See a book on general topology for details.

53 The intervals.

54 The ball is both a closed and open set. Take a smaller ball inside it. It is also closed and open, so that the complement of the smaller ball in the bigger ball is a closed and open set. This gives us the decomposition we want.

55 Suppose a set S contains both x and another point y; we will show S cannot be connected. To simplify the notation, let $r = |x - y|$. To show S is disconnected, we need to find the sets U_1 and U_2 in the definition. Remember that balls are clopen. For U_1 we take the open ball of radius $r/2$ around x; this contains x and not y. For U_2 we take the complement of U_1, which is open because U_1 is closed; this contains y but not x. The union of U_1 and U_2 is the whole space, so this does what we want.

56 First of all, the trivial case: the empty set and all of $\mathbb{Q}$ are both clopen sets. But there are other clopen sets in $\mathbb{Q}$. For example, consider the set of rational numbers a/b whose square is less than two:

$$S = \left\{\frac{a}{b} : \left(\frac{a}{b}\right)^2 < 2\right\} = \left\{\frac{a}{b} : -\sqrt{2} < \frac{a}{b} < \sqrt{2}\right\}.$$

This is clearly open, but it is also closed (can you check that?). It's not hard to see that this means that $\mathbb{Q}$ is totally disconnected also with respect to the usual absolute value.

On the other hand, there are no nontrivial clopen sets in $\mathbb{R}$, and $\mathbb{R}$ is connected; prove it.

57 We showed in a previous problem that the closed unit ball around 0 was the disjoint union of $p - 1$ open balls of radius 1. Scaling and translating, we see that the same will be true for any closed ball. For open balls, we can use a dirty trick. Again, look at the open ball of radius 1 around 0; for the reasons explained in Problem 50, this is *equal* to the closed ball of radius $1/p$ around 0; by the argument above, this is the disjoint union of open balls! This proves our claim for the open ball of radius 1 around 0, but again we can scale and translate to get the general result.

The fact that we had to use the special property that the values of the p-adic absolute value are all of the form p^n with n an integer should be a hint that this will *not* work in general. Nevertheless, it's pretty hard (at this point) to come up with a counter-example (the algebraic closure of $\mathbb{Q}$ will work, but defining an absolute value on that field is a non-trivial task).

58 (Very much a set of hints rather than a solution.) To show that $\mathcal{O}$ is a subring, we need to show that it contains 0 and 1 and is closed under addition, multiplication, and change of sign. This is all easy, and you only need to use the non-archimedean property to show closure under addition. (Do it.) To show that $\mathfrak{P}$ is an ideal, we need to check that it is closed under addition (easy), contains 0 (clear), and that if $x \in \mathcal{O}$ and $y \in \mathfrak{P}$, then $xy \in \mathfrak{P}$. The two assumptions say that $|x| \leq 1$ and $|y| < 1$; since $|xy| = |x||y|$, it follows that $|xy| < 1$, i.e., $xy \in \mathfrak{P}$. If $x \in \mathcal{O}$ but $x \notin \mathfrak{P}$, then we must have $|x| = 1$. It then follows that $|1/x| = 1$, so that $1/x \in \mathcal{O}$, which means that x is invertible in $\mathcal{O}$. Finally, any ideal strictly containing $\mathfrak{P}$ would have, by what we have just shown, to contain an invertible element, and would therefore be all of $\mathcal{O}$; this shows $\mathfrak{P}$ is maximal.

59 The jazzy proof: we have an injective homomorphism $\mathbb{Z} \hookrightarrow \mathbb{Z}_{(p)}$, and it maps $p\mathbb{Z}$ into $p\mathbb{Z}_{(p)}$ (do you see that?). By the usual hocus-pocus, this gives an injective map

$$\mathbb{Z}/p\mathbb{Z} \longrightarrow \mathbb{Z}_{(p)}/p\mathbb{Z}_{(p)}$$

(it's injective because an integer a maps to zero only if $a/1 \in p\mathbb{Z}_{(p)}$, which happens only if $p|a$). To see that it is also onto, we use an argument involving congruences: if $p \nmid b$, then there exists an integer b_1 such that $bb_1 \equiv 1$ (mod p). Then for any a/b in $\mathbb{Z}_{(p)}$, the integer ab_1 maps to the class of a/b in $\mathbb{Z}_{(p)}/p\mathbb{Z}_{(p)}$.

The non-jazzy proof is not all that different.

60 This is very similar to the calculations for $\mathbb{Q}$, and we leave it for the reader to puzzle over at leisure. Hints: the valuation rings all look like sets of rational functions with restrictions on the numerators and denominators, and the residue fields are often (but not always) equal to F itself.

61 The open ball is the coset $a + \mathfrak{P}$ of the ideal $\mathfrak{P}$ in $\mathcal{O}$. (Are other balls also cosets?) Problem 49 gets translated to the statement that the residue field is finite.

62 Yes it is always the case in the examples we considered, but no, it is not always true. This question is really closely related to the fact that in all our examples we have a very restricted range of values for our absolute value function.

63 Checking that $v(x)$ is a valuation is an easy exercise in logs and inequalities. For the other three statements:

$i)$ If $v_p(x) = n$, then $|x| = p^{-n}$, so that $v(x) = n \log p$. Hence v and v_p differ by multiplication by a constant, $\log p$. The image of v is $\log p \cdot \mathbb{Z}$, i.e., the real numbers which are integral multiples of $\log p$. (It's easy to see that this is a subgroup of $\mathbb{R}$, and that it is isomorphic to $\mathbb{Z}$.)

$ii)$ If the value group is discrete, look at the element x with smallest nonzero $v(x)$. It's not too hard to prove that it must be a generator of $\mathfrak{P}$. Conversely, if $\mathfrak{P}$ is principal check that the valuation of a generator must be the minimal nonzero element of the value group, which must then be discrete.

$iii)$ This takes some work. We showed in (ii) that in this case $\mathfrak{P}$ is a principal ideal, but we still need to show that every other ideal is too. See [Ser74] for a detailed discussion of what hypotheses are necessary and for the proofs.

64 As the hint suggests, choose any $x_0 \in \Bbbk$, $x_0 \neq 0$, such that $|x_0|_1 < 1$. Then (ii) says that $|x_0|_2$ is also less than 1, so that there exists a positive real number α such that $|x_0|_1 = |x_0|_2^{\alpha}$. (Just take logs on both sides to find α; why is it important to choose $|x_0|_1 < 1$? What if no such x_0 exists?) This gives us our α.

Now choose any other $x \in \Bbbk$, $x \neq 0$. If $|x|_1 = |x_0|_1$, then we must also have $|x|_2 = |x_0|_2$, because otherwise either x/x_0 or x_0/x would have $|\ |_2$ less than 1 and (ii) would be violated. So in this case the equation $|x|_1 = |x|_2^{\alpha}$ holds. Also, if $|x|_1 = 1$, then we must have (by (ii) applied to either x or $1/x$)

that $|x|_2 = 1$ also, so that the equation $|x|_1 = |x|_2^\alpha$ holds trivially. Finally, notice that the equality for some x implies the equality for any power of that x; in particular, we know that $|x_0^n|_1 = |x_0^n|_2^\alpha$ for any integer n.

So we may assume that $|x|_i \neq 1$ and $|x|_i \neq |x_0|_i$ for $i = 1, 2$. As before, choose β such that $|x|_1 = |x|_2^\beta$; again, this means that we also have $|x^n|_1 = |x^n|_2^\beta$ for all integers n. In particular, we can assume that $|x|_1 < 1$ (otherwise replace it with $1/x$), which of course also implies that $|x|_2 < 1$.

What we want to do is show that α and β must be equal. Since we want to use (ii), the natural way to proceed is to show that if they are not equal, then we can manufacture an element that has $|\ |_1 < 1$ but $|\ |_2 > 1$, which would contradict (ii). The reader should fiddle with this idea for a while to convince himself that it is not very easy to carry it through. Faced with that, we have no choice but to take a more roundabout and more devious route.

Let n and m be any two positive integers. Then we have

$$|x|_1^n < |x_0|_1^m \iff \left|\frac{x^n}{x_0^m}\right|_1 < 1 \iff \left|\frac{x^n}{x_0^m}\right|_2 < 1 \iff |x|_2^n < |x_0|_2^m.$$

Taking logs of the first and last equations, we get

$$n \log |x|_1 < m \log |x_0|_1 \iff n \log |x|_2 < m \log |x_0|_2,$$

which we can write as

$$\frac{n}{m} < \frac{\log |x_0|_1}{\log |x|_1} \iff \frac{n}{m} < \frac{\log |x_0|_2}{\log |x|_2}.$$

This says that the set of fractions which is smaller than the first quotient of logs is exactly the same as the set of fractions which is smaller than the other; since there are fractions as close as we like to any real number, this means that the two numbers must be equal (otherwise, some fraction will be bigger than one but smaller than the other). Thus, we get

$$\frac{\log |x_0|_1}{\log |x|_1} = \frac{\log |x_0|_2}{\log |x|_2},$$

and therefore

$$\frac{\log |x_0|_1}{\log |x_0|_2} = \frac{\log |x|_1}{\log |x|_2}.$$

But plugging in $|x_0|_1 = |x_0|_2^\alpha$ shows that the first quotient equals α, and similarly the second quotient equals β. This shows $\alpha = \beta$, and we are finally done!

65 This only requires a straightforward reading of the definition.

66 The point is that saying $|x| = 1$ is equivalent to saying that both $|x|$ and $|1/x|$ are ≤ 1. Then do the obvious thing.

67 According to the Lemma, equivalent absolute values differ by raising to a positive power. Since both $1^\alpha = 1$ and $0^\alpha = 0$ for any α, anything equivalent to the trivial absolute value is itself trivial. We would only need to change the definition of "nontrivial" if there were more absolute values that were equivalent to the trivial one. Since there aren't, we can let "trivial" mean "trivial" and have done with it.

68 Easy: $|p|_p < 1$ but $|p|_q = 1$ whenever p and q are two different primes.

69 Let A be the image of $\mathbb{Z}$ in $\Bbbk$, i.e., the elements of $\Bbbk$ which are integral multiples of 1. We showed that $|\ |$ was non-archimedean if and only if we had $|a| \le 1$ for all $a \in A$. Now use Problem 66.

70 First of all, if $n = 1$ we're OK, since $|1| = 1 = 1^\alpha$. If $1 < n < n_0$, the estimate $|n| \le Cn^\alpha$ would still be true, since in this case $|n| = 1$ (remember that we chose n_0 to be the smallest integer with absolute value more than 1). Furthermore, when we go on consider n^N we will certainly get to integers bigger than n_0, so the proof doesn't need to consider this case separately.

71 If we assume that the real numbers are known, this is easy: just take a sequence of rational numbers like

$$1, \quad 1.4, \quad 1.41, \quad 1.414, \quad 1.4142, \quad \text{etc.}$$

which get closer and closer to $\sqrt{2}$. This is clearly Cauchy and has no limit in $\mathbb{Q}$.

If we want to do this within $\mathbb{Q}$, we just need to be a bit more careful. For example, we might use Newton's method to generate a sequence of rational numbers which approximate $\sqrt{2}$.

72 Let $\Bbbk = \mathbb{R}$, and let $|\ |$ be the usual absolute value. The famous example is

$$x_n = 1 + \frac{1}{2} + \frac{1}{3} + \cdots + \frac{1}{n}.$$

For this sequence, we have $x_{n+1} - x_n = 1/(n+1)$, which clearly tends to zero as $n \to \infty$. The sequence is increasing, so that if it has a limit its terms must all be bounded (they are all less than any number which is bigger than the limit). However, a standard argument that the reader will very likely have seen in her calculus class shows that $x_{2^k} \ge (k + 2)/2$, so that the sequence cannot be bounded, hence cannot have a limit.

The infinite sum

$$1 + \frac{1}{2} + \frac{1}{3} + \cdots + \frac{1}{n} + \cdots$$

is called the *harmonic series*, and it is a staple of calculus courses because it shows that a series can get infinitely large even though its summands get closer and closer to zero.

73 The first one is clear: if a field contains $\mathbb{Q}$ and is complete, it must contain the limit of any Cauchy sequence made up of elements of $\mathbb{Q}$. For the second, you need to show that any real number can be arbitrarily well approximated by rational numbers. Can you prove that?

74 No, because $\mathbb{Q}$ is already complete with respect to the trivial absolute value: since the only possible absolute values are 0 and 1, a sequence will be Cauchy only if $|x_m - x_n| = 0$ for all large enough m and n. But this means that $x_m = x_n$ for all large enough m and n, and of course any such sequence converges (because it just stops).

75 Use the approach we described to construct a sequence tending to a cube root of 3. The main point is to show that one can always get a solution modulo 2^{n+1} from a solution modulo 2^n, and this is done just as in the other case.

76 The sum is easy, since

$$(x_n + y_n) - (x_m + y_m) = (x_n - x_m) + (y_n - y_m).$$

For the product, use the identity

$$x_n y_n - x_m y_m = x_n(y_n - y_m) + y_m(x_n - x_m),$$

plus the fact that x_n and y_m cannot get arbitrarily big as n and m grow.

77 The zero element is the sequence

$$0, 0, 0, 0, 0, 0, \ldots$$

The unit element is

$$1, 1, 1, 1, 1, 1, \ldots$$

A sequence (x_n) is invertible exactly when the x_n are bounded away from zero (i.e., there exists a bound b such that $|x_n| > b$ for all n; in particular $x_n \neq 0$ for all n).

We need to know they are bounded away from 0 rather than simply non-zero, because otherwise the "inverse sequence" might not be Cauchy! (Make sure you understand this one: what's an example of a Cauchy sequence of non-zero rational numbers x_n such that the sequence given by $y_n = 1/x_n$ is not Cauchy?) On the other hand, things are less bad than they seem: you should be able to show that if a Cauchy sequence does not tend to zero, then it is bounded away from zero.

78 Here's an example: sequence one is

$$0, p, 0, p^2, 0, p^3, 0, p^4, \ldots$$

and sequence two is

$$p, 0, p^2, 0, p^3, 0, p^4, \ldots$$

79 In any Cauchy sequence (x_n), the terms x_n are bounded (if this is not immediately clear, you should write down a proof). Hence, if $y_n \to 0$, then also $x_n y_n \to 0$, which is what we want to prove.

80 As it says, just follow the proof through. The argument you used in Problem 77 to show that the inverse of an invertible Cauchy sequence with terms bounded away from zero is itself a Cauchy sequence will work for "almost inverses" too.

81 If $\lambda = 0$, then $(x_n) \in \mathcal{N}$, so that $x_n \to 0$, so that $|x_n|_p \to 0$, so $|\lambda|_p = 0$, which is only reasonable. On the other hand, if $\lambda \neq 0$, then Lemma 3.2.10 says that the sequence $|x_n|_p$ is constant for sufficiently large n, which means it certainly has a limit.

82 If the difference tends to zero, the absolute value of the difference tends to zero, so that the difference of the absolute values tends to zero.

83 Once you do remember, there is nothing left to prove.

84 Problem 83 handled one part. What remains to be shown are the multiplicativity and the non-archimedean property. Both are easy: just write down the known properties of $|\ |_p$ for the terms of any sequence, and take the limit. For example, if λ is represented by (x_n) and μ is represented by (y_n), then the product $\lambda\mu$ is represented by $(x_n y_n)$. Now, for each n we have $|x_n y_n|_p = |x_n|_p |y_n|_p$. taking the limit gives $|\lambda\mu|_p = |\lambda|_p |\mu|_p$. Something similar works for the addition.

85 Yes, this is essentially obvious.

86 Lemma 3.2.10 says it all.

87 $<$ becomes $\leq$ because it's perfectly possible for a sequence to tend to a certain value while remaining consistently smaller than that value. We need to decrease ε slightly to guarantee that y doesn't end up in the *closed* ball of radius ε.

88 Follow the steps! This is mostly a question of keeping your cool when dealing with sequences of sequences.

89 This shouldn't be too hard, given all the hints. The point of number one is that if two continuous functions agree on a dense set, then they are equal. Number two is quite direct, and number three follows from the fact that, as suggested, the absolute value function is continuous, because

$$\big|\, |x| - |y| \,\big| \leq |x - y|.$$

(Can you prove this? It holds for an arbitrary absolute value. Do you see why it says that the absolute value is continuous?)

90 For something to be determined "really uniquely," it must not only be unique up to isomorphism, but up to unique isomorphism. An example is the process of forming an algebraic closure of a field. Given a field k, its algebraic closure is unique up to isomorphism, but has a great many automorphisms, so that given two algebraic closures there are a great many *different* isomorphisms between them. What this means is that the algebraic closure is not really canonically determined. (One should always speak of *an* algebraic closure, but one *can* speak of *the* completion.)

Another example of the same thing happens in linear algebra: all the vector spaces of dimension n are isomorphic, but it is still unwise to simply identify them all, because given any two there are *many* different ways to establish the isomorphism, and why should we favor any one of them over the others?

91 This is easy: since $\mathbb{Z}_p$ is the closed unit ball with center 0, it is an open set containing 0, hence a neighborhood of 0. Since multiplication by p sends open sets to open sets, this means that for every n the set $p^n\mathbb{Z}_p$ is a neighborhood of zero. That

$$\bigcup_{n \in \mathbb{Z}} p^n \mathbb{Z}_p = \mathbb{Q}_p$$

is clear from the first statement in the Corollary; to see that they are a fundamental system of neighborhoods we need to show that any open set containing zero contains a $p^n\mathbb{Z}_p$, and this is clear (because any open ball containing 0 contains a closed ball of smaller radius, which will be one of the $p^n\mathbb{Z}_p$).

92 There are lots, of course; an example would be the family consisting of the open intervals $(-1/n, 1/n)$ plus the open intervals $(-n, n)$, where n ranges through the positive integers.

93 The proof we sketched for Problem 91 pretty much shows this already.

94 This has all been done, albeit in different terms. Multiplication by p^n is injective because $\mathbb{Z}_p$ is contained in $\mathbb{Q}_p$, which is a field. The existence of the map from $\mathbb{Z}_p$ to $\mathbb{Z}/p^n\mathbb{Z}$ is part of what was proved in Proposition 3.3.3; that it is surjective is obvious because $\mathbb{Z}_p$ contains $\mathbb{Z}$. To check that the kernel is correct, we need to look once again at the proof of the Proposition, which shows this pretty clearly.

95 This is pretty silly if we just remember that $\mathbb{Z}_p$ is contained in $\mathbb{Q}_p$, which is a field: if $nx = 0$ in $\mathbb{Z}_p$, then $nx = 0$ in $\mathbb{Q}_p$, and then $n = nx\dfrac{1}{x} = 0$. To do it without referring to $\mathbb{Q}_p$ seems a bit perverse, but it follows from two facts: first, if $p \nmid n$, then n is invertible in $\mathbb{Z}_p$; second, multiplication by p^n is injective.

96 You'll probably need to look some of these up. Just some comments: (i) is standard but pretty hard; (ii) just uses the fact that the inverse image of an open set by a continuous function is again open; (iii) takes some thought to come up with a way to use the condition about covering sets; for (iv), to show that any compact set will have these two properties is not too hard (use (iii) for the first one), but the converse takes some work.

97 A closed interval is compact, and is a neighborhood of any of its interior points, so it's enough to note that any point is in the interior of some closed interval; e.g., $x \in [x - 1, x + 1]$.

98 The hint pretty much proves everything.

99 Because any ball in $\mathbb{Q}_p$ is equal to a ball of radius p^n, and any ball in $\mathbb{Z}_p$ with radius greater than or equal to one will simply be all of $\mathbb{Z}_p$.

100 To reproduce the argument we gave for $\mathbb{Z}_p$, we only need to check that the other quotients $\mathcal{O}/\mathfrak{P}^n$ are also finite, since $\mathcal{O}$ is always the closed unit ball in $\Bbbk$. To see this, we look at the obvious map $\mathcal{O}/\mathfrak{P}^n \longrightarrow \mathcal{O}/\mathfrak{P}$; its kernel is $\mathfrak{P}^n/\mathfrak{P}$. If we show the kernel is finite, we will be done (because the assumption is that the image is too). Can you do that?

The "do we really need" questions are both pretty hard. (The answer is likely to be "yes" in both cases.)

101 If $x_n \in \mathbb{Z}$ for all n, then $|x_n| \le 1$ for all n. Now if $x_n \to x$, then there is some n such that $|x - x_n| < 1$. But then

$$|x| = |x_n + (x - x_n)| \le \max\{|x_n|, |x - x_n|\} \le 1$$

so that $x \in \mathbb{Z}_p$.

102 This is not too hard, but does rely on the reader being comfortable with topology and with infinite products. That the map is an injective homomorphism is not too hard to show, because any element in $\mathbb{Z}_p$ is the limit of its associated coherent sequence. For details on how to construct $\mathbb{Z}_p$ from this point of view, see chapter 2 of [Ser73].

103 Given such a family of maps $f_n : R \longrightarrow A_n$, and given $r \in R$, the sequence $(f_n(r))$ is a coherent sequence. By the previous problem, we can find an element of $\mathbb{Z}_p$ corresponding to this sequence. Taking this as the image of r gives a map $R \longrightarrow \mathbb{Z}_p$ which does what we want.

104 If you are comfortable with the algebraic concepts involved, this should not be very hard. The point is that any integer between 0 and $p^n - 1$ is congruent to a unique element of the form $a_0 + a_1 p + \cdots + a_{n-1} p^{n-1}$ with the a_i chosen from our set X of coset representatives. This can be proved, for example, by induction on n. Once that is known, just repeat the proof in the text.

105 We need $x = a_0 + a_1 p + \cdots$ with $a_0 \neq 0$.

106 $\mathbb{Z}^{\times} = \{\pm 1\}$, $\mathsf{F}[t]^{\times} = \mathsf{F}^{\times}$, and $\mathbb{C}[[t]]^{\times}$ is all power series with nonzero initial term, i.e., of the form $a_0 + a_1 t + \cdots$ with $a_0 \neq 0$.

107 Let x_n be a sequence of elements of $\mathbb{Z}_p$. We want to pick out a subsequence that converges. To do this, use the following iterative procedure:

(i) There are only p possible choices for the zeroth coefficient in the p-adic expansion of the x_n. Hence there must be infinitely many x_n all of which have the same initial term a_0. Choose n_1 such that x_{n_1} is one of these.

(ii) For each of the infinitely many x_n whose p-adic expansions start with a_0, there are p choices for the first coefficient. Hence there must be infinitely many x_n all of whose p-adic expansions start with $a_0 + a_1 p$. Choose n_1 so that x_{n_1} is one of these.

Now keep going. Why does this procedure fail for sequences in $\mathbb{Q}_p$?

108 This can be done in many ways. Notice that, however we do it, the Taylor expansion is *finite*, so we don't need to worry about convergence questions.

Here's the jazziest proof I can think of: the field generated by $\mathbb{Q}$ and the coefficients of $F(X)$ can be embedded in $\mathbb{C}$, and the theorem is clearly true for polynomials with complex coefficients. (Talk about overkill...)

109 Just follow what was done to go from α_1 to α_2.

110 In Newton's method, we start with an initial guess x_0 and then compute what we hope are better and better approximations using the formula

$$x_{n+1} = x_n - \frac{f(x_n)}{f'(x_n)}$$

In our setup, we found α_{n+1} by setting it equal to $\alpha_n + p b_n$, and computed b_n by setting $F(\alpha_n) = px$ and $b_n = -x(F'(\alpha_n))^{-1}$. Plugging everything in gives

$$\alpha_{n+1} = \alpha_n - p\frac{F(\alpha_n)}{p}(F'(\alpha_n))^{-1} = \alpha_n - \frac{F(\alpha_n)}{F'(\alpha_n)}$$

In other words, it is exactly the same formula!

There are differences, of course. First of all, we checked that this procedure never leaves $\mathbb{Z}_p$ (in other words, the division in the formula can always be performed in $\mathbb{Z}_p$). Next, we checked that in the p-adic case the method *always* works, provided only that $F'(\alpha_1) \not\equiv 0 \pmod{p}$. This is far from true in the classical case. Finally, we get an extra bit of information, $\alpha \equiv \alpha_1 \pmod{p}$, which can be read as saying that the root we get is not too far from the initial estimate (this too is not true in the classical case).

111 If $F'(\alpha_1)$ is divisible by p, it is not invertible in $\mathbb{Z}_p$, so that we can't pick the b_1 in the computation, and the proof falls through. Indeed, the polynomial $X^2 - 3$ has roots modulo 2 but no roots in $\mathbb{Q}_2$.

112 See [Cas86] for a full proof. The method is very similar to the one we used. An example where the stronger result is necessary is the equation $X^2 - 17$, which does have roots in $\mathbb{Q}_2$.

113 If α_1 exists, then its image in $\mathbb{Z}/p\mathbb{Z}$ is an element of order dividing m in the cyclic group $(\mathbb{Z}/p\mathbb{Z})^\times$ of order $p-1$. It follows that $\gcd(m, p-1) \neq 1$ unless $\alpha_1 \equiv 1 \pmod{p}$. Furthermore, the least exponent m with this property must be a divisor of the gcd, and hence must be a divisor of $p - 1$. Conversely, in a cyclic group of order $p - 1$ there must certainly be elements of any order dividing $p - 1$ (if x is a generator, $x^{(p-1)/d}$ is of order d).

114 It's basically straight Hensel's lemma. For the two loose ends, note that if there is an m-th root of unity, then it must be in $\mathbb{Z}_p$ (because its absolute value must equal one); furthermore, it will be congruent to an *integer* α_1 with $\alpha_1^m \equiv 1 \pmod{p}$ (just take the first term in its p-adic expansion). Now use the previous problem and the uniqueness part of Hensel's Lemma.

115 The roots of unity are exactly the elements of $\mathbb{Z}_p^\times$ that satisfy $x^m = 1$ for some power m. It is easy to see that the set of such elements in any abelian group always forms a subgroup. To see that there are $p - 1$ roots, note that the numbers $1, 2, 3, \ldots, p - 1$ are all solutions of $X^{p-1} \equiv 1 \pmod{p}$, and are all incongruent modulo p. Applying Hensel's lemma gives $p - 1$ roots which are all incongruent modulo p, and in particular are all different. Since a polynomial can only have as many roots as its degree, these must be all the roots. Since, by the previous problem, any root of unity must be a root of this polynomial, these must be all the roots of unity in $\mathbb{Q}_p$. Finally, any finite subgroup of any field is cyclic. (To see that the group of roots of unity is cyclic in a more direct way, apply the reasoning above to all polynomials

$X^d - 1$ as d ranges through the divisors of $p - 1$, and count to show that some $(p - 1)$-st root of unity must exist which is not a root of any of these. Such a root of unity will be a generator.)

116 The first assertion is a straight application of the stronger form of Hensel's lemma. For the second, write the 2-adic unit in the form $1 + 2x$, and square. The conclusion follows by considering $\mathbb{Z}/8\mathbb{Z}$.

117 Polynomials that are quite different in $\mathbb{Z}_p[X]$, such as $X + 1$ and $X + (p + 1)$, are identical modulo p, so being relatively prime modulo p is a more restrictive condition than being so over $\mathbb{Z}_p$.

118 We use the notation in the proof, and focus mostly on the question about the final twist. Since $g_1(X) = X$ and $h_1(X) = 1$, the obvious solution for $a(X)g_1(X) + b(X)h_1(X) \equiv 1 \pmod{p}$ is $a(X) = 0$, $b(X) = 1$, which yields $\tilde{s}_1(X) = 0$ and $\tilde{r}_1(X) = X^2 + 1$. But now if we simply set $g_2(X) = g_1(X) + p\tilde{r}_1(X)$ and $h_2(X) = h_1(X) + p\tilde{s}_1(X)$, we end up with $g_2(X) = 2X^2 + X + 2$ and $h_2(X) = 1$, which yields a factorization, all right, but a rather unsurprising one!

If we do it right, we get $r_1(X) = 1$ (the remainder of dividing $X^2 + 1$ by X) and $s_1(X) = X$, which gives $g_2(X) = X + 2$ and $h_2(X) = 1 + 2X$, and all is well. The reader should go through at least one more iteration herself.

119 Just follow what we did in our proof.

120 If you did the previous problem, this should be easy.

121 For $p \neq 2, 17$ one can do this by a straight application of Hensel's lemma, as follows: if neither 2 nor 17 are squares modulo p, then their product must be a square modulo p; then use Hensel's lemma. (To see why the product of two quadratic non-residues must be a square modulo p, remember that $(\mathbb{Z}/p\mathbb{Z})^\times$ is cyclic, so that being a square means being an even power of the generator; the product of two odd powers of the generator must be an even power of the generator!)

For $p = 2$, note that 17 is a square in $\mathbb{Q}_2$. For $p = 17$, note that $6^2 \equiv 2 \pmod{17}$, so that (Hensel's lemma!) 2 is a square in $\mathbb{Q}_{17}$. For $p = \infty$, there are clearly lots of roots. And there are clearly no rational roots.

122 This isn't too hard if one uses more advanced tools such as biquadratic reciprocity. An elementary (but not easy) proof can be found in [Cas86], page 57, and one using algebraic number theory is outlined in [Cas91], page 88.

123 All sorts of polynomials are irreducible over $\mathbb{Q}$ and reducible over some $\mathbb{Q}_p$ (think of $X^2 + 1$ for example), so the "only if" part is bunk. The "if" part works, because if a polynomial were reducible over $\mathbb{Q}$ it would certainly be reducible over all the $\mathbb{Q}_p$. Of course, the "if" part is not very interesting... How about this: is it true that a polynomial will be irreducible over $\mathbb{Q}$ if and only if it is irreducible over *some* $\mathbb{Q}_p$? (In other words, given a polynomial that is irreducible over $\mathbb{Q}$, can I find a prime p such that the polynomial is irreducible over $\mathbb{Q}_p$?) If so, this proves the statement in the exercise with "irreducible" replaced by "reducible."

124 We have

$$ax^2 + by^2 + cz^2 = a'n^2x^2 + by^2 + cz^2 = a'(nx)^2 + by^2 + cz^2,$$

which establishes the correspondence. We are interested in deciding when there are roots in the rational numbers (or p-adic numbers, or integers modulo $p^n = m$ for some m), and the correspondence shows that (in each case) the equation with a will have a root if and only if the equation with a' does. So we might as well work with a'. Doing the same for b and c, we see that we can assume that all three coefficients are square-free.

125 By the previous problem, we may assume a, b, and c are square-free and have no common factors, and we do. Let $k = \gcd(a, b)$, which we assume is greater than 1. Notice that k must be square-free. Then we can set $a = ka'$, $b = kb'$, and we know that k is relatively prime to c. Suppose that $ax^2 + by^2 + cz^2 = 0$. If we look at the last equation, we see that k must divide cz^2, since it divides the other two terms. Since it is prime to c, it divides z^2. Since it is square-free, it must divide z. So write $z = kz'$, plug in, divide by k, and continue from there.

126 For $n = 0$, we get the sum of p ones, which is p, hence is $\equiv 0 \pmod{p}$. Instead of trying to give a general proof, here's the proof for $n = 1$: choose and fix a number a, $2 \le a \le p-1$ (this is possible, since p is odd). I want to compare the two sums

$$\sum_{x=0}^{p-1} x \quad \text{and} \quad \sum_{x=0}^{p-1} ax.$$

It is not hard to show that the numbers 0, a, $2a$, $\ldots (p-1)a$ are all noncongruent modulo p. (Do it!) Since there are p of them, they must be congruent, in some order, to 0, 1, $\ldots p-1$. (In other words, modulo p the list of the ax is just a permutation of the list of x.) This means that the two sums are congruent:

$$\sum_{x=0}^{p-1} x \equiv \sum_{x=0}^{p-1} ax \pmod{p}.$$

We can rewrite this as

$$0 \equiv \sum_{x=0}^{p-1} x - \sum_{x=0}^{p-1} ax \pmod{p}$$
$$\equiv (1-a) \sum_{x=0}^{p-1} x$$

and, since $1 - a \not\equiv 0 \pmod{p}$, the conclusion follows.

Reorganizing this to take care of more general exponents is not too hard: what we need is to show that we can choose our a so that $a^n \not\equiv 1 \pmod{p}$. If so, the same proof will work!

127 What we need to check is that the polynomial $f(X) = aX^2 + by_0^2 + cz_0^2$ satisfies the conditions in Hensel's Lemma. But that's easy: $f(x_0) \equiv 0$ is our assumption, and $f'(x_0) = 2ax_0 \not\equiv 0 \pmod{p}$ because p is odd and does not divide a or x_0.

128 Well, certainly in the application of Hensel's Lemma (we need to know that whichever of x_0, y_0, or z_0 is not divisible by p has a coefficient next to it which is not divisible by p). But presumably also in the Proposition: where? (Hint: suppose one of a, b or c is divisible by p; is the Proposition still true?)

129 Of course, the idea is to use the result in Problem 112, using Hensel's Lemma as we did in Problem 127. The difficulty is that, since $p = 2$, there is no doubt that the derivative will be divisible (once) by p. If we look hard at the conditions in Problem 112, we see that we need to find an initial solution (x_0, y_0, z_0) such that $ax_0^2 + by_0^2 + cz_0^2 \equiv 0 \pmod{8}$.

Very well, we know that the sum of two of the coefficients, say a and b, is divisible by 4: $a + b \equiv 0 \pmod{4}$. Now there are two possibilities:

- if $a + b \equiv 0 \pmod{8}$, then we can choose $x_0 = y_0 = 1$ and $z_0 = 0$, and all is well;

- if not, we will have $a + b \equiv 4 \pmod{8}$; choosing $x_0 = y_0 = 1$ and $z_0 = 2$ will then do what we want (check!).

In either case, we are in business, and Hensel's Lemma gives the solution in $\mathbb{Q}_2$.

130 Suppose a is even, b and c are odd, and $ax^2 + by^2 + cz^2 = 0$. As before, we can assume that at least one of x, y and z is a 2-adic unit, and that all three are in $\mathbb{Z}_2$. There are two cases to consider:

- x is in $2\mathbb{Z}_2$. Then clearly ax^2 is divisible by 8, and it is easy to see that y and z must be 2-adic units. Since the square of a 2-adic unit is always in $1 + 8\mathbb{Z}_2$, it follows that

$$0 = ax^2 + by^2 + cz^2 \equiv b + c \pmod{8}.$$

- x is a 2-adic unit. Then y and z again must be 2-adic units (if, say, $y \in 2\mathbb{Z}_2$, then $ax^2 + by^2$ would be divisible by 2, and therefore cz^2 would be divisible by 2; but we know c is odd, so z would have to be in $2\mathbb{Z}_2$; but then $by^2 + cz^2$ would be in $4\mathbb{Z}_2$, hence so would ax^2...). Once again, the square of a 2-adic unit is always in $1 + 8\mathbb{Z}_2$, and we get

$$a + b + c \equiv 0 \pmod{8}.$$

The converse is once again an application of the generalized form of Hensel's Lemma (but it's actually easier this time, because we have information about a, b, and c modulo 8).

131 Necessity is easy: since $p|a$, we have $by^2 + cz^2 \equiv 0 \pmod{p}$, and it's not hard to see that both y and z will have to be p-adic units. Hence, we can rewrite the equation as $b + (y/z)^2 c \equiv 0 \pmod{p}$, and it is now a matter of showing that if a p-adic unit fits into this equation, then we can find an integer that does (and that is easy).

The sufficiency is Hensel's Lemma again, of course.

132 The magic words here are "quadratic reciprocity." Have a chat with the local number theorist, who is likely to wax poetic over this one.

133 $n!$ converges to zero, n and $1/n$ diverge, p^n converges to zero, and $(1 + p)^{p^n}$ converges to 1.

134 We've pretty much already proved this. Look at Lemma 3.2.10.

135 Exactly the same proof that works over $\mathbb{R}$ works here also. Basically, the fact that $|x + y| \leq |x| + |y|$ says that when the sequence of partial sums of $\sum |a_n|$ is Cauchy, then so is the sequence of partial sums of $\sum a_n$.

136 If $\sum a_n = 0$, the inequality is vacuous. If not, for any partial sum, we have

$$\left| \sum_{n=0}^{N} a_n \right| \leq \max_{0 \leq n \leq N} |a_n|$$

by the non-archimedean property. Now note that for large enough N we have

$$\max_{0 \leq n \leq N} |a_n| = \max_n |a_n|$$

because the a_n tend to zero, and

$$\left| \sum_{n=0}^{\infty} a_n \right| = \left| \sum_{n=0}^{N} a_n \right|$$

by Lemma 3.2.10.

137 Examples in $\mathbb{R}$:

$$\sum \frac{1}{n} \qquad \sum \frac{1}{n \log n} \qquad \sum_{p \text{ prime}} \frac{1}{p}$$

are all divergent.

138 We used the ultrametric inequality a number of times, but basically in two ways:

i) to conclude that a series converges when we know that its terms tend to zero,

ii) to conclude that a sum is less than ε when each of the summands is less than ε.

Both uses are crucial, so we don't expect that this result remains true over $\mathbb{R}$. Can you construct a counterexample? There are theorems of this sort that are true over $\mathbb{R}$, but in that case the crucial property is absolute convergence. See, for example, section 8.21 of [Apo74].

139 This is true in the classical setting (i.e., over $\mathbb{R}$), and the same proof works here. But it's not hard to do: work with partial sums. We have

$$\sum_{n=0}^{N} a_n + \sum_{n=0}^{N} b_n = \sum_{n=0}^{N} c_n$$

because these are all *finite* sums; now take the limit.

140 Again, this is a classical result, and the same proof works in our setting. The proof is similar to the one in the previous exercise, but one has to be a little careful because the product of the N-th partial sums is *not* equal to the N-th partial sum of the product series, and one has to deal with the extra terms. It might be fun to go through it to see if working p-adically makes it any easier.

141 Well, the basic content of the intermediate value theorem is that the image of an interval under a continuous function is an interval. This is a special case of a general fact, true in any metric space: the image of a connected set by a continuous function is a connected set. This is true in the p-adic context, but is kind of silly, since the only connected sets are those which consist of exactly one point!

A more interesting question is this: suppose $f(X)$ is a polynomial (might as well choose an easy function to work with). What can you say about the range of values of $f(x)$ as x runs through $\mathbb{Z}_p$?

142 It's what you think it should be!

143 The easiest version is locally constant everywhere except at zero. Define $f(0) = 0$ and make $f(x)$ be constant on annuli around zero of smaller and smaller radius. Then arrange the values so that the derivative at $x = 0$ exists and equals 0. For a more "natural" construction, see Problem 348.

144 The chain rule is true because the usual proof works quite well. Once that is known, it's easy to see that one can make "almost constant" functions by taking one such and composing with any other non-constant differentiable function. That'll yield a great many examples!

145 In each case, one has to look at how the p-adic valuation of the general term changes as $n \to \infty$. For (i), note that $v_p(p^n x^n) = n + n v_p(x) = n(1 + v_p(x))$; if $v_p(x) > -1$, this will tend to infinity with n, so that

$$|p^n x^n|_p = p^{-n(1+v_p(x))}$$

will tend to zero, and the series will converge. Otherwise, the series will diverge. So the radius of convergence is given by $v_p(x) > -1$, or $|x|_p < p$. (ii) is very similar. The hardest one is (iii): we want to compute $v_p(n! x^n) = v_p(n!) + n v_p(x)$. The difficulty is to estimate $v_p(n!)$. This will be done later in the chapter, but give it a go now. If you can show that $v_p(n!)$ grows faster than linearly in n, then the series will converge for all x. Does it?

146 All that needs to be checked is that the definition of the sum and product power series agrees with the sum and product series in problems 139 and 140.

147 Clearly the formula is

$$c_n = \sum_{m=1}^{n} a_m d_{m,n},$$

where

$$g(X)^m = \sum_{n=m}^{\infty} d_{m,n} X^n.$$

So what's needed is a formula for the coefficient $d_{m,n}$ of degree n in $g(X)^m$. That can be gotten by induction from the definition of the product of two power series. If you can't find it by yourself, look further down in this section!

148 It's easy to see that the extra condition doesn't hold, since $g(1) = 0$ is certain to be smaller than the terms of the series. The first few terms of $h(X)$ look like this:

$$h(X) = 1 - 2x + 4x^2 - \frac{16}{3}x^3 + \frac{20}{3}x^4 - \frac{104}{15}x^5 + \frac{304}{45}x^6 + \ldots$$

Since $g(X)$ is just a binomial, it isn't hard to write out the general term of $h(X)$. (Do it!)

Provided we can show the estimate on the coefficients, it's clear that $h(1) = 1 - 2 +$ multiples of 4, so that $h(1) \equiv 3 \pmod 4$. So it all boils down to showing that the estimate $v_2(a_n) \geq 1 + n/4$ holds for $n \geq 2$. This is actually rather hard, but you might want to give it a try. For a (rather sophisticated) reference, see [Lan89], Chapter 14, section 2, which will also give a clue about why this particular power series is interesting.

149 This is straightforward manipulation of formal series. (Or can you think of a smarter way to prove these?)

150 Imitate the classical proof.

151 The phrase assumes that the roles of f and g in the proposition are symmetric, that is, that if we start with g and construct a new series as specified, the result will be f. Can you check that?

152 This is a matter of writing out $g(x)$ and using Proposition 4.1.4 to reorganize it into a power series in x. Say $f(X) = \sum c_n X^n$. Since $|a| = 1$ and $|b| < \rho$, we have

$$|x| < \rho \Longleftrightarrow |ax + b| < \rho$$

and

$$g(x) = f(ax + b) = \sum_{n=0}^{\infty} c_n (ax + b)^n = \sum_{n=0}^{\infty} \sum_{k=0}^{n} c_n \binom{n}{k} a^k x^k b^{n-k}.$$

Now check that we can reorganize that to get

$$g(x) = \sum_{k=0}^{\infty} \left(\sum_{n=k}^{\infty} \binom{n}{k} c_n a^k b^{n-k} \right) x^k.$$

153 First, the region of convergence of a power series is either an open or a closed ball, hence is an open set. Hence, if $x_m \to x$ and $f(x)$ and $g(x)$ converge, we can conclude that $f(x_m)$ and $g(x_m)$ converge for large enough m. Now use Proposition 4.4.2 to reduce the problem to the case where $x = 0$.

154 Just use the proposition repeatedly (equivalently, use an induction proof where the step is provided by the proposition).

155 Well, the formula for a_k suggests that it's $f^{(k)}(x)/k!$ that's the interesting quantity, and notice that the formula says that if the a_n are in $\mathbb{Z}_p$ then so are the coefficients of $f^{(k)}(x)/k!$. That's kind of neat.

Actually, in certain cases one wants to consider whether the "quasi-derivative" defined by

$$f^{[k]}(x) = \sum_{n \geq k} \binom{n}{k} a_n (x - \alpha)^{n-k}$$

has nice enough properties to replace the derivative. This is relevant, for example, if we are working over a field of characteristic p and we have $k > p$.

156 We need to consider $c_{nj} = a_n x^j \alpha^{n-1-j}$. Since both x and α are in $\mathbb{Z}_p$, we get $|c_{nj}| \leq |a_n| \to 0$, which gives one of the conditions we need to check. For the other, note that $c_{nj} = 0$ if $j \geq n$.

157 If we put $f(X) = \sum a_n X^n$, and assume it converges on $p^m \mathbb{Z}_p$, then, as in the proof of the Corollary, We have to look at the series $\sum a_n p^{mn} X^n$. We need to find N such that

$$|p^{mN} a_N| = \max_n |p^{nm} a_n| \quad \text{and} \quad |p^{mn} a_n| < |p^{mN} a_N| \text{ for } n > N$$

Then $f(X)$ has at most N zeros on $p^m \mathbb{Z}_p$.

158 The first series converges for $|x| < p$, hence for $x \in \mathbb{Z}_p$, and since $|p^n| = p^{-n}$ is strictly decreasing, we have $N = 0$, so that there are no roots in $\mathbb{Z}_p$. (In fact, we have

$$\sum p^n x^n = \frac{1}{1 - px}$$

and this is clearly never equal to zero.) The second series converges on $p^2 \mathbb{Z}_p$, and changing variables as above gives $N = 0$ again. (What is the sum?) The third one is again the hardest; to count the roots in $\mathbb{Z}_p$, one needs to find the last n such that $n!$ is not divisible by p, which gives $N = p - 1$. Thus, the series has at most $p - 1$ roots in the unit disk. If you managed to determine the precise radius of convergence, can you say anything about other possible roots?

159 Since $v_p(n)$ is the largest m such that p^m divides n, it's clear that $v_p(n) \leq \log n / \log p$. But then $v_p(n)/n \leq \log n / n \log p$, which tends to zero as $n \to \infty$, which gives what we want.

160 If $p = 2$, then $-1 = 1 - p \in B$, so that $\log_2(-1)$ makes sense. On the other hand, we must have $2 \log_2(-1) = \log_2(-1)^2 = \log_2(1) = 0$, so that $\log(-1) = 0$. Writing out the series for $\log(1 - 2)$ gives

$$-\left(2 + \frac{2^2}{2} + \frac{2^3}{3} + \frac{2^4}{4} + \cdots + \frac{2^n}{n} + \cdots \right)$$

and saying that this converges to zero in $\mathbb{Q}_2$ amounts to saying that its partial sums get more and more divisible by 2. This was the claim in chapter one. For an estimate of the power of 2 dividing a partial sum, we might write

$$\left(2 + \frac{2^2}{2} + \frac{2^3}{3} + \frac{2^4}{4} + \cdots + \frac{2^N}{N}\right) + \left(\frac{2^{N+1}}{N+1} + \cdots\right) = 0,$$

which shows that

$$\left|2 + \frac{2^2}{2} + \frac{2^3}{3} + \frac{2^4}{4} + \cdots + \frac{2^N}{N}\right|_2 = \left|\frac{2^{N+1}}{N+1} + \cdots\right|_2 \leq \max_{n>N}\{|2^n/n|_2\}.$$

Thus we need to estimate $|2^n/n|_2$, or $v_2(2^n/n)$, for large n. Now, $v_2(2^n/n) = n - v_2(n) \geq n - \log n/\log 2$, so a lower bound for the exponent will be given by the least value of $n - \log n/\log 2$ for $n > N$. Now use some calculus.

161 Define a power series by $f(X) = \log(1 + pX)$, which will converge for $x \in \mathbb{Z}_p$. We need to find the last N for which the coefficient a_N has the maximum absolute value. Writing down the series explicitly (do it!), one sees that $N = 1$ if $p \neq 2$ and $N = 2$ if $p = 2$, which gives us the answer we want.

162 Just use the previous result for the first part: if $x \in 1 + p\mathbb{Z}_p$ satisfied $x^p = 1$, then clearly $\log_p(x) = 0$, and the previous problem says this can only happen if $x = 1$. For the second statement, notice that if $x^p = 1$, then $|x| = 1$, so that any such root must be in $\mathbb{Z}_p$. Reducing modulo p gives an element $\overline{x}$ of $\mathbb{Z}/p\mathbb{Z}$ whose p-th power is one, which implies that $\overline{x} = 1$ in $\mathbb{Z}/p\mathbb{Z}$, i.e., that $x \in 1 + p\mathbb{Z}_p$. In a nutshell, a p-th root of unity must be in $1 + p\mathbb{Z}_p$, and Strassman's estimate says that there are no nontrivial ones there. Thus, there are no nontrivial p-th roots of unity in $\mathbb{Q}_p$.

163 This is very similar to, but easier than, the previous theorem.

164 Write out the expression of $n!$ as a product, and work out how many numbers are multiples of p, how many of p^2, etc.

165 What the hint says.

166 Not serious ones!

167 Rewrite the inequality as $v \leq 1+p+\cdots+p^{v-1}$ and prove it by induction.

168 Since $\log_2(-1) = 0$ and the terms of the series are non-zero, there's no chance that the condition $|g(x)| \geq |a_m x^m|$ is going to be satisfied. This points out a general fact: whenever $g(x) = 0$, we'll have trouble applying Theorem 4.3.3.

169 This is very similar to what we did in the text for the exponential. The regions of convergence will, of course, be the same as those for the exponential function. (Why "of course"?) The "p-adic trig functions" won't be periodic, because of Corollary 4.4.10.

170 The elements of $\mathbb{Z}/n\mathbb{Z}$ can be represented by the integers between 1 and n. It's easy to see that if a is invertible in $\mathbb{Z}/n\mathbb{Z}$, then $\gcd(a,n)=1$ (can you prove it?). For the converse, use the fact that if $\gcd(a,n)=1$ then we can find integers r and s such that $ra+sn=1$, and reduce modulo n.

171 $(1+qx)(1+qy) = 1 + q(x+y+qxy)$, and

$$\frac{1}{1+qx} = 1 - qx + q^2x^2 - q^3x^3 + \cdots$$

which converges because $x \in \mathbb{Z}_p$. Similarly with p instead of q.

172 We already know that there are exactly $(p-1)$ roots of unity in $\mathbb{Z}_p$, by a combination of Hensel's Lemma (Problem 115) and Strassman's Theorem (Problem 162). Further, Hensel's Lemma already tells us that no two of the $(p-1)$-st roots of unity are congruent modulo p. (For $p=2$, one needs to change this slightly; see problems 116 and 163.) Can you come up with a more direct argument?

173 Since π gives an injective homomorphism between V and $(\mathbb{Z}/q\mathbb{Z})^\times$, and these groups have the same number of elements, π must in fact be an isomorphism. Now let $u \in \mathbb{Z}_p^\times$, and suppose $\pi(u) = \overline{n} \in (\mathbb{Z}/q\mathbb{Z})^\times$. Choose $\zeta \in V$ such that $\pi(\zeta) = \overline{n}$. Then $u_1 = u\zeta^{-1} \in U_1$. The map

$$u \mapsto (\zeta, u_1)$$

gives the isomorphism between $\mathbb{Z}_p^\times$ and $V \times U_1$.

174 Not much needs to be changed. V is not the image of $\mathbb{F}_2^\times$ (which has order one, after all), but we can still define ω as the projection on V, and make the resulting notational changes.

175 We know $x = \omega(x)\langle x\rangle$. Since $\omega(x)^{p-1} = 1$, we have $\omega(x)^p = \omega(x)$; taking p-th powers over and over, we see that $\omega(x)^{p^n} = \omega(x)$ for any n.

On the other hand, $\langle x\rangle = 1 + qy$ for some y. Taking p-th powers,

$$\langle x\rangle^p = (1+qy)^p = 1 + pqy + \text{multiples of } q^2,$$

so that $\langle x\rangle^p \in 1 + p^2\mathbb{Z}_p$. Repeating, we see that $\langle x\rangle^{p^n} \in 1 + p^{n+1}\mathbb{Z}_p$, so that $\langle x\rangle^{p^n}$ tends to 1 as $n \to \infty$. Putting these together gives what we want.

176 If $|x| = 1$, we want to look at the sequence $\binom{\alpha}{n}$ as n tends to infinity. If it tends to zero, then the series converges for $|x| = 1$; if not, not, and the radius of convergence is 1. Can you decide? The answer may very well depend on α!

177 This one is much easier: if $|\alpha| > 1$, then $|\alpha - i| = |\alpha|$ (because "all triangles are isosceles"). Putting this together with our various estimates on $|n!|$ should allow you to get an answer.

178 When α is a *positive* integer, this is obvious, since

$$\mathbf{B}(\alpha, x) = (1 + x)^\alpha = \sum_{i=0}^{\alpha} \binom{\alpha}{i} x^i$$

is actually a polynomial. For negative integers, all we need to notice is that we have $\mathbf{B}(\alpha, x)^{-1} = (1 + x)^{-\alpha}$.

179 According to Koblitz, this is a special case of a theory due to Bombieri. It takes quite a bit of work, though some bits aren't hard. For example, we already know that $\mathbf{B}(1/2, x)$ converges if $|x| < 1$, that is, if $x \in p\mathbb{Z}_{(p)}$ (since x is rational). So we need to know for which a/b is it true that $(a/b)^2$ is a one-unit. Now $(a/b)^2 - 1 \in p\mathbb{Z}_p$ means that p divides $a^2 - b^2 = (a+b)(a-b)$ (which is the numerator), and hence that it divides either $a + b$ or $a - b$. This shows the "if" in (i). For the converse, we need to know that the series does not converge if $x \notin p\mathbb{Z}_p$. The rest is similar.

180 See any book on real analysis or general topology.

181 The easiest way is to exploit the proposition that follows this problem in the text: a function on $\mathbb{Z}$ that cannot be extended to $\mathbb{Z}_p$ will work. Say, choose an element $\alpha \in \mathbb{Z}_p$ which is not in $\mathbb{Z}$ (say, any a/b with $p \nmid b$ and $b > 1$), and define $f(x) = 1/(x - \alpha)$ for any $x \in \mathbb{Z}$. Then f is continuous on $\mathbb{Z}$ but not uniformly continuous (check!).

182 The main point is that $a_n - a_m$ small implies $f(a_n) - f(a_m)$ small by the uniform continuity. It's pretty much a direct check in $\varepsilon - \delta$ style.

183 If b_k is another sequence tending to x, then $a_k - b_k$ tends to zero; by the boundedness and uniform continuity, it follows that $f(a_k) - f(b_k)$ tends to zero, which is what we want.

184 This is very similar to the other two problems: easy $\varepsilon - \delta$ stuff.

185 Basically, all that needs to be done is to run through the argument and check that the only result we needed was that $\mathbb{Z}_p$ was compact and that $\mathbb{Z}$ was a dense subset. Hence the argument works for any compact subset of $\mathbb{Q}_p$ and any dense subset of that. (Even for $\mathbb{Z}_p$, that's an advantage. For example, if $p \neq 2$ the even integers are dense in $\mathbb{Z}_p$, and we can interpolate from them to all of $\mathbb{Z}_p$.)

186 This one's really pretty hard. Here's one way. First check that each term in the series for $\mathbf{B}(\alpha, x)$ is continuous as a a function of α. This is easy, since $\binom{\alpha}{n}$ is a polynomial in α. Then check that the series converges uniformly (as a series of functions of α—we mentioned this property in the proof of Prop. 4.4.4; see a book on real analysis for more information). This implies (exactly as in the classical case) that the sum is a continuous function of α.

187 This is simpler than it seems. To begin with, in $\mathbb{Z}_2$ there is a good notion of "even," since 2 is not invertible. Hence if we define $(-1)^\alpha = 1$ is 2 divides α and $= -1$ if not, everything works. On the other hand, if $p \neq 2$, then 2 is invertible, and there's clearly no good way to extend the whole function.

If we want to do the interpolation "in pieces," as in the text, then it works. Take $p = 3$, so that $p - 1 = 2$, and the two choices for α_0 are 0 and 1. In fact, f_0 and f_1 are pretty easy to work out: $f_0(\alpha) = 1$ for all α and $f_1(\alpha) = -1$ for all α. Then f_0 interpolates $(-1)^\alpha$ for $\alpha \equiv 0 \pmod 2$, i.e., for even α, and f_1 does the same for odd α. A dumb example, but maybe it sheds some light on what is going on...

What happens if $p = 5$?

188 Easy: do the same as you did in Problem 42.

189 Again, this is a repeat of Problem 43.

190 The hardest property to check is (ii). For the sup-norm, even that one comes easily: we want to check that $\|v + w\| \leq \|v\| + \|w\|$. Let $v = a_1 v_1 + \cdots + a_n v_n$ and $w = b_1 v_1 + \cdots + b_n v_n$; the inequality translates into

$$\max_i |a_i + b_i| \leq \max_i |a_i| + \max_i |b_i|.$$

But that follows easily from the fact that $|a_i + b_i| \leq |a_i| + |b_i|$ for each i.

For the r-norms, it's a little harder to get (ii); in fact, it may be worth looking it up in books on functional analysis (where it's done in much greater generality). If you'd like to give it a try, here is an outline of the standard proof.

First of all, it's relatively easy to prove the triangle inequality if $r = 1$ or $r = 2$, so we'll concentrate on providing hints for the rest. (Actually, the

proof we sketch works fine for $r = 2$.) Next, for each $r > 1$, let $r' > 1$ be the real number such that
$$\frac{1}{r} + \frac{1}{r'} = 1.$$

We sometimes call r and r' a dual pair.[1] A lot of the proof depends on the duality between the r-norm and the r'-norm. The first lemma is the following:

• Let α and β be positive real numbers, and let r and r' be as above. Then we have
$$\alpha\beta \leq \frac{\alpha^r}{r} + \frac{\beta^{r'}}{r'}.$$

To prove that, plot the function $y = x^{r-1}$, the lines $x = \alpha$ and $y = \beta$, and try to locate in your picture the various quantities that appear in the inequality.

Now the next step: prove the *Holder Inequality*. Let $v_1, v_2, \ldots, v_n$ be a basis, and take two elements
$$v = a_1 v_1 + a_2 v_2 + \cdots + a_n v_n$$

and
$$w = b_1 v_1 + b_2 v_2 + \cdots + b_n v_n.$$

Then show that
$$\sum_{i=1}^{n} |a_i|\,|b_i| \leq \|v\|_r \|w\|_{r'}$$

For the proof, apply the previous inequality with $\alpha = |a_i|/\|v\|_r$ and $\beta = |b_i|/\|w\|_{r'}$ for each $i = 1, 2, \ldots, n$, and add the results. (For $r = 2$, this should be a familiar formula—is it?)

Finally, use the Holder Inequality to prove that $\|v + w\|_r \leq \|v\|_r + \|w\|_r$. Here's the idea: start with the sum whose r-th root is the norm:

$$\sum_{i=1}^{n} |a_i + b_i|^r = \sum_{i=1}^{n} |a_i + b_i|^{r-1}|a_i + b_i|$$
$$\leq \sum_{i=1}^{n} |a_i + b_i|^{r-1}|a_i| + \sum_{i=1}^{n} |a_i + b_i|^{r-1}|b_i|$$

(where we've just used the triangle inequality for the absolute value), and now apply Holder's inequality to both summands and the pair (r, r').

This was a hard one!

---

[1]Notice that if $r = 2$, then $r' = 2$; this is what makes the case $r = 2$ special. On the other hand, if $r = 1$, the only sensible choice for r' is $+\infty$. What norm would that correspond to?

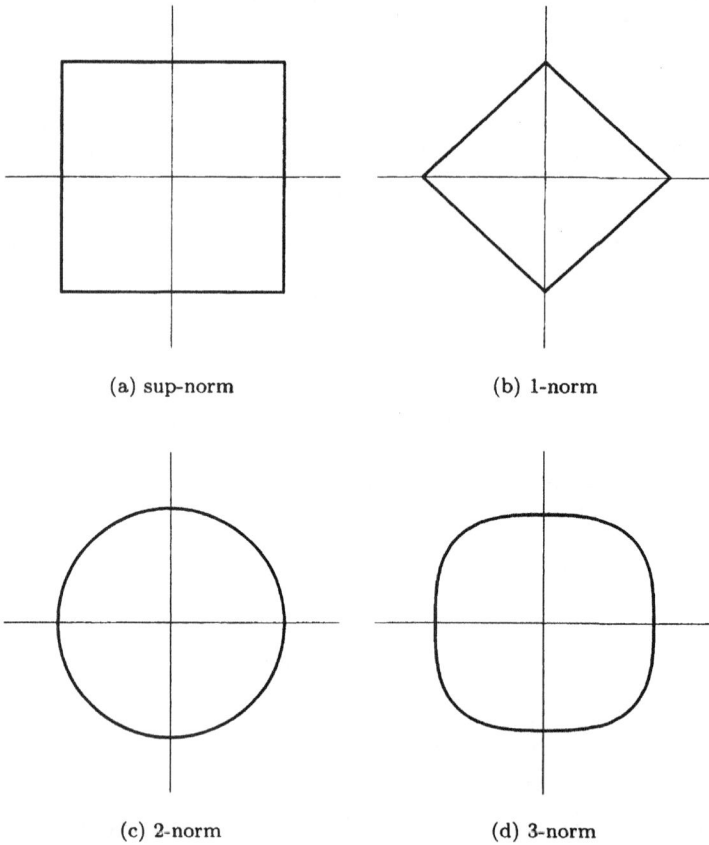

(a) sup-norm   (b) 1-norm

(c) 2-norm   (d) 3-norm

Figure A.1: Unit balls for various norms

191 See figure A.1.

192 Try a vector space of dimension one.

193 Well, $\|(1, -1)\| = 0$ kind of messes things up. (On the other hand, the other two conditions are satisfied; is that significant?)

194 To prove that equivalent norms define the same topology, it's enough to show that an open ball with respect to one norm is an open set with respect to the other. Since this is a vector space with a norm, it's enough to prove this for one ball, say, the open unit ball. So let $B = \{x \in V : \|x\|_1 < 1\}$. If $x \in B$, then let $r = \|x\|_1$. Choose $R < (1 - r)/C$; it's easy to see that the set $N = \{y \in V : \|y - x\|_2 < R\}$, which is an open ball with respect to $\| \ \|_2$, is contained in B. This shows B is open with respect to $\| \ \|_2$, and, since everything is symmetric, proves what we want.

For the converse, we can be direct or we can be fancy. For a direct approach, show that if the two topologies are equivalent, the closed unit ball with respect to one norm must contain a closed unit ball with respect to the other. (For example, argue that the open unit ball for $\| \ \|_1$ is open with respect to $\| \ \|_2$, and hence contains an open $\| \ \|_2$-ball around zero, which contains a closed $\| \ \|_2$-ball—of slightly smaller radius—around zero.) Then look closely at what this means to get one of the inequalities we want.

A fancier approach would be this: consider the identity map $\iota : V \longrightarrow V$, so that $\iota(v) = v$. We give the "first" V the norm $\| \ \|_1$ and we give the "second" V $\| \ \|_2$. Since this yields the same topology on "both" V's, both ι and its inverse are continuous linear transformations. Unwinding the continuity yields the inequalities we want.

195 The two inequalities in the definition of equivalence can be restated as

$$\frac{1}{D}\|\mathbf{v}\|_2 \leq \|\mathbf{v}\|_1 \leq C\|\mathbf{v}\|_2$$

for any $\mathbf{v} \in V$. This clearly translates to what we said about closed balls.

196 Easy.

197 The sketches make it clear that any ball with respect to one of the norms, say $\| \ \|_1$, both contains and is contained in balls with respect to the other norms, and this translates directly into the existence of C and D.

198 $\max\{|a|, |b|\} \leq |a| + |b| \leq 2\max\{|a|, |b|\}$.

199 This is mostly straightforward if we do it in the usual "circle of implications" way. To see, for example, that (i) implies (ii), suppose that f is continuous at $\mathbf{0}$. Then given any $\varepsilon > 0$ there exists a $\delta > 0$ such that $\|\mathbf{v}\| \leq \delta$ implies $\|f(\mathbf{v})\| \leq \varepsilon$. Making δ smaller if necessary, we can find an element $x \in K$ such that $|x| = \delta$. But then we have

$$\|\mathbf{v}\| \leq 1 \Longrightarrow \|x\mathbf{v}\| \leq |x| = \delta \Longrightarrow \|f(x\mathbf{v})\| \leq \varepsilon$$
$$\Longrightarrow \|xf(\mathbf{v})\| \leq \varepsilon \Longrightarrow \delta\|f(\mathbf{v})\| \leq \varepsilon$$
$$\Longrightarrow \|f(\mathbf{v})\| \leq \frac{\varepsilon}{\delta},$$

so that the sup is finite. The other implications are similar.

For an added challenge, show that all of these conditions are also equivalent to the assertion that there exists some $\mathbf{v} \in V$ and some positive $r \in \mathbb{R}$ such that f is bounded on the closed ball of radius r around $\mathbf{v}$.

200 This is a standard example in functional analysis (these spaces are sometimes known as $\ell_\infty(K)$ and $\ell_1(K)$). If necessary, look it up.

201 See Chapter 6!

202 The picture in two dimensions is like this: the unit ball with respect to the sup-norm is a "square" defined by $|a_1| \leq 1$ and $|a_2| \leq 1$, where a_1 and a_2 are the coordinates with respect to our basis. (In $\mathbb{R}^2$, this is the square given by $-1 \leq x \leq 1$ and $-1 \leq y \leq 1$.) What we are doing is partitioning the sides of the square into pieces of radius less than ε and using this partition to cut the "square" up into lots of "rectangles." Then we show that the rectangles do the job. *Now* draw the picture.

203 Checking that 2 is not a square in $\mathbb{Q}_5$ is just a matter of seeing that it is not a square modulo 5, which is easy. For the norm, we can try

$$\|a + b\sqrt{2}\| = \sqrt{|a|^2 + |b|^2}$$

This gives the 5-adic norm when $b = 0$, i.e., on $\mathbb{Q}_5$, but is not an absolute value on $\mathbb{Q}_5(\sqrt{2})$—why not?

204 The point is that one of the cube roots of two is real, and the other two are complex. The field obtained by adjoining the real root is contained in $\mathbb{R}$, hence can't be equal to its image under an automorphism mapping $\sqrt[3]{2}$ to a complex cube root.

205 Either $\sigma(\sqrt{D}) = \sqrt{D}$ or $\sigma(\sqrt{D}) = -\sqrt{D}$, because those are the only two roots of $X^2 - D$. But any field that contains $\sqrt{D}$ contains $-\sqrt{D}$.

206 Any σ must map ζ to another root of $X^2 + X + 1$; the roots are ζ and ζ^2, so we're OK. Similarly the image of $\sqrt[3]{2}$ must be a cube root of two; there are three: $\sqrt[3]{2}$, $\zeta\sqrt[3]{2}$, and $\zeta^2\sqrt[3]{2}$, and they are all in K.

207 The point is that the minimal polynomial of α is the characteristic polynomial of the matrix.

208 Follow the hints!

209 If we remember that $K = F(\alpha)$ is isomorphic to the quotient of $F[X]$ by the ideal generated by $f(X)$, it's not hard. Let $\mathbf{C}$ be an algebraically closed field containing K. For any root α' of $f(X)$, consider the map $K[X] \longrightarrow \mathbf{C}$ mapping X to α'; pass to the quotient to get a map from $K = F(\alpha)$ to $\mathbf{C}$, whose image must be K, by normality. To get the final conclusion, write $f(X)$ as a product of linear factors.

210 If K/F is normal, but K is not equal to $F(\alpha)$, then just use Problem 208.

211 Does taking the product in the normal closure work?

212 A general quadratic extension works exactly like the example in the text. For the second half, it can be easier or harder depending on the elements you choose to work with; I'd try ζ, $\sqrt[3]{2}$, and $\zeta + \sqrt[3]{2}$. The first two are easy; the determinant method is tempting for the last one, but we'll have to compute a six-by-six determinant...

213 No big deal. The only real point: given x and y in the algebraic closure, the field $\mathbb{Q}_p(x, y)$ is a finite extension of $\mathbb{Q}_p$; hence the norm we have defined gives an absolute value on $\mathbb{Q}_p(x, y)$. It follows that $|x + y| \le \max\{|x|, |y|\}$ and that $|xy| = |x|\,|y|$, which is what we needed to prove.

214 Use the same strategy as in the Lemma, i.e., reduce modulo p after making sure that everything is in $\mathbb{Z}_p$.

215 Suppose $f(X)$ factors in $\mathbb{Q}_p[X]$; by the lemma, it also factors in $\mathbb{Z}_p[X]$. Since $f(X)$ is monic, the top coefficients of each of the factors must be invertible in $\mathbb{Z}_p$ (yes?), and therefore are non-zero modulo p. If we now reduce modulo p we get a non-trivial factorization in $\mathbb{F}_p[X]$.

216 The argument will still work if we assume that the top coefficient of $f(X)$ is invertible. Otherwise, the reduction modulo p of $f(X)$ will have degree *smaller* than the degree of $f(X)$, and things begin to get weird.

217 Well, modulo p an Eisenstein polynomial looks like X^n. If we factor that as $X^r \cdot X^s$, the factors are not relatively prime, so we can't apply the Lemma. If we factor as $X^n \cdot 1$, we can, but the factorization will be as the product of a polynomial of degree n and a polynomial of degree zero, which means it will be the trivial factorization.

218 Yes, and this is proved in Chapter 6. In fact, proving this first would allow us to simplify many of the proofs in this section.

219 Let $f(X) \in \mathbb{Z}_p[X]$ be a monic polynomial of degree n such that $\bar{f}(X)$ is irreducible in $\mathbb{F}_p[X]$ and whose roots generate an extension $\mathbb{F}$ of degree n. Let K be the extension of $\mathbb{Q}_p$ obtained by adjoining a root of $f(X)$. We know both K and $\mathbb{F}$ are extensions of degree n, and we've arranged things so that $\mathbb{F}$ is a normal extension. Now: (1) use Hensel's Lemma to show that $f(X)$ has n roots in K, then (2) conclude that K is a normal extension of $\mathbb{Q}_p$. Use the fact that automorphisms preserve absolute values to show (3) that every automorphism of $K/\mathbb{Q}_p$ induces an automorphism of $\mathbb{F}/\mathbb{F}_p$. This gives a map from $\mathrm{Gal}(K/\mathbb{Q}_p)$ to $\mathrm{Gal}(\mathbb{F}/\mathbb{F}_p)$. Then it's a matter of showing this map is injective (and is therefore an isomorphism).

220 Eliminate one prime at a time from the denominator.

221 Yes, of course: a polynomial in $\mathbb{Z}[X]$ which satisfies the conditions in the Eisenstein criterion for some prime p is irreducible in $\mathbb{Q}_p[X]$, and *a fortiori*[2] irreducible in $\mathbb{Q}[X]$.

222 Yes. Can you prove it?

223 Let's do the first one;

$$\mathbf{N}_{F_1/\mathbb{Q}_5}(1 + 3\sqrt{2}) = (1 + 3\sqrt{2})(1 - 3\sqrt{2}) = 1 - 18 = -17,$$

so

$$v_5(1 + 3\sqrt{2}) = \frac{1}{2}v_5(-17) = 0.$$

The others are similar, but keep in mind that what we know about valuations still works. For example, it's easy to see that $v_5(\sqrt{2}) = 0$, by computing the norm, and then it follows that $v_5(5\sqrt{2}) = 1$; hence ("all triangles are isosceles") $v_5(1 + 5\sqrt{2}) = 0$.

224 The nicest one of these is $x = \sqrt{5}$:

$$\mathbf{N}_{F_2/\mathbb{Q}_5}(\sqrt{5}) = (\sqrt{5})(-\sqrt{5}) = -5,$$

so $v_5(\sqrt{5}) = 1/2$. As well it should be!

225 Let's try $x = 1 - \zeta$:

$$\mathbf{N}_{F_3/\mathbb{Q}_3}(1 - \zeta) = \mathbf{N}_{\mathbb{Q}_3(\zeta)/\mathbb{Q}_3}\left(\mathbf{N}_{F_3/\mathbb{Q}_3(\zeta)}(1 - \zeta)\right)$$

$$= \mathbf{N}_{\mathbb{Q}_3(\zeta)/\mathbb{Q}_3}(1 - \zeta)^2.$$

(Remember that F_3 is an extension of degree 2 of $\mathbb{Q}_3(\zeta)$.) To compute the norm, take $\{1, \zeta\}$ as a basis for $\mathbb{Q}_3(\zeta)$ over $\mathbb{Q}_3$. The matrix of multiplication by $1 - \zeta$ is

$$\begin{pmatrix} 1 & 1 \\ -1 & 2 \end{pmatrix}$$

(remember that $1 + \zeta + \zeta^2 = 0$ for that one), so the norm (which is the determinant) is 3. It follows that

$$\mathbf{N}_{F_3/\mathbb{Q}_3}(1 - \zeta) = 9$$

---

[2]Run for the dictionary! What does he mean, *a fortiori*?

and then that

$$v_3(1 - \zeta) = \frac{1}{4}v_3(9) = \frac{1}{2}.$$

Try some of the others.

226 It's not too hard to see that the answers must be $e = 1, 2, 2$, respectively. But how would a proof go?

227 For F_1, $\pi = 5$ will do, and for F_2, $\pi = \sqrt{5}$. For F_3, the computation is problem 225 helps: $\pi = 1 - \zeta$ does the job.

228 Follow the hints; this is mostly straightforward. For example, to show that $\mathfrak{p}_K$ is principal, just note that

$$\begin{aligned}
x \in \mathfrak{p}_K &\implies v_p(x) > 0 \implies v_p(x) \geq 1/e \\
&\implies v_p(\pi^{-1}x) \geq 0 \implies \pi^{-1}x \in \mathcal{O}_K \\
&\implies x \in \pi\mathcal{O}_K.
\end{aligned}$$

This is already enough to show π generates $\mathfrak{p}_K$.

The only non-trivial bit in the remainder is showing that the elements of $\mathcal{O}_K$ are exactly the elements of K which are roots of monic polynomials with coefficients in $\mathbb{Z}_p$. In one direction, it's easy: if α is the root of such a polynomial, then its norm is (up to sign) a power of the zeroth coefficient, which is in $\mathbb{Z}_p$. Hence, $v_p(\alpha) = \frac{1}{n}v_p(N\alpha) \geq 0$. For the converse, look at Lemma 5.3.6.

229 For F_1, we get $\mathcal{O} = \mathbb{Z}_5[\sqrt{2}]$, and $\Bbbk = \mathbb{F}_5[\sqrt{2}]$ is a field of order 25. For F_2, $\mathcal{O} = \mathbb{Z}_5[\sqrt{5}]$ and $\Bbbk = \mathbb{F}_5$. For F_3, $\mathcal{O} = \mathbb{Z}_3[\zeta, \sqrt{2}]$ and $\Bbbk$ is a field of order 9.

230 Routine, but important routine. Make sure you understand how both portions of the proof work—for the most part, it's a question of keeping track of what is divisible by what. Can you come up with a more conceptual proof?

231 $X^2 - 5$, of course.

232 We've done all the work already, when we computed the norm of $1 - \zeta$, which is 3. It follows that $v_3(1 - \zeta) = 1/2$, and the extension (which is of degree 2) is totally ramified. The Eisenstein polynomial for $1 - \zeta$ is $X^2 - 3X + 3$ (just square $1 - \zeta$ and see what coefficients work, or use the fact that $\zeta^2 + \zeta + 1 = 0$). Notice that in this case there is another uniformizer, $\sqrt{-3}$, since it's easy to see that

$$\zeta = \frac{-1 + \sqrt{-3}}{2}$$

(since it is a root of $X^2 + X + 1$).

233 Easy: *exactly* the same proof works—just replace all the p's by π's.

234 The obvious reformulation works, and again the proof is the same.

235 Well, if you solved problem 112, then your solution solves this one too.

236 F_1 contains the 24-th roots of unity; for F_2, there's no new information (only that it contains the 4-th roots of unity, which $\mathbb{Q}_5$ already does). F_3 contains the 8-th roots of unity (the degree is 4, and $e = 2$, so $f = 2$). As for other roots of unity, F_3 certainly contains the cube roots of unity, by construction. (Notice that the cube roots of unity are 1-units, since $\zeta - 1$ is a uniformizer. That means they are "invisible" from the Hensel's Lemma side, and therefore not predicted by the Corollary.) What about the other two fields?

237 The first is really easy: $x^m = 1$ implies $|x| = 1$. For the second part, if $x^m = 1$ and m is prime to $p^f - 1$, then look at the image of x in $\mathbb{k}^\times$, and remember that this last is a cyclic group of order $p^f - 1$.

238 Write $x = 1 + \pi u$ and raise to the p-th power, remembering that π is a divisor of p. The version for general r requires, of course, an easy induction argument.

239 Just replace 1 by the other m-th root of unity and repeat exactly the same argument.

240 Expanding $(1 - x_1)^\ell - 1 = 0$ shows that

$$\ell x_1 + \sum_{i=2}^{\ell} \binom{\ell}{i} x_1^i = 0.$$

Dividing by x_1 and rearranging shows that

$$|\ell| = \left| \sum_{i=2}^{\ell} \binom{\ell}{i} x_1^{i-1} \right|.$$

But the left hand side is 1, since $\ell \neq p$, and every term of the right hand side is in $\mathfrak{p}_K$, which is a contradiction.

241 The quotients are clearly abelian, and problem 238 shows that $x \in U_n$ implies $x^p \in U_{n+1}$, so that every element of the quotient is of order p. Now: why is the quotient a finite group? (An idea: fix a uniformizer, and consider the map $U_n \to \mathcal{O}_K$ given by $1 + \pi^n x \mapsto x$; what properties does this have?)

242 To see that they are both unramified, it's enough to check that both $X^2 - 2$ and $X^2 - 3$ are irreducible over $\mathbb{F}_5$. That they are the same simply means that one can express $\sqrt{2}$ as $a + b\sqrt{3}$ with a and b in $\mathbb{Q}_5$. That's rather hard to imagine, isn't it? But notice that 6 is a square in $\mathbb{Q}_5$ by Hensel's lemma! So if $\gamma \in \mathbb{Q}_5$ satisfies $\gamma^2 = 6$, we have $\sqrt{2}\sqrt{3} = \pm\gamma$, and there we are:

$$\sqrt{2} = \pm\frac{\gamma}{\sqrt{3}} = \pm\frac{\gamma\sqrt{3}}{3}.$$

As to the p-adic expansion of a 24-th root of unity, we need to choose our "digits" first. Since the residue field is $\mathbb{F}_5[\sqrt{2}]$, we might take coefficients from the nonzero elements of the set $\{a + b\sqrt{2} : 0 \le a,\ b \le 4\}$. Then, to find the expansion, we need to determine, first, the reduction modulo p. That'll have to be an element of order 24 in $\mathbb{F}_5[\sqrt{2}]$. Once you find one, use the procedure in Hensel's Lemma to get closer and closer to the real thing.

243 F_3 is an extension of degree four, and is ramified (in fact, e=2). Hence, the subfield $\mathbb{Q}_3(\sqrt{2})$, which is unramified, must be the maximal unramified subfield.

244 Here's a fancy proof that uses the uniqueness we have just proved. Suppose the extension is unramified. Then it is equal to the unique unramified extension of degree 3. Now consider the extension $K' = \mathbb{Q}_3(\zeta\sqrt[3]{2})$ obtained by adjoining another cube root of 2 (here, as before, $\zeta^3 = 1$, $\zeta \ne 1$). If K is unramified, then so is K', since they are clearly isomorphic. If they are both unramified, then they are equal, by the uniqueness. If $K = K'$, then, since both $\sqrt[3]{2}$ and $\zeta\sqrt[3]{2}$ are in K, we must have $\zeta \in K$ and, in fact, $\mathbb{Q}_3(\zeta) \subset K$, which is impossible: extensions of degree 3 can't contain subextensions of degree 2! Hence, K must be ramified, and since e must divide the degree, we must have $e = 3$.

Of course, that's a very fancy chain of reasoning! (It's the one that the author followed, though...) Can you exhibit directly an element of K whose minimal polynomial over $\mathbb{Q}_3$ is Eisenstein?

245 The point is that for any such m one can find an r such that m divides $p^r - 1$, and the $(p^r - 1)$-st roots of unity are in $\mathbb{Q}_p^{unr}$.

246 The image of v_p is still $\mathbb{Z}$, since there has been no ramification. The residue field is the algebraic closure of $\mathbb{F}_p$.

247 The residue field is the same (it can't very well become any bigger), but the image of v_p on $\overline{\mathbb{Q}}_p$ is $\mathbb{Q}$.

248 Let $\alpha \in K$ be a uniformizer; since K is totally ramified, the minimal polynomial for α is an Eisenstein polynomial, so $\alpha^e + a_{e-1}\alpha^{e-1} + \cdots + a_1\alpha + a_0 = 0$ with $p|a_i$ for all i and $p^2 \nmid a_0$. Now rearrange the equation to get

$$\alpha^e + a_0 = -(a_{e-1}\alpha^{e-1} + \cdots + a_1\alpha).$$

Every term on the right hand side is divisible by $p\alpha$, so that $v_p(\alpha^e - a_0) \geq 1 + 1/e$. This suggests that the pu in the problem will be $-a_0$, but to make it work we have to show that we can pass from the "approximate root" α to a real root. The obvious way to do this is Hensel's Lemma, but that method doesn't work: if we put $f(X) = X^e - pu$, our estimates give $v_p(f(\alpha)) \geq 1 + 1/e$ and $v_p(f'(\alpha)) = 1 - 1/e$, which isn't enough to use problem 235.

Here's a more direct method (taken from [Kob84]) that avoids that dilemma by using some analysis: we have $|\alpha^e - pu| \leq p^{-1}p^{-1/e} = |pu|p^{-1/e}$. Dividing through by $|pu|$ gives $|(\alpha^e/pu) - 1| \leq p^{-1/e}$. In other words, α^e/pu is a 1-unit. Using the binomial series, we can raise any 1-unit to any p-adic integer. Since $p \nmid e$, we have $1/e \in \mathbb{Z}_p$; use the binomial series to get $(\alpha^e/pu)^{1/e}$, and then use that to get a root.

249 Yes, of course. Just replace p by a uniformizer π everywhere.

250 This is mostly a matter of time and patience. There should be no difficulty in checking that everything still works as before.

251 This is hard! Clearly, the expansion begins with the term in π^{p-1}. The coefficient will be the reduction mod p of the quotient p/π^{p-1}. Let's see...

We know that π is a root of the polynomial

$$\Phi_p(X+1) = X^{p-1} + \cdots + p$$

(we've shown that everything in the dots is divisible by p, but we don't really know exactly what the coefficients are). Plugging in gives

$$\pi^{p-1} + a_{p-2}\pi^{p-2} + \cdots + a_1\pi + p = 0$$

dividing by p and rearranging,

$$\frac{\pi^{p-1}}{p} = -\frac{a_{p-2}\pi^{p-2} + \cdots + a_1\pi}{p}.$$

Now, we can certainly determine this for any specific p, but it's hard to see what to do to get a general result. And that's just for the first coefficient!

252 Just use $x_1 = 1$ as your initial root.

253 Well, if I could, I would very likely have put the simpler one in the text... Can you?

254 If $K = \mathbb{Q}_p(\zeta_p)$ contained *both* a $(p-1)$-st root of $-p$ and a $(p-1)$-st root of p, then it would contain an element ξ such that $\xi^{p-1} = -1$. Such a thing would be a $2(p-1)$-st root of unity, and we already know that K (a totally ramified extension of $\mathbb{Q}_p$) contains exactly as many prime-to-p roots of unity as $\mathbb{Q}_p$ does, and hence can't contain ξ.

255 Show first that the minimum will occur at a power of p.

256 This is really an open-ended project, which should be fun to play with. Is the extension still totally ramified? Can we get higher-order roots of $-p$? How about the logarithm function?

257 To use Krasner's Lemma, what we want to do is find a generator x of $\mathbb{Q}_p(\zeta_p)$ and a $(p-1)$-st root a of $-p$ such that $|x - a|$ is less than $|a - a'|$ whenever a' is some other $(p-1)$-st root of $-p$. (The a' are the conjugates of a over $\mathbb{Q}_p$, since the polynomial $X^{p-1} + p$ is irreducible.) Understanding the $|a - a'|$ part isn't hard: any a' must be of the form ξa, where $\xi^{p-1} = 1$, and hence, $|a - a'| = |a - \xi a| = |a|\,|1 - \xi| = p^{-1/(p-1)}$. (Remember that we proved before that $|1 - \xi| = 1$ when $\xi \neq 1$ is a root of unity of order prime to p.) The other bit—finding the x—is a little harder. One idea is to repeat the proof in the text to get

$$(1 - \zeta)^{p-1} \cdot (\text{a } 1\text{-unit}) = -p.$$

Use this equation to choose a appropriately so that we get

$$(1 - \zeta) \cdot (\text{a } 1\text{-unit}) = a.$$

(In other words, we want to "take the $(p-1)$-st root of both sides of the equation;" of course, $-p$ has many $(p-1)$-st roots, and what the equation does is tell us how to choose the right one.) This gives

$$|(1 - \zeta) - a| < p^{-1/(p-1)},$$

which gives what we want.

(Notice that the real work of the proof ends up being the same. Krasner's Lemma just replaces Hensel's Lemma as the trump card.)

258 Suppose we can show that the function

$$\phi : (a_0, a_1, \ldots, a_{n-1}, b_0, b_1, \ldots, b_{n-1}) \mapsto D$$

is continuous. Then notice that $\phi(a_0, a_1, \ldots, a_{n-1}, a_0, a_1, \ldots, a_{n-1}) = 0$ (since in that case the λ's and the μ's are the same). By continuity, it will follow that we can make D as small as we like by choosing the b's close enough to the a's, which proves Claim 2.

It remains to show that ϕ is indeed continuous. In fact, it's not hard to see that ϕ is a *polynomial* in $a_0, a_1, \ldots, a_{n-1}, b_0, b_1, \ldots, b_{n-1}$. This is a very classical fact, and one that you may have met before; if not, try to come up with a proof.

259 Solving this one is, at least at a first stage, a matter of reading through proofs carefully to see what fails if we drop any of the hypotheses. It's rather clear, for example, that characteristic zero plays a minor role (though we might need to add a separability condition if we drop it). Will the theorems work in complete archimedean fields? (That's not much fun, because they will be talking about roots of polynomials with real coefficients, not a very mysterious topic.) Will they work if we drop completeness? That would be quite interesting, since we would then have a choice of absolute values to work with.

260 Well, we've come close to proving this one in the proof of the Corollary 5.7.3, since there we obtained the conclusion by showing that some root of $g(X)$ was very close to some root of $f(X)$. See if you can push it through to get this result. If you can't, check [Kob84, Section III.3].

261 What happens in either if we add a term $b_m X^m$ to $g(X)$ where m is very large and b_m is very small? How do the roots of

$$b_0 + b_1 X + \cdots + b_{10} X^{10}$$

relate to the roots of

$$b_0 + b_1 X + \cdots + b_{10} X^{10} + p^{100000} X^{100000}?$$

(See the section on Newton polygons, in Chapter 6, for further light on this one.)

262 Yes. (Prove it.)

263 The condition $\mathbb{Q}_p(\zeta_{i-1}) \subset \mathbb{Q}_p(\zeta_i)$ will hold if m_{i-1} is a divisor of m_i, so we have to make sure that holds. But we need to remember a little more about unramified extensions to get this right. First of all, remember that we get an unramified extension of degree f by adjoining the $(p^f - 1)$-st roots of unity. So we might as well assume that $m_i = p^{f_i} - 1$. You should first show that if $f_{i-1} | f_i$ then $p^{f_{i-1}} - 1 | p^{f_i} - 1$. The divisibility condition in the m's, then, translates to a divisibility condition in the f's. Next, suppose we have a tower of fields $\mathbb{Q}_p \subset \mathbb{Q}_p(\zeta_{i-1}) \subset \mathbb{Q}_p(\zeta_i)$. Then $[\mathbb{Q}_p(\zeta_i) : \mathbb{Q}_p] = f_i$ and $[\mathbb{Q}_p(\zeta_{i-1}) : \mathbb{Q}_p] = f_{i-1}$, which shows that $[\mathbb{Q}_p(\zeta_i) : \mathbb{Q}_p(\zeta_{i-1})] = f_i/f_{i-1}$. Now it should not be hard to find the appropriate choice for the f's.

264 Elements of $\mathbb{Q}_p^{\mathrm{unr}}$ still have p-adic expansions, since p is still a uniformizer. The coefficients in such an expansion will be chosen from a set of lifts of elements of the residue field, and the roots of unity we have used are precisely such a set of lifts. It's not clear that this clarifies anything. Note, however, that there are constructions of transcendental elements in $\mathbb{R}$ which proceed by constructing an appropriate decimal expansion. Is there any analogy?

265 This should be clear. There is nothing in our constructions that depends explicitly on the field being $\mathbb{Q}$: any field with a non-archimedean valuation will clearly do.

266 The definitions follow this problem in the text. The residue field is the algebraic closure of $\mathbb{F}_p$; the valuation ideal is not principal (there is no smallest positive rational number!), and therefore there is no uniformizer.

267 Suppose $x \in \mathbb{C}_p$ and $v_p(x) = r = a/b$. Choose a root π of $X^b - p^a$ in $\overline{\mathbb{Q}}_p$; it's fair to say that π is a "fractional power of p," and it's also clear that $v_p(\pi) = a/b$. Then $y = x/\pi$ is clearly a unit. Go on from there.

268 Follow the outline. (This makes a nice longer project.)

269 One would need to show that the closed unit ball is not compact. To do that, you need to exhibit a covering of the closed unit ball by open sets which has no finite subcovering. Can you find one? (The closed unit ball is just the valuation ring $\mathfrak{O}$. Consider the image of $\mathfrak{O}$ under reduction modulo $\mathfrak{p}$; how many elements are in the image? Now translate back to topological language.)

270 Mostly routine. The point about ρ is simply that there are enough different possible radii for balls in $\mathbb{C}_p$ (any p^r with $r \in \mathbb{Q}$ is allowed, and this is a dense subset of $\mathbb{R}$).

271 It was true in $\mathbb{Q}_p$ because the ideal in question was a principal ideal. This isn't true over $\mathbb{C}_p$.

272 Imitate the proof in Chapter 3.

273 Given the caution about choosing δ appropriately, it's just a matter of repeating the original proof.

274 Since every polynomial with coefficients in $\mathbb{F}$ has a root, having a common factor is the same as having a common root.

275 Yes, because we know that $g(X)$ is monic.

276 Parts (i) and (iv) are clear, (ii) follows by writing out the sum and applying the ultrametric inequality coefficient by coefficient, and (v) is a straight application of the ultrametric inequality (and was done in the text just before the statement of the theorem).

277 Clear, since the absolute value of each of the coefficients is independent of the field we think it belongs to.

278 The inequality is very easy to get: just use the non-archimedean property directly. Over $\mathbb{C}_p$, the equality holds, but this takes some proving. Let's do it in the special case where $c = 1$. In this case, after multiplying by a constant if necessary, we can assume $\|f(X)\|_1 = 1$, so that all the coefficients are in $\mathfrak{O}$ and at least one is a unit. Then reduce it modulo $\mathfrak{p}$ to get a polynomial with coefficients in $\mathbb{F}$; the fact that $f(X)$ has a coefficient that is a unit means that the reduced polynomial is non-zero. Since $\mathbb{F}$ is an infinite field, there must be an element $\alpha \in \mathfrak{O}$ such that $\bar{f}(\bar{\alpha}) \neq 0$ in $\mathbb{F}$. Then it's clear that $|f(\alpha)| = 1$, which proves the equality.

Can you generalize to arbitrary c?

279 Basically, we just replace things like $f(X) \equiv g_1(X)h_1(X) \pmod{\mathfrak{P}}$ with their translation (in this case, $\|f(X) - g_1(X)h_1(X)\|_1 < 1$). As to a version for the $\| \ \|_c$, can you decide? (Take a look, for example, at the proof of Lemma 6.2.2 and the statement of Proposition 6.2.3.)

280 Finding α uses a trick we have used before: if $c = p^r$ and $r = a/b$, we choose α to be a root of the polynomial $X^b - p^{-a}$, which exists because $\mathbb{C}_p$ is algebraically closed. Proving that $\|f(X)\|_c = \|\phi(f(X))\|_1$ is a matter of writing out the definitions. What this tells us is that all of these norms should have similar properties, since the equality allows us to transfer theorems about one to the other. In fancier terms, the theorem gives an isometric isomorphism between two normed rings.

281 One would need to be a lot more careful, the problem being that it is no longer clear that an α exists. What we would need to do is to further restrict c. To be precise, the argument still works over a field whose ramification index is e if we restrict c to be a real number of the form p^r where $r \in \frac{1}{e}\mathbb{Z}$.

282 Not really, since no α is available. But look at the proof of Lemma 6.2.2.

283 If we have a sequence of polynomials of bounded degree, we might as well think of them as being all of the same degree (by padding the top terms with zeros if necessary). So let $f_i(X)$ be a sequence of polynomials of degree n. The first requirement is to dig out a candidate for the limit, and the obvious thing works: consider each of the coefficients and note that they form a Cauchy sequence themselves. Then it's a matter of showing that the polynomial just obtained is the limit we want.

As to why the boundedness is essential, the simplest example makes the point: take $c = 1$ and look at the sequence

$$f_0(X) = 1$$
$$f_1(X) = 1 + pX$$
$$f_2(X) = 1 + pX + p^2X^2$$
$$\cdots$$
$$f_i(X) = 1 + pX + p^2X^2 + \cdots + p^iX^i$$

This is clearly Cauchy, and clearly its limit cannot possibly be a polynomial.

284 Because we know something about its N-th coefficient. Fill in the details.

285 Just notice that the inequality holds at every step of the inductive construction of $g(X)$.

286 Start from the fact that $\|f(X) - g(X)\|_c < \|f(X)\|_c$. This says, in particular, that $|a_N - b_N|c^N < |a_N|c^N$, which implies, via "all triangles are isosceles," that $|b_N| = |a_N|$. Since we know that $\|g(X)\|_c = \|f(X)\|_c = |a_N|c^N$, the claim follows.

287 Just from the proof we can see that $g_i(X)$ converges to $g(X)$ at least as fast as δ^i converges to zero. That already says that the convergence is quite good. It may be, of course, that a more delicate analysis shows that the convergence is in fact faster than that.

288 If $c = 1$, then we can multiply $f(X)$ by a constant to assume that $\|f(X)\|_1 = 1$. In that case, the assumption reduces to saying that a_N is a p-adic unit and that $a_j \in \mathfrak{p}$ if $j > N$. The reduction of $f(X)$ modulo $\mathfrak{p}$ is then of degree N, and, after multiplying by another (unit) constant if necessary, we can assume the reduction is monic. This gives a congruence $f(X) \equiv g_1(X) \cdot 1 \pmod{\mathfrak{p}}$. Now apply Theorem 6.1.2.

289 What we need to prove is

- the sum of two series in A_c belongs to A_c,

- the product of two series in A_c belongs to A_c (this implies that the product of a series by a scalar does too, of course).

But both statements follow directly from Proposition 4.3.2.

290 This follows from the fact that absolute values are independent of the field in which we place ourselves.

291 As long as we are thinking of the closed ball in $\mathbb{C}_p$, this is clear. First, if $|a_n|c^n \to 0$, then clearly the series converges for any x such that $|x| \leq c$. For the converse, we just need to note that in $\mathbb{C}_p$ we can always find an x whose absolute value is exactly equal to c. Convergence at that x implies that $|a_n|c^n \to 0$.

Notice that it is important to work over $\mathbb{C}_p$. It is easy to come up with a series that converges in the closed ball of radius $c = p^{-1/100}$ in $\mathbb{Q}_p$ but which is not in A_c, simply because the closed ball of radius c in $\mathbb{Q}_p$ is exactly the same as the closed ball of radius p^{-1}.

If c is not a rational power of p, then there are no x such that $|x| = c$, so the closed ball of radius c is the same as the open ball of radius c. Can you go on from there?

292 Since $0 \leq |a_n|c_2^n < |a_n|c_1^n$, $|a_n|c_1^n \to 0$ implies $|a_n|c_2^n \to 0$.

293 This is identical to problem 276, except for the fact that we've replaced the equality in (iii) with an inequality. But that makes (iii) much easier.

294 No. Can you prove it? (Here's a strategy: handle $c = 1$ first, by exactly the same method, which is feasible because the reduction modulo $\mathfrak{p}$ is still a polynomial. Then use the usual tricks to handle other values of c.)

295 Same as before.

296 Again, the same argument as was used for polynomials works here, and shows that the map is an isometric isomorphism between the two spaces. (It's even a ring isomorphism.)

297 The map will be continuous if we can find a constant M with the property that
$$\|f(X)\|_{c_1} \leq 1 \implies \|f(X)\|_{c_2} \leq M.$$
Try to decide whether such a constant exists. (Thinking about the norms as sup-norms on appropriate balls may help.)

298 We have $g(X) = b_0 + b_1 X + \cdots + b_N X^N$ and $|b_N| = \max |b_n|$. Dividing through by b_N gives a monic polynomial whose coefficients all have absolute value less than or equal to 1. So what we want to prove is this: if $g(X) = b_0 + b_1 X + \cdots + b_{N-1} X^{N-1} + X^N$ satisfies $|b_i| \leq 1$ for all i and α is a root of $g(X)$, then $|\alpha| \leq 1$. To prove it, plug α into $g(X)$ to get
$$\alpha^N + b_{N-1}\alpha^{N-1} + \cdots + b_1 \alpha + b_0 = 0,$$

and rewrite this as

$$\alpha^N = -(b_{N-1}\alpha^{N-1} + \cdots + b_1\alpha + b_0).$$

It follows that

$$|\alpha|^N \le \max_{0 \le i \le N-1}\{|b_i||\alpha|^i\},$$

and, since $|b_i| \le 1$, it follows that

$$|\alpha|^N \le \max_{0 \le i \le N-1}\{|\alpha|^i\}.$$

But this clearly implies $|\alpha| \le 1$.

299 Games with two indices are always a little tricky, but a careful walk through the proof should convince you that all is well.

300 No. It's also not Cauchy.

301 If $c = p^r$ for some $r \in \mathbb{Q}$, then it's easy to see that the answer is yes: just use the trick we've been using over and over. How would you handle general c?

302 Is this obvious? If $a_n \to 0$, then given $\varepsilon > 0$ we can find N such that $|a_n| < \varepsilon$ if $n > N$. But then $\max_{n>k}|a_n| < \varepsilon$ as soon as $k > N$.

303 It certainly should.

304 Do write out the details. The proof follows blow-by-blow the proof of Proposition 6.2.3, so there should be no difficulty in putting it together. But doing so will help you understand what's going on.

305 If $\|h(X) - 1\|_1 < 1$, then whenever $|x| \le 1$ we have $|h(x) - 1| < 1$, which implies $h(x) \ne 0$.

306 By Proposition 154, we can rewrite $f(X)$ as a power series in $(X - x)$, and in this case the fact that n exists becomes obvious. To show that the two definitions of the multiplicity agree, write $f(X)$ as a product as in the Weierstrass Preparation Theorem, and then take derivatives. The advantage of Cassels' definition is, of course, that it doesn't depend on the theorem.

307 It's the usual thing: change variables so as to translate from $\| \ \|_c$ to $\| \ \|_1$. For more general values of c, it's a little harder—see below.

308 This particular game should be routine by now. Just follow the usual outline.

309 If ζ is a p^m-th root of unity, then $f(\zeta - 1) = \log_p(\zeta) = 0$. Hence, $f(X)$ has infinitely many zeros in the open unit ball. How can that be?

310 Write $g(X)$ as a product of linear factors and rearrange as necessary.

311 The roots of $g_0(X)$ (in $\mathbb{C}_p$) are the roots of $f(X)$ in the closed unit ball, counted with multiplicities. The roots of $g_1(pX)$ are the roots of $f(X)$ in the closed ball of radius p, counted with multiplicities. Hence, every root of $g_0(X)$ (in $\mathbb{C}_p$) is also a root of $g_1(pX)$, with the right multiplicities. Therefore, $g_0(X)$ must be a divisor of $g_1(pX)$.

312 If you are at all hesitant, it might be helpful to write out a detailed proof.

313 We've clearly done enough to prove convergence with respect to $\| \ \|_1$. For $c = p^r$ with $r \in \mathbb{Q}$, we then use the usual dodge: change variables, and note that this transforms entire functions into entire functions. What should you do with other c? (Hint: how does convergence with respect to $\| \ \|_c$ relate to convergence with respect to $\| \ \|_{c'}$ when $c > c'$?)

314 The point is that "convergent in the closed ball of radius c" and "convergent with respect to $\| \ \|_c$" are equivalent.

315 As the next problem suggests, one can make such a function by using an infinite product. Something like

$$f(X) = \prod_{i=1}^{\infty}(1 - p^i X)$$

will work. It's also easy to arrange it directly in a power series, something like

$$\sum_{i=0}^{\infty} p^{n(i)} X^i,$$

by making $n(i)$ grow fast enough.

316 This is basically routine. What we are asking for, in a way, is a p-adic version of the general theory of infinite products. This would make a nice project. For example: we've proved that p-adic infinite series converge whenever their general term tends to zero; is it true that p-adic infinite products converge whenever their general term tends to 1?

317 See figures A.2 and A.3. Note that in (*iii*), we don't really need to divide by 3, because it is a 5-adic unit (dividing by a unit doesn't change any of the valuations).

(i)

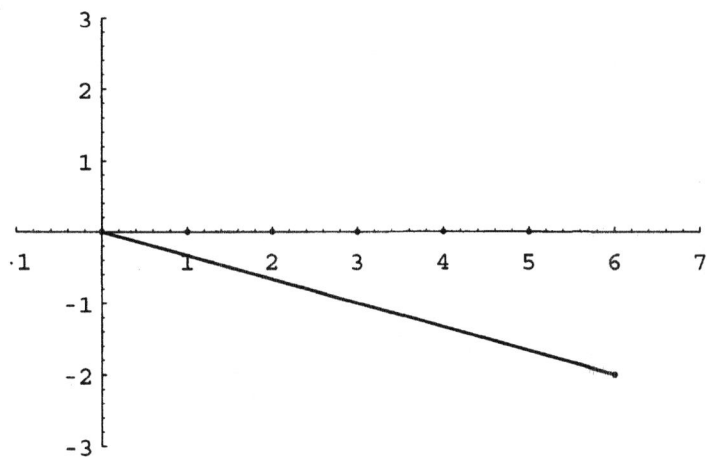

(ii)

Figure A.2: The first two Newton polygons for problem 317

(iii)

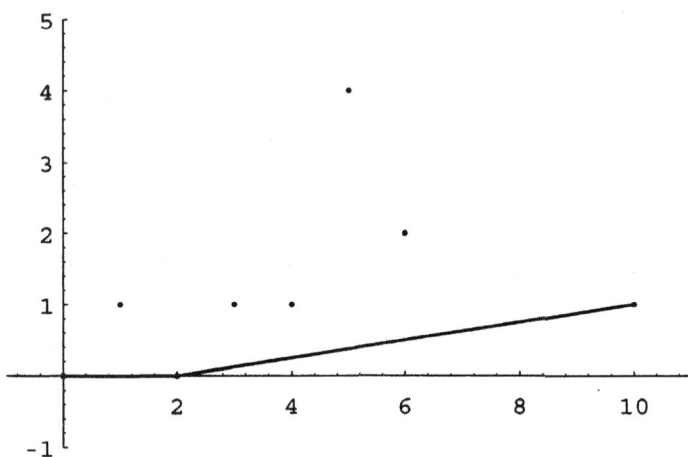

(iv)

Figure A.3: Two more Newton polygons for problem 317

318 If the polynomial has degree n, the polygon has only one line segment, of slope $-1/n$. (In the language we'll introduce below, Eisenstein polynomials of degree n over $\mathbb{Q}_p$ are "pure of slope $-1/n$.")

319 We would have to start with a vertical line beginning at $(0, v_p(a_0))$ and use that point as the initial center for the rotation, but otherwise all would be the same. The polygons for $f(X)$ and $af(X)$ will be the same except for a vertical translation.

320 This is just a matter of sorting through the definitions. $h(X)$ will be pure of slope m if and only if $v_p(a_i) \geq mi$ for all i and $v_p(a_n) = mn$; translating this to absolute values gives what we want.

321 Routine, but worth writing up carefully. It's mostly a matter of translating Proposition 6.2.3 to the language of Newton polygons.

322 Use the fact that $\|f(X)g(X)\|_c = \|f(X)\|_c \|g(X)\|_c$ and the characterization of pure polynomials in problem 320.

323 It's just a question of translating inequalities for v_p to the language of absolute values.

324 Write

$$f(X) = 1 + a_1 X + \cdots + a_n X^n = (1 - \lambda_1 X)(1 - \lambda_2 X) \cdots (1 - \lambda_n X)$$

and work out valuations. We can start by noting that $v_p(a_n) = nm = \sum v_p(\lambda_i)$, and go on from there. (This is much trickier than it looks!)
   A more sophisticated method would be to compute $\| \ \|_{p^m}$ of each term in the product, and then to try to use the characterization of pure polynomials in terms of the norm. Does that work?

325 By induction, we can work at the j-th break, assuming that it happens at a point (x_j, y_j) and that the next point is $(x_j + i_{j+1}, y_j + m_{j+1}i_{j+1})$. It's then a matter of translating the assertion that this is the $(j+1)$-th segment of the polygon into valuations and absolute values.

326 Yes. Can you prove it?

327 The polygon of $h(X)$ is obtained from the polygon of $f(X)$ by removing the segment corresponding to λ, i.e., a segment of slope m and x-length one, and then translating the resulting polygon to the origin. This clearly follows from our analysis of the roots, but it could be used as the starting point for that analysis if we could prove it directly. See [Kob84] for a direct proof.

328 They tell us exactly what sorts of roots each polynomial has:

i) five unit roots, one root of valuation −2;

ii) six roots of valuation 2;

iii) two unit roots, four roots of valuation −1/2;

iv) two unit roots, eight roots of valuation −1/8.

Notice that the fact that the last two polynomials are congruent modulo 5 makes them have the same number of unit roots, but that this says very little about the other roots.

329 One polygon has a segment of slope 0 and a segment of slope 3; the other has a segment of slope 0 and a segment of slope 1. So even though the polynomials are "close," their polygons—and therefore their roots—look quite different. In other words, even if $\|f(X) - g(X)\|_1$ is very small, the root distribution of $f(X)$ and $g(X)$ outside the unit ball can be completely different. (*Inside* the unit ball, they will of course have exactly the same number of zeros of each valuation—check!)

330 The obvious thing to try is to require that $f(X)$ and $g(X)$ be close with respect to the *c*-norm. Does that work?

331 Yes. Find an example.

332 A segment of slope −1 and length 1, followed by an infinite horizontal line. The radius of convergence doesn't change, of course, since we've only made a finite number of changes.

333 Just use the same idea: compare the polygon to a line of slope b, remembering that by assumption the segments in the polygon all have slope less than b. (Does this work when $b = m$ but the sup is not attained? The segments will still "all have slope less than b" in that case... What if the sup *is* attained?)

334 Draw a picture if necessary, and refer to the previous solution.

335 The conclusion is that the radius of convergence is 0. The proofs still work.

336 One can't do much better than the obvious: the series will converge on the closed ball if the points in the polygon get farther and farther above (vertical distance) the last segment.

337 The first looks like a parabola, and defines an entire function. The second is a horizontal line, and so is the third (most coefficients will have valuation zero). The second converges on the closed ball of radius 1, the third on the open ball. The fourth is tricky: the polygon connects the points $(0,0)$, $(p, -2)$, $(p^2, -4)$,... The series converges on the open ball of radius 1.

338 We'll leave these to the reader, who has certainly got the point by now.

339 It's just a matter of putting together all the information we already have. To show there are no zeros of smaller absolute value, consider a line through the $(k-1)$-st break point of slope smaller than m_k, and so on.

340 Yes it is. Can you prove it?

B   A Brief Glance at the Literature

As we end our promenade, it is important to point out to our reader where to go for further adventure in the p-adic realm. We will limit ourselves to a brief outline of the major sources of information (of which we are aware), and invite the reader to explore at will. The comments are, of course, personal opinion.

B.1   Texts

The first category of books are those which are intended as basic textbooks covering the fundamentals of the theory. In general, these are aimed at graduate students, but they should be accessible to anyone who has managed to read this book. Many of these books were major sources of information during the preparation of this book.

p-adic Numbers, p-adic Analysis, and Zeta-functions, by Neal Koblitz, [Kob84], is probably the closest in spirit and subject matter to this book. Koblitz includes an introduction to p-adic numbers and p-adic analysis, and then goes on to discuss p-adic interpolation, the construction of the p-adic zeta-function, and several other related topics. The culmination of this book is an exposition of Dwork's proof of the rationality of the zeta-function of a hypersurface over a finite field, which is one of the landmarks of modern number theory. While the introductory portion of Koblitz's book has much in common with this book, Koblitz goes much further than we have.

Local Fields, by J. W. S. Cassels, [Cas86], is a much broader book that contains a great deal that is interesting. Its treatment of the fundamentals had a lot of influence on our choices when we were writing this book, but once again Cassels goes much further than we do. A particularly interesting characteristic of this book is the large number of examples of honest-to-goodness applications of p-adic methods to the rest of mathematics.

A Course in p-adic Analysis, by A. Robert, [Rob00], is a new book that overlaps this one at the beginning, but then goes much further into p-adic analysis. Robert also includes some unusual material, such as a way of constructing a topological model of the p-adic numbers inside $\mathbb{R}^n$.

Les Nombres p-adiques, by Y. Amice, [Ami75], is another elementary introduction to p-adic numbers and p-adic methods, a little brief but very useful. Readers who read French might enjoy looking through her book, which is slanted towards functional analysis and rationality theorems.

Ultrametric Calculus, by W. H. Schikof, [Sch84], despite its unprepossessing title, is quite an advanced book. Again, the focus is largely on p-adic analysis, and the author assumes that his reader has a good knowledge of classical analysis as a starting point.

Introduction to p-Adic Numbers and Valuation Theory, by G. Bachman, [Bac64], starts with a basic introduction to p-adic numbers, and then veers off into a discussion of valuation theory in general. It is an interesting book, with very little intersection with this one.

Introduction to p-adic Numbers and Their Functions, by Kurt Mahler, [Mah73], is rather hard to classify. While it presents itself as an introduction (and does develop the theory from scratch), it is really focused on a rather sophisticated account of continuous and differentiable functions on $\mathbb{Q}_p$ with a special focus on their interpolation properties. This material is very different in flavor from the topics we have discussed, and the book is well worth the effort.

Finally, there is *Primeiros Passos p-ádicos*, [Gou89], the seed from which this book grew. This one the reader, even if fluent in Portuguese, can safely disregard, since the only things it contains that were not incorporated into this version are the errors, which have been replaced by new and subtler errors.

B.2   Software

It's not really clear to me that software is "literature," but it *is* quite clear that sophisticated software is becoming an integral part of "doing mathematics." The reader who enjoys working with computers may, in fact, find that supplementing the problems in the text with actual computation proves illuminating. It seems good, then, to have a look at the currently available programs that can do p-adic arithmetic.

The natural choice here is the GP-PARI system, due to H. Cohen, C. Batut, D. Bernardi, and M. Olivier. This exists both as an interactive "calculator" called GP and as a library of C routines that one can use in one's own programs. In GP, one works with p-adic numbers by using "big-oh notation:" p-adic numbers are represented as approximations to their p-adic expansions, if the given expansion is to be treated as correct up to the term in p^{99}, one indicates this by adding +O(p^100) at the end of it. Here's an example in which we compute (an approximation to) the square root of 2 in $\mathbb{Q}_7$:

```
? a=2+O(7^30)
%1 = 2 + O(7^30)
? sqrt(a)
%2 = 4 + 5*7 + 4*7^2 + 5*7^4 + 4*7^5 + 5*7^6 + 4*7^7 + 2*7^8 +
4*7^11 + 5 *7^12 + 5*7^13 + 6*7^14 + 4*7^15 + 5*7^16 + 5*7^17 +
2*7^18 + 5*7^20 + 3 *7^21 + 4*7^22 + 3*7^25 + 7^26 + 7^27 +
3*7^29 + O(7^30)
```

The GP-PARI package is freely distributable and runs on many different kinds of computers. See http://www.parigp-home.de/ for more information.

Of course, it is perfectly possible to implement p-adic arithmetic in any of the big "computer algebra" packages. I am aware of such an implementation in MAPLE, and I'm sure it has also been done on MATHEMATICA, MAGMA, and other packages. Of course, one can simulate p-adic arithmetic by working with congruences modulo high powers of p, but this can be ungainly. For example, when working modulo p^n, most packages return an error if one tries to divide an expression by p. For a true p-adic approach, the division should be done, and the result treated as correct modulo p^{n-1}.

B.3   Other Books

There are many other books that either deal with specific aspects of the theory or contain material that relates to one or another topic covered in this book. Here are the ones I like best, in no particular order.

I guess I'll mention first the book that started it all: *Theorie der Algebraischen Zahlen*, by Kurt Hensel, [Hen08]. Hensel had introduced p-adic numbers in various journal articles, but this is their first appearance in a book. For people who read German, this is worth a look, particularly to note the differences between Hensel's point of view and the one we have taken.

Numbers, by a crowd of people headed by H.-D. Ebbinghaus, [EHH+91], is a delightful book about number systems in general. Its first part is called "From the Natural Numbers, to the Complex Numbers, to the p-adics." It is written in a very compressed style, and the various chapters can only survey the basics of each of the number systems, but reading them still is quite an enjoyable ride. There are two other parts: about real division algebras and about Conway's "surreal numbers." This one is definitely worth a look.

Exercises in Number Theory, by the fictitious D. P. Parent ([Par84], which is a translation of the 1978 French original) is a problem book which is really much more ambitious than its title suggests. Each chapter gives a compact introduction to one of the major areas of modern number theory and then presents the reader with problems (full solutions are included). The final chapter is called "p-adic Analysis."

The best elementary treatment of the Hasse-Minkowski theorem is probably the one in J.-P. Serre's *A Course in Arithmetic*, [Ser73]. This book includes a chapter on the basic structure of the p-adic numbers, from the point of view of "coherent sequences," and then goes on to develop the theory of quadratic forms and prove the Hasse-Minkowski theorem over $\mathbb{Q}$. The second half of the book focuses on Analytic Number Theory, and may serve as an introduction to the (classical, rather than p-adic) theories of L-functions and modular forms.

The functional-analytic side of p-adic analysis is the focus of *Analyse Non-Archimedienne*, by A. F. Monna, [Mon70], and of *Non-Archimedian Functional Analysis*, by A. C. M. van Rooij, [vR78]. By contrast, Koblitz's *p-adic Analysis: a Short Course on Recent Work*, [Kob80], focuses on material which has played a big role in arithmetical algebraic geometry: the p-adic L-functions, Gamma function, regulators, and the various theorems relating them.

Rigid Analytic Geometry is a difficult subject, but we should mention *Non-Archimedean Analysis*, [BGR84], by S. Bosch, U. Güntzer, and R. Remmert, which tries to lay down a solid foundation for the subject. This is a very advanced book, but one which contains an enormous amount of information.

These, of course, only scratch the surface, since an enormous amount of research has focused on p-adic methods and their application to number theory. Much of it has in fact not yet been put into books. Other parts have been published as monographs focusing on very specific material, such as Iwasawa's *Lectures on p-adic L-functions*, [Iwa72], *Schottky Groups and Mumford Curves*, by L. Gerritzen and M. van der Put, [GvdP80], B. Dwork's *Lectures on p-adic Differential Equations*, [Dwo82], Volovich and Vladimirov's *p-adic Analysis and Mathematical Physics*, [VV94], A. Khrennikov's *p-adic Valued Distributions in Mathematical Physics*, [Khr94], or my own[1] *Arithmetic of p-adic Modular Forms*, [Gou88]. There are also several books that collect papers on related themes, such as [AST92], [BP92], or [MS94]. A good recent overview of the whole area is Mazur's article "The theme of p-adic variation" in [AALM99]. Our adventurous reader will have no trouble finding more and more to learn, and may soon be in the position to teach us something herself.

---

[1]Could I resist a chance like this?

Bibliography

[Ahl79]   L. V. Ahlfors. *Complex Analysis*. McGraw-Hill, New York, third edition, 1979.

[Ami75]   Y. Amice. *Les Nombres p-adiques*. Presses Universitaires de France, Paris, 1975.

[Apo74]   T. M. Apostol. *Mathematical Analysis*. Addison-Wesley, 1974.

[AST92]   A. Adolphson, S. Sperber, and M. Tretkoff, editors. *p-adic Methods in Number Theory and Algebraic Geometry*, volume 133 of *Contemporary Mathematics*. American Mathematical Society, 1992.

[AALM99] V. Arnold, M Atiyah, P. Lax, and B. Mazur, editors. *Mathematics: Frontiers and Perspectives*. American Mathematical Society, 1999.

[Bac64]   G. Bachman. *Introduction to p-Adic Numbers and Valuation Theory*. Academic Press, New York and London, 1964.

[BGR84]   S. Bosch, U. Güntzer, and R. Remmert. *Non-Archimedian Analysis*. Springer-Verlag, Berlin, Heidelberg, New York, 1984.

[BP92]   A. Baker and R. Plymen, editors. *p-adic methods and their applications*. Oxford University Press, 1992.

[BS96]   E. B. Burger and T. Struppeck. Does $\sum_{n=0}^{\infty} \frac{1}{n!}$ really converge? Infinite series and p-adic analysis. *Amer. Math. Monthly*, 103:565–577, 1996.

[Car95]   H. Cartan. *Elementary theory of analytic functions of one or several complex variables*. Dover Publications, 1995.

[Cas86]   J. W. S. Cassels. *Local Fields*. Cambridge University Press, Cambridge, 1986.

[Cas91]   J. W. S. Cassels. *Lectures on Elliptic Curves*. Cambridge University Press, Cambridge, 1991.

[CT91]   J. Coates and M. J. Taylor, editors. *L-functions and Arithmetic*, volume 153 of *London Mathematical Society Lecture Note Series*. Cambridge University Press, Cambridge, 1991.

[Die44]   J. Dieudonné. Sur les fonctions continues p-adiques. *Bull. Sci. Math.*, 68:79–95, 1944.

[Dwo82]   B. Dwork. *Lectures on p-Adic Differential Equations*. Springer-Verlag, 1982.

[EHH⁺91]   H.-D. Ebbinghaus, H. Hermes, F. Hirzebruch, et al. *Numbers*. Springer-Verlag, Berlin, Heidelberg, New York, 1991.

[Gou88]   F. Q. Gouvêa. *Arithmetic of p-adic Modular Forms*, volume 1304 of *Lecture Notes in Mathematics*. Springer-Verlag, Berlin, Heidelberg, New York, 1988.

[Gou89]   F. Q. Gouvêa. *Primeiros Passos p-ádicos*. IMPA-CNPq, Rio de Janeiro, 1989.

[GvdP80]   L. Gerritzen and M. van der Put. *Mumford Groups and Schottky Curves*, volume 817 of *Lecture Notes in Mathematics*. Springer-Verlag, Berlin, Heidelberg, New York, 1980.

[Has80]   H. Hasse. *Number Theory*. Springer-Verlag, Berlin, Heidelberg, New York, 1980.

[Hen08]   Kurt Hensel. *Theorie der Algebraischen Zahlen*. Teubner, Leipzig and Berlin, 1908.

[Iwa72]   K. Iwasawa. *Lectures on p-adic L-functions*, volume 74 of *Annals of Mathematics Studies*. Princeton University Press, Princeton, 1972.

[Khr94]   A. Khrennikov. *p-adic Valued Distributions in Mathematical Physics*. Kluwer Academic, 1994.

[Kob80]   N. Koblitz. *p-adic Analysis: a Short Course on Recent Work*, volume 46 of *London Mathematical Society Lecture Note Series*. Cambridge University Press, Cambridge, 1980.

[Kob84]   N. Koblitz. *p-adic Numbers, p-adic Analysis, and Zeta-functions*. Springer-Verlag, Berlin, Heidelberg, New York, second edition, 1984.

[Lan89]   S. Lang. *Cyclotomic Fields I and II*. Springer-Verlag, Berlin, Heidelberg, New York, 1989.

[Mah73]   K. Mahler. *Introduction to p-adic numbers and their functions*. Cambridge University Press, Cambridge, 1973.

[Mon70]   A. F. Monna. *Analise Non-Archimedienne.* Springer-Verlag, Berlin, Heidelberg, New York, 1970.

[MS94]   B. Mazur and G. Stevens, editors. *p-adic Monodromy and the Birch and Swinnerton-Dyer Conjecture,* volume 165 of *Contemporary Mathematics.* American Mathematical Society, 1994.

[Par84]   D. P. Parent. *Exercises in Number Theory.* Spriger-Verlag, 1984.

[Rob95]   A. Robert. Le théorème des acroissements finis p-adique. *Ann. Math. Blaise Pascal,* 2(1):245–258, 1995.

[Rob00]   A. Robert. *A Course in p-adic Analysis* Springer-Verlag, 2000.

[Rud76]   W. Rudin. *Principles of Mathematical Analysis.* McGraw-Hill, New York, third edition, 1976.

[Sch84]   W. H. Schikhof. *Ultrametric Calculus.* Cambridge University Press, Cambridge, 1984.

[Ser73]   J.-P. Serre. *A Course in Arithmetic.* Springer-Verlag, Berlin, Heidelberg, New York, 1973.

[Ser74]   J.-P. Serre. *Local Fields.* Springer-Verlag, Berlin, Heidelberg, New York, 1974.

[Tat71]   J. T. Tate. Rigid analytic spaces. *Inv. Math.,* 12:257–289, 1971.

[Vit89]   A. G. Vitushkin, editor. *Several Complex Variables I,* volume 7 of *Encylopaedia of the Mathematical Sciences.* Springer Verlag, 1989.

[vR78]   A. C. M. van Rooij. *Non-Archimedian Functional Analysis.* M. Dekker, 1978.

[VV94]   V. S. Vladimirov and I. V. Volovich. *p-adic Analysis & Mathematical Physics.* World Scientific, 1994.

[Wig64]   E. P. Wigner. The unreasonable effectiveness of mathematics in the natural sciences. *Communications in Pure and Applied Mathematics,* 1964.

[ZS75]   O. Zariski and P. Samuel. *Commutative Algebra.* Springer-Verlag, 1975.

Index

Universitext

图书在版编目（ＣＩＰ）数据

p 进数导论 : 第 2 版 = P-adic Numbers : An Introduction Second Edition :
英文 / （美）F. Q. 戈维亚（F. Q. Gouvêa）著 . — 影印本 . —
北京 : 世界图书出版有限公司北京分公司 , 2017.4
ISBN 978-7-5192-2612-1

Ⅰ . ① p… Ⅱ . ① F… Ⅲ . ① p 进数域—英文 Ⅳ . ① O156.2

中国版本图书馆 CIP 数据核字 (2017) 第 068721 号

中文书名	p 进数导论 第 2 版
英文书名	p-adic Numbers: An Introduction Second Edition
著　　者	Fernando Q. Gouvêa
责任编辑	刘 慧　　高 蓉
装帧设计	任志远
出版发行	世界图书出版有限公司北京分公司
地　　址	北京市东城区朝内大街 137 号
邮　　编	100010
电　　话	010-64038355（发行）　64037380（客服）　64033507（总编室）
网　　址	http://www.wpcbj.com.cn
邮　　箱	wpcbjst@vip.163.com
销　　售	新华书店
印　　刷	北京建宏印刷有限公司
开　　本	711 mm × 1245 mm　1/24
印　　张	13.5
字　　数	259 千字
版　　次	2017 年 8 月第 1 版　　2021 年 3 月第 3 次印刷
版权登记	01-2015-3452
国际书号	ISBN 978-7-5192-2612-1
定　　价	48.00 元